Potential Health Benefits of Fruits and Vegetables III

Potential Health Benefits of Fruits and Vegetables III

Guest Editors

Luca Mazzoni
Franco Capocasa
Maria Teresa Ariza Fernández

MDPI

Basel • Beijing • Wuhan • Barcelona • Belgrade • Novi Sad • Cluj • Manchester

Guest Editors
Luca Mazzoni
Department of Agricultural, Food and Environmental Sciences
Università Politecnica delle Marche
Ancona
Italy

Franco Capocasa
Department of Agricultural, Food and Environmental Sciences
Università Politecnica delle Marche
Ancona
Italy

Maria Teresa Ariza Fernández
Instituto Andaluz de Investigación y Formación Agraria, Pesquera, Alimentaria y de la Producción Ecológica
IFAPA-Centro de Churriana
Málaga
Spain

Editorial Office
MDPI AG
Grosspeteranlage 5
4052 Basel, Switzerland

This is a reprint of the Special Issue, published open access by the journal *Applied Sciences* (ISSN 2076-3417), freely accessible at: https://www.mdpi.com/journal/applsci/special_issues/917MY11T38.

For citation purposes, cite each article independently as indicated on the article page online and as indicated below:

Lastname, A.A.; Lastname, B.B. Article Title. *Journal Name* **Year**, *Volume Number*, Page Range.

ISBN 978-3-7258-8293-9 (Hbk)
ISBN 978-3-7258-8294-6 (PDF)
https://doi.org/10.3390/books978-3-7258-8294-6

Contents

applied sciences

MDPI

Editorial

Potential Health Benefits of Fruits and Vegetables III

Luca Mazzoni [1,*], Franco Capocasa [1,*] and Maria Teresa Ariza Fernández [2,*]

[1] Agricultural, Food and Environmental Sciences, Università Politecnica delle Marche, Via Brecce Bianche 10, 60131 Ancona, Italy
[2] Breeding and Biotechnology, Instituto Andaluz de Investigación y Formación Agraria y Pesquera (IFAPA), IFAPA-Centro de Churriana, Cortijo de la Cruz, s/n, 29140 Málaga, Spain
* Correspondence: l.mazzoni@staff.univpm.it (L.M.); f.capocasa@staff.univpm.it (F.C.); mariat.ariza@juntadeandalucia.es (M.T.A.F.)

Over the past few decades, there has been a marked increase in public awareness about the connection between nutrition and the prevention of chronic diseases [1,2]. Plant-based foods, particularly fruits and vegetables, have become essential components of a healthy diet due to their rich content of bioactive phytochemicals [3,4]. These natural compounds have been widely recognized for their role in supporting health and reducing the risk of disease [5]. A broad spectrum of scientific research has focused on profiling the phytochemical constituents of various fruits and vegetables [6,7], as well as exploring the biological mechanisms through which they exert beneficial effects [8,9]. This Special Issue aims to present cutting-edge findings on the chemical composition of these foods and to explore their health-enhancing properties, drawing from evidence gathered through a range of experimental approaches, including in vitro studies, animal models, and clinical trials.

This Special Issue features 12 original papers, including nine research articles and three review papers. A common starting point in this field of research is the assessment of antioxidant properties and the profile of bioactive compounds in fruits and vegetables. This is typically achieved by examining their composition under various conditions, such as genotype, environmental factors, farming practices, and post-harvest handling.

In line with this focus, almost half of the research articles (four out of nine) explore the phytochemical content of plant-based foods—specifically, one study on vegetables and three on fruits. Starting from the manuscript that explores the quality of a vegetable matrix, the article examines how cryogenic grinding—a process in which materials are ground at extremely low temperatures using liquid nitrogen—affects the nutritional and antinutritional components of rapeseed cake, a by-product of oil extraction (Contribution 1). Traditionally, rapeseed cake has been used in animal feed; however, its potential for broader applications is limited by the presence of antinutritional factors, such as glucosinolates, tannins, and phytic acid. These compounds can interfere with digestion and the absorption of nutrients. The study finds that cryogenic grinding significantly reduces these unwanted substances, improving the nutritional profile of the final product. Additionally, because cryogenic grinding prevents the heat buildup typical of conventional grinding methods, it helps preserve sensitive nutrients, such as proteins, amino acids, and lipids. This means the ground rapeseed cake retains more of its original nutritional value. The resulting powder also has a finer and more uniform particle size, which improves bioavailability and enhances its functional properties, making it more suitable not only for animal feed but also as a possible ingredient in human food products. In conclusion, the article highlights cryogenic grinding as an effective technique to enhance both the nutritional quality and functional potential of rapeseed cake, opening new opportunities for its use in the food and feed industries. The other three studies on the topic of nutritional quality have highlighted the promising role of fruit pomaces and medicinal plant materials as natural sources of enriching nutritional and bioactive compounds. One noteworthy example is the use of black chokeberry (*Aronia melanocarpa* L.) pomace in wheat bread, which has shown significant potential to enhance both the nutritional value and antioxidant capacity of bakery products (Contribution 2). Specifically, the addition of chokeberry pomace at a level of 3% relative to flour content resulted in a marked increase in dietary fiber and total phenolic content, leading to improved antioxidant activity in the final product. However, higher concentrations negatively affected dough rheology and bread volume, as well as intensified the dark coloration of the crumb, factors that can influence consumer acceptability. These results suggest that low-level incorporation strikes a balance between nutritional enhancement and sensory quality. In a similar context, the application of fruit pomaces (including chokeberry, cherry, and blackcurrant) in the production of gluten-free extruded snacks has demonstrated substantial improvements in both nutritional and functional attributes (Contribution 3). For example, the incorporation of 20% chokeberry pomace into corn-based formulations significantly boosted the total phenolic and anthocyanin content, with increases up to tenfold compared to controls, without compromising essential technological parameters such as expansion ratio and texture. These findings reinforce the potential of fruit-derived by-products as effective natural enhancers of antioxidant activity and dietary quality in gluten-free processed foods. Extending beyond processed food systems, a comparative study on milk thistle (*Silybum marianum*) fruits from organic and conventional farming revealed nuanced differences in phytochemical composition and microbiological quality (Contribution 4). While the overall silymarin and oil content were comparable across both cultivation systems, organic samples were notably richer in silydianin and silychristin compounds known for their hepatoprotective properties. Interestingly, despite showing higher microbial counts (including mesophilic bacteria and molds), the organic fruits remained within acceptable safety limits. Moreover, extracts derived from organic milk thistle exhibited greater antimicrobial activity, particularly against *Staphylococcus aureus*, suggesting additional therapeutic potential associated with cultivation practices. Taken together, these studies illustrate the multi-faceted benefits of integrating fruit pomaces and phytochemical-rich plant materials into food systems. Beyond enhancing nutritional density and antioxidant capacity, such ingredients contribute to the functional, sensory, and microbial profile of food products, offering sustainable strategies for waste valorization and health-oriented product innovation.

Beyond compositional studies, a further step in assessing the health benefits of bioactive compounds in fruits and vegetables involves investigating additional biological activities, such as antimicrobial, bactericidal, and anticancer properties. In this Special Issue, three of the research articles specifically address these functional aspects, further supporting recent investigations that have underscored the biomedical and nutraceutical relevance of fruit-derived phytochemicals and their role in modulating cellular and microbial processes. The hydroethanolic (50%) extract of *Plinia cauliflora* fruit has demonstrated dual bioactivity of significant interest (Contribution 5). In vitro assays revealed a potent inhibitory effect against the single-stranded DNA-binding protein (SSB) of *Klebsiella pneumoniae*, a key factor in bacterial DNA replication and repair, with an IC_{50} of 73 ± 8 μg/mL, suggesting potential antibacterial applications through interference with the replication machinery. Simultaneously, the same extract exerted pronounced cytotoxic effects on B16F10 murine melanoma cells, as evidenced by a dose-dependent reduction in cell viability, decreased migratory capacity, and activation of apoptotic pathways, indicating its promise as a multifunctional antitumor agent that acts through mechanisms involving redox imbalance and pro-apoptotic signaling. In the context of nutrient bioavailability, a comparative analysis of L-citrulline uptake from watermelon flesh, rind, and peel using the Caco-2 human intestinal epithelial cell monolayer model revealed differential permeability profiles (Contribution 6). While the flesh possessed the highest intrinsic citrulline concentration, peel extracts exhibited superior intestinal transport rates, potentially due to interactions with dietary fiber or coexisting polyphenolic compounds that influence paracellular transport and active uptake mechanisms. These findings highlight the potential of underutilized anatomical fractions of watermelon as functional ingredients with improved bioaccessibility profiles, thereby supporting sustainable valorization strategies. Further supporting the health-protective properties of fruit-derived compounds, ethanolic extracts of strawberry (*Fragaria* × *ananassa*) and kiwifruit (*Actinidia deliciosa*) were evaluated for their radioprotective capacity in vitro (Contribution 7). Both extracts significantly mitigated DNA damage and oxidative stress in irradiated models, correlating strongly with their high total phenolic content, antioxidant activity (as measured by DPPH and FRAP assays), and ascorbic acid levels. These effects were attributed to synergistic interactions between flavonoids, phenolic acids, and vitamin C, which collectively scavenge free radicals and stabilize cellular redox balance. Together, these studies highlight the multifaceted bioactivity of fruit-derived extracts, ranging from antibacterial and anticancer mechanisms to enhanced nutrient bioavailability and radioprotection, underscoring their potential applications in the development of functional foods, dietary supplements, and adjunct therapeutic formulations.

The final stage in assessing the potential health benefits of fruits and vegetables involves in vivo testing, encompassing studies on both animal models and human participants; in this Special Issue, two research articles contribute to this phase by reporting findings from in vivo experiments conducted on mice, investigating the protective effects of plant-derived extracts in two distinct models of inflammation and tissue injury. In a DSS-induced rat model of inflammatory bowel disease (IBD), supplementation with fermented onion notably improved intestinal health compared to fresh onion or probiotic controls (Contribution 8). Specifically, high-dose fermented onion (20 g/kg) significantly preserved colonic crypt architecture, increased goblet cell density, and augmented mucosal thickness, while also restoring short-chain fatty acid (SCFA) levels—particularly butyrate, acetate, and propionate—and reducing systemic IgA/IgG titers and pro-inflammatory cytokines (IL-6, TNF-α, IL-1β). These results suggest that fermentation enhances the conversion of FODMAPs and enriches bioactive compounds, such as quercetin, thereby mitigating inflammatory damage more effectively than fresh onion. In a parallel study utilizing a mouse model of particulate matter ($PM_{2.5}$)-induced pulmonary injury, oral administration of immature Asian pear (*Pyrus* spp.) extract yielded dose-dependent protective effects (Contribution 9). Treatment inhibited oxidative stress markers, including decreased glutathione (GSH), superoxide dismutase (SOD), and catalase (CAT) activity, while reducing lipid peroxidation (MDA) and downregulating inflammatory mediators, such as TNF-α, IL-6, CXCL1, and CXCL2. Mechanistically, the extract suppressed the activation of the NF-κB, p38 MAPK, and PI3K/Akt signaling pathways and preserved the expression of the tumor suppressor PTEN, demonstrating efficacy comparable to dexamethasone in mitigating $PM_{2.5}$-triggered subacute lung injury.

The remaining three contributions in this collection are review articles, each offering an in-depth examination of topics related to the health-promoting properties of fruits and vegetables from nutritional, agricultural, and environmental perspectives. One review critically evaluates the benefits and limitations of a whole-foods approach to increasing dietary nitrate intake, highlighting the need to move beyond the current emphasis on beetroot juice toward a more diverse and sustainable strategy involving a variety of nitrate-rich plant foods (Contribution 10). Another review examines the role of organic horticultural practices in simultaneously mitigating climate change and enhancing the nutritional quality of fruits and vegetables, with a focus on phytochemical enrichment and soil health (Contribution 11). The third review revisits the functional properties of red beetroot (*Beta vulgaris*), emphasizing its continued relevance as a nutrient-dense superfood, rich in nitrates, betalains, phenolic compounds, and saponins, which contribute to its antioxidant, anti-inflammatory, and cardioprotective effects (Contribution 12). Collectively, these articles underscore the importance of integrating whole-food dietary strategies, sustainable agricultural systems, and bioactive-rich crops to support both human health and ecological resilience.

In summary, this Special Issue offers a comprehensive and multifaceted contribution to the expanding body of research on the health-promoting properties of fruits and vegetables. The collection spans detailed compositional analyses and innovative processing approaches, as well as functional assessments conducted through in vitro and in vivo models, culminating in broad, integrative reviews that contextualize these findings within larger nutritional, agricultural, and environmental frameworks. Together, these contributions not only advance our understanding of the bioactive compounds present in plant-based foods but also illuminate the mechanisms by which they may exert protective effects against various chronic conditions. Importantly, this volume highlights the importance of interdisciplinary approaches, which bridge food science, nutrition, pharmacology, and agronomy, to support both human and planetary health. Building on the foundation laid by the previous Special Issues, "Potential Health Benefits of Fruits and Vegetables I" and "Potential Health Benefits of Fruits and Vegetables II", this third installment further consolidates the scientific rationale for promoting fruit and vegetable consumption as a key component of preventive health strategies. Moreover, it sets the stage for the forthcoming fourth edition, which will continue to explore emerging trends, novel methodologies, and translational insights into the role of plant-based diets in disease prevention and health optimization. The collective findings presented here reaffirm the critical importance of fruits and vegetables not only as nutritional staples but also as dynamic sources of bioactive agents that can contribute meaningfully to sustainable health solutions.

Funding: This research was funded by the PP.AVA23.INV2023.005 and PP.TRA23.TRA2023.010 projects, co-financed by FEDER funds (80%) within the Andalusian Operational Programme 2021–2027, and by the Breeding Value project funding from the European Union's Horizon 2020 research and innovation programme under grant agreement No. 101000747.

Conflicts of Interest: The authors declare no conflicts of interest.

List of Contributions:

1. Kraljić, K.; Škevin, D.; Čukelj Mustač, N.; Benković, M.; Drakula, S.; Balbino, S.; Mandura Jarić, A.; Mamilović, K.; Ramljak, I.; Ćurić, D. Influence of Cryogenic Grinding on the Nutritional and Antinutritional Components of Rapeseed Cake. *Appl. Sci.* **2023**, *13*, 5841. https://doi.org/10.3390/app13105841.
2. Cacak-Pietrzak, G.; Dziki, D.; Gawlik-Dziki, U.; Parol-Nadłonek, N.; Kalisz, S.; Krajewska, A.; Stępniewska, S. Wheat Bread Enriched with Black Chokeberry (*Aronia melanocarpa* L.) Pomace: Physicochemical Properties and Sensory Evaluation. *Appl. Sci.* **2023**, *13*, 6936. https://doi.org/10.3390/app13126936.
3. Gumul, D.; Berski, W.; Zięba, T. The Influence of Fruit Pomaces on Nutritional, Pro-Health Value and Quality of Extruded Gluten-Free Snacks. *Appl. Sci.* **2023**, *13*, 4818. https://doi.org/10.3390/app13084818.
4. Sadowska, K.; Andrzejewska, J.; Ligocka, A.; Korczyk-Szabo, J.; Haban, M. Bioactive Substances and Microbiological Quality of Milk Thistle Fruits from Organic and Conventional Farming. *Appl. Sci.* **2023**, *13*, 8536. https://doi.org/10.3390/app13148536.
5. Lin, E.-S.; Huang, C.-Y. The Inhibition Activities of the Fruit Extract of Plinia Cauliflora against Melanoma Cells and the Single-Stranded DNA-Binding Protein (SSB) from Klebsiella Pneumoniae. *Appl. Sci.* **2023**, *13*, 11061. https://doi.org/10.3390/app131911061.

6. Gu, I.; Balogun, O.; Brownmiller, C.; Kang, H.W.; Lee, S.-O. Bioavailability of Citrulline in Watermelon Flesh, Rind, and Skin Using a Human Intestinal Epithelial Caco-2 Cell Model. *Appl. Sci.* **2023**, *13*, 4882. https://doi.org/10.3390/app13084882.
7. Ribeiro, M.D.S.; Sebastià, N.; Montoro, A.; García-Martínez, E. Strawberry (*Fragaria* × *ananassa*) and Kiwifruit (*Actinidia deliciosa*) Extracts as Potential Radioprotective Agents: Relation to Their Phytochemical Composition and Antioxidant Capacity. *Appl. Sci.* **2023**, *13*, 8996. https://doi.org/10.3390/app13158996.
8. Sangpreecha, N.; Chanmuang, S.; Park, K.-H.; Sangar, M.; Sharma, D.; Song, D.; Park, Y.-J.; Sung, H.-M.; Promyo, K.; Ham, K.-S. Effects of Fermented Onion on Gut Health in Dextran Sodium Sulfate (DSS)-Induced Inflammatory Bowel Disease (IBD) Rats. *Appl. Sci.* **2023**, *13*, 1590. https://doi.org/10.3390/app13031590.
9. Kim, M.-R.; Lee, J.-H.; Ku, M.-U.; Kim, K.-Y.; Shin, S.; Hong, E.-J.; Ku, S.-K.; Choi, J.-S. Effectiveness of Immature Asian Pear Extract on Pulmonary Injury Caused by Particulate Matter in Mice. *Appl. Sci.* **2023**, *13*, 9578. https://doi.org/10.3390/app13179578.
10. Griffiths, A.; Alhulaefi, S.; Hayes, E.J.; Matu, J.; Brandt, K.; Watson, A.; Siervo, M.; Shannon, O.M. Exploring the Advantages and Disadvantages of a Whole Foods Approach for Elevating Dietary Nitrate Intake: Have Researchers Concentrated Too Much on Beetroot Juice? *Appl. Sci.* **2023**, *13*, 7319. https://doi.org/10.3390/app13127319.
11. Manyaku, A.; Witbooi, H.; Laubscher, C.P. The Significance of Organic Horticulture in Mitigating Climate Change and Promoting the Production of Healthier Fruits and Vegetables. *Appl. Sci.* **2024**, *14*, 4966. https://doi.org/10.3390/app14124966.
12. Sentkowska, A.; Pyrzynska, K. Old-Fashioned, but Still a Superfood—Red Beets as a Rich Source of Bioactive Compounds. *Appl. Sci.* **2023**, *13*, 7445. https://doi.org/10.3390/app13137445.

References

1. Tufail, T.; Fatima, S.; Bader Ul Ain, H.; Ikram, A.; Noreen, S.; Rebezov, M.; AL-Farga, A.; Saleh, R.; Shariati, M.A. Role of Phytonutrients in the Prevention and Treatment of Chronic Diseases: A Concrete Review. *ACS Omega* **2025**, *10*, 12724–12755. [CrossRef] [PubMed]
2. Sun, J.; Luo, S.; Deng, J.; Yang, H. Phytochemicals in Chronic Disease Prevention. *Nutrients* **2023**, *15*, 4933. [CrossRef] [PubMed]
3. Josa, E.; Barril, G.; Ruperto, M. Potential Effects of Bioactive Compounds of Plant-Based Foods and Medicinal Plants in Chronic Kidney Disease and Dialysis: A Systematic Review. *Nutrients* **2024**, *16*, 4321. [CrossRef] [PubMed]
4. Alvarez-Suarez, J.M.; Giampieri, F.; Gasparrini, M.; Mazzoni, L.; Forbes-Hernández, T.Y.; Afrin, S.; Battino, M. Guava (*Psidium guajava* L. Cv. *Red Suprema*) Crude Extract Protect Human Dermal Fibroblasts against Cytotoxic Damage Mediated by Oxidative Stress. *Plant Foods Hum. Nutr.* **2018**, *73*, 18–24. [CrossRef] [PubMed]
5. Koyama, S.; Joseph, P.V.; Shields, V.D.C.; Heinbockel, T.; Adhikari, P.; Kaur, R.; Kumar, R.; Alizadeh, R.; Bhutani, S.; Calcinoni, O.; et al. Possible Roles of Phytochemicals with Bioactive Properties in the Prevention of and Recovery from COVID-19. *Front. Nutr.* **2024**, *11*, 1408248. [CrossRef] [PubMed]
6. Ňorbová, M.; Vollmannová, A.; Fedorková, S.; Musilová, J.; Lidiková, J. The Forgotten Fruit (*Cydonia oblonga* Mill.) and Its Chemical Composition: A Review. *Eur. Food Res. Technol.* **2024**, *250*, 2093–2102. [CrossRef]
7. Li, Y.; Zhao, H.; Xiang, K.; Li, D.; Liu, C.; Wang, H.; Pang, W.; Niu, L.; Yu, R.; Sun, X. Factors Affecting Chemical and Textural Properties of Dried Tuber, Fruit and Vegetable. *J. Food Eng.* **2024**, *365*, 111828. [CrossRef]
8. Del Rio, D.; Rodriguez-Mateos, A.; Spencer, J.P.E.; Tognolini, M.; Borges, G.; Crozier, A. Dietary (Poly)Phenolics in Human Health: Structures, Bioavailability, and Evidence of Protective Effects Against Chronic Diseases. *Antioxid. Redox Signal.* **2013**, *18*, 1818–1892. [CrossRef] [PubMed]
9. Liu, Y.; Deng, J.; Zhao, T.; Yang, X.; Zhang, J.; Yang, H. Bioavailability and Mechanisms of Dietary Polyphenols Affected by Non-Thermal Processing Technology in Fruits and Vegetables. *Curr. Res. Food Sci.* **2024**, *8*, 100715. [CrossRef] [PubMed]

applied sciences

Article

Effects of Fermented Onion on Gut Health in Dextran Sodium Sulfate (DSS)-Induced Inflammatory Bowel Disease (IBD) Rats

Neeracha Sangpreecha [1,†], Saoraya Chanmuang [1,†], Kyung-Hee Park [1], Madhuri Sangar [1], Divya Sharma [1], Doyoung Song [1], Yun-Ja Park [2], Hea-Mi Sung [3], Kitipong Promyo [4] and Kyung-Sik Ham [1,*]

1 Department of Food Engineering and Solar Salt Research Center, Mokpo National University, Muan 58554, Republic of Korea
2 Jangsoofood Co., Naju-si 58307, Republic of Korea
3 Jeonnam Bioindustry Foundation Food Research Center, Naju-si 58307, Republic of Korea
4 Institute of Agricultural Technology, School of Food Technology, Suranaree University of Technology, Nakhon Ratchasima 30000, Thailand
* Correspondence: ksham@mokpo.ac.kr; Tel.: +82-61-450-2425; Fax: +82-61-450-1521
† These authors contributed equally to this work.

Featured Application: This paper is the first report to compare the effects of fresh and fermented onions on gut health using rats. These results could be useful for developing healthy foods using onions.

Abstract: Onion is a well-known health-beneficial vegetable. However, fresh onion is high in FODMAPs (fermentable oligosaccharides, disaccharides, monosaccharides, and polyols) which may be problematic for IBD. Fermentation of onion may help to lower FODMAP problems and increase the availability of bioactive compounds, especially quercetin. We investigated the effect of fermented onion on DSS-induced IBD in rats. Rats were divided into six groups and treated orally with saline as a control and negative control (DSS), probiotics, low and high doses of fermented onion, or fresh onion extract for 3 weeks. After two weeks, rats were given drinking water containing 0.2% DSS for 5 days, except for the control followed by two days of regular water. The colonic histomorphology, immunity, oxidative stress, short-chain fatty acids, and biochemical analysis showed improved IBD conditions in the fermented onion groups. In contrast, the consumption of fresh onion appeared to exacerbate the IBD condition. These results suggest that the consumption of a high dose of fermented onion can ameliorate IBD symptoms.

Keywords: inflammatory bowel disease; dextran sodium sulfate; fermented onion; onion

1. Introduction

Inflammatory bowel disease (IBD) is a chronic inflammation of the digestive tract. Its incidence is increasing worldwide at an unprecedented rate [1]. IBD can be caused by many factors. In particular, oxidative stress causes protein, lipid, and DNA damage, which leads to the pathogenesis of intestinal disorders [2]. The most important factor of IBD is intestinal homeostasis, which can occur through the equilibrium between the immune system and intestinal microbiota. Controlling the invasion of harmful microorganisms is one of the processes for maintaining this balance. Therefore, to prevent the adhesion and invasion of harmful microorganisms, the intestinal mucosa is furnished with protective mechanisms including the mucus layer, a building complex, and an effective mucosal barrier, which is formed by mucins produced by intestinal goblet cells [3]. Currently, therapeutic drugs for IBD such as steroids, immunomodulators, and antibodies are now available. However, most of them have limitations due to their side effects including increased risk of infection, pulmonary toxicity, congestive heart failure, congenital disabilities, myelosuppression, liver toxicity, pancreatitis, and malignancy [4]. Since eating habits play an important role in overall health, changes in diet and food consumption are used to improve, control, or even cure IBD. Numerous studies have shown that dietary components impact gut microbiota and may also have an effect on gut homeostasis directly [5].

Onion (*Allium cepa*), a monocotyledon, is one of the oldest fructan-containing vegetables. Onions are known to be high in polyphenols, the most abundant of which is quercetin [6]. According to research, quercetin has antihypertensive, antibacterial activities, and anti-inflammatory properties [7]. Furthermore, studies have shown that quercetin helps to suppress lipid peroxidation, inhibit the initial process of inflammation, and improve the immune system [8].

Onions are high in fructan, one of the FODMAPs (fermentable oligosaccharides, disaccharides, monosaccharides, and polyols) that has been shown to have negative effects on IBD due to its poor absorption in the small intestine and is rapidly fermented by bacteria in the large intestine, leading to symptoms of irritable bowel syndrome (IBS) in IBD patients. On the other hand, the fermentation of onions enhances antioxidant activity, and the availability of bioactive compounds such as flavonoids and short-chain fatty acids (SCFAs) [9,10]. A recent study showed that a diet containing a high content of fermented food increased microbiome diversity and decreased inflammatory signals and activity [11]. Therefore, fermentation of onion may reduce onion's FODMAP characteristics and improve IBD. Thus, this study aims to evaluate the effect of fermented onion on gut health in DSS-induced IBD rats.

2. Materials and Methods

2.1. Materials and Chemicals

A commercial probiotic that was approved by the Ministry of Food and Drug Safety of Korean government was purchased from Seoyun family Co., Ltd. (Seoul, Republic of Korea) and was used as a positive control in this study. Thiobarbituric acid (TBA), malondialdehyde (MDA), hydroxylamine hydrochloride, 5,5′-dithiobis (2-nitrobenzoic acid) (DTNB), 2-4-dinitrophenylhydrazine (DNPH), guanidine hydrochloride, acetic acid (AA), n-butyric acid (BA), propionic acid (PA), and anhydrous methanol were purchased from Sigma-Aldrich Chemical (St. Louis, MO, USA). Dextran sodium sulfate (DSS) was purchased from MP Biomedicals (Santa Anna, CA, USA). The antibody for interleukin 1 beta (IL-1β) was purchased from Abcam (Waltham, CA, USA). Antibodies for nuclear factor Kappa-B (NF-κB) p65, interleukin-6 (IL-6), tumor necrosis factor-alpha (TNF-α), β-actin, GAPDH and DAPI solution (1 mg/mL) were procured from Themo Fisher Scientific Inc. (Waltham, MA, USA). The antibody for horseradish peroxide (HRP)-conjugated goat anti-rabbit was obtained from Jackson ImmunoResearch Inc. (West Grove, PA, USA). Formic acid was obtained from Daejung chemicals (Siheung-si, Republic of Korea). Tri-chloroacetic acid (TCA) was purchased from Wako Pure Chemical, Inc. (Kyoto, Japan).

2.2. Selection of a Lactic Acid Bacteria for the Fermentation of Onion

Fresh onion (100 g) was mixed with 200 g of water. After boiling for 2 h at 90 °C, the mixture was inoculated with 109 cells of 16 different lactic acid bacteria (*Lactococcus lactis subsp. Cremoris* (KCCM-40699), *Leuconostoc mesenteroides* (KCCM-13374), *Lactobacillus brevis* (KCCM-40017), *Lactobacillus cellobiosus* (KCCM-40983), *Lactobacillus plantarum subsp. Plantarum* (KCCM-13093), *Lactobacillus plantarum* (KCCM-11322), *Lactobacillus casei* (KCCM-12452), *Lactobacillus rhamnosus* (KCCM-32405), *Lactobacillus acidophilus* (KCCM-32820), *Lactobacillus delbrueckii subsp. Bulgaricus* (KCCM-35465), *Lactobacillus casei subsp. Casei* (KCCM-35465), *Lactococcus lactis subsp. Lactis* (KCCM-40104), *Lactobacillus acidophilus* (KCCM-40265), *Lactobacillus helveticus* (KCCM-40989), *Lactobacillus paracasei subsp. Paracasei* (KCCM-41276), *Streptococcus salivarius* (KCCM-41580)), which were obtained from Korea Culture Center of Microorganisms (KCCM). The growth of lactic acid was monitored by counting the number of lactic acid bacteria in fermented onion using a 3M Petri film lactic acid bacteria count plate (3M Korea Co., Seoul, Republic of Korea). Sensory evaluation of fermented onion was also carried out.

Based on the growth of lactic acid bacteria and sensory evaluation (data not shown), *Lactobacillus casei* (KCCM-12452) was selected for the fermentation of onion. Furthermore, *L. casei* has been known as a probiotic that has been reported to balance the gut microbiota, relieve gastrointestinal dysfunction, prevent infection, modulate inflammatory and immunological responses, and alleviate IBS symptoms [12,13].

2.3. *Preparation of Fermented Onion for Animal Experiment*

Onion (80 kg) was ground and mixed with water (160 kg), and then heated for 2 h at 90 °C. Then, the mixture was inoculated with *Lactobacillus casei*, and incubated at 37 °C for 48 h without agitation. The culture broth was filtered and sterilized by boiling. The boiled filtrate was used for the animal experiment.

2.4. *Animal Studies*

Wistar rats were purchased from Orient Bio Inc. (Seongnam, Republic of Korea). The animals were housed under standard housing conditions of temperature (23 ± 2 °C), relative humidity (55 ± 10%), and light (12/12 h light/dark cycle). The animals had ad libitum access to a chow diet and water. Ethical approval for this study was obtained from the Institutional Animal Care and Use Committee (IACUC) of Mokpo National University (Muan, Republic of Korea). All animal experiments were performed according to the National Institutes of Health (NIH) guidelines for the care and use of laboratory animals and the guidelines of the IACUC.

2.5. *Experimental Design*

An inflammatory bowel disease (IBD) model was established through the oral administration of 2% DSS with a molecular weight of 36,000–50,000 purchased from MP Biomedicals (Santa Anna, CA, USA). Male Wistar rats (5 weeks old) were adapted for 1 week. After acclimatization, rats were divided into six groups (n = 8/group) and were treated orally with a different solution: control, saline; PC, positive control, 0.44 g of commercial probiotic/kg rat); LO, low dose of fermented onion (equivalent to 10 g of fresh onion/kg rat); HO, high dose of fermented onion (equivalent to 20 g fresh onion/kg rat); and FO, fresh onion solution (10 g/kg rat) for 3 weeks. Two weeks after starting the experiment, every group except the control was given 2% DSS in drinking water for 5 days, followed by two days of regular water. At the end of the experiment (week 3), animals were anesthetized with isoflurane and euthanized by draining blood by means of cardiac puncture. Colon tissue was surgically removed, then 3 cm of the distal colon containing feces was removed and stored in Carnoy's fixative (60% ethanol, 30% chloroform, 10% glacial acetic acid) overnight, and the remaining tissue was flushed with 0.9% NaCl, followed by storage at −80 °C. A 0.5 cm segment was excised from the end of the proximal colon and soaked in a 10% neutral, buffered formalin solution (Sigma).

2.6. *Colonic Histomorphology*

2.6.1. Hematoxylin and Eosin Staining

The colon samples, after being fixed for 2 days in 10% neutral formalin, were paraffin-embedded and sectioned transversely. The cross-sections of colon tissues (5 μm in thickness) were prepared for rehydration and passed through a series of xylene–alcohol mixtures (100%, 95%, and 70%). The slides were dyed with hematoxylin for 6–8 min and then washed with tap water for 5 min, then placed in eosin solution for 1 min. The slides were dehydrated with a series of alcohol concentrations (70%, 95%, and 100%). Finally, slides were placed in xylene and mounted using Canada balsam.

2.6.2. Alcian Blue and Nuclear Fast Red Staining

The cross-sections of the colon after rehydration in the xylene–alcohol mixtures were stained with alcian blue (AB). Slides were dipped in acetic acid for 3 min and stained with AB dye for 30 min and then stained with Nuclear Fast Red for 5 min. After staining, all slides were dehydrated with a series of alcohol–xylene mixtures (70%, 95%, and 100% ethanol). Finally, the slides were placed in xylene and mounted using Canada balsam. For each stained section, at least 30 bright-field images were captured by a digital camera, under 20–40× magnification, using an Olympus BX43 microscope (Olympus, Tokyo, Japan). Histological damage and goblet cell number were evaluated. The Image J software was used to determine the goblet cell number.

2.6.3. Mucus Layer Thickness Using Periodic Acid-Schiff (PAS) Staining

One-centimeter-long segments of the colon, together with fecal content, were carefully collected and dipped in a water-free methanol-Carnoy's fixative (60% dry methanol, 30% chloroform, and 10% glacial acetic acid) and fixed overnight. The tissues were then washed in methanol and embedded in paraffin, and sections 5 μm in thickness were cut using a microtome and placed on glass slides. Slides underwent deparaffinization with xylene and rehydration with 100% and 95% ethanol. Thereafter, the slides were stained using the periodic acid-Schiff procedure. Briefly, slides were oxidized for 5 min in 0.5% periodic acid solution. After rinsing, tissues were placed in Schiff reagent for 15 min, washed in warm water, and assessed by a light microscope. For each stained section, at least 10 bright-field images were captured by a digital camera, under 20–40× magnification, using an Olympus BHS microscope (Tokyo, Japan). The Image J 1.53g software was used to determine mucus layer thickness.

2.7. Oxidative Stress Parameters

2.7.1. Determination of Protein Carbonyl Content (PCOs)

Protein carbonyl content was measured using the modified method of Reznick and Packer [14]. An aliquot (1 mL) of 10 mM DNPH in 2 M HCl was added to the reaction mixture (2 mg tissue protein). The sample was incubated for 1 h in the dark at room temperature and vortexed at intervals of 15 min. Then, 1 mL of cold trichloroacetic acid (TCA) (10%, *w*/*v*) was added to each reaction mixture which was centrifuged at 3000× *g* for 10 min. The protein pellet was washed three times with 2 mL of ethanol/ethyl acetate (1:1, *v*/*v*) and dissolved in 1.5 mL of guanidine hydrochloride (6 M), incubated for 10 min at 37 °C. The absorbance was measured at 595 nm. The carbonyl content was calculated in nmol of carbonyl groups per mg of protein based on the molar extinction of DNPH (e = 2.1 × 104 $mL^{-1} \cdot cm^{-1}$).

2.7.2. Determination of Thiobarbituric Acid Reactive Substances (TBARS)

The lipid peroxidation content in colon tissue was determined by the TBARS method. Briefly, colon tissue (0.15 g) was homogenized with 1 mL of 0.01 M tris buffer (pH 7.4). Homogenate was centrifuged at 5000× *g* at 4 °C for 15 min. The supernatant (0.1 mL) was mixed with 0.3 mL of 10 mM KH_2PO_4 solution and 1 mL of 0.15 M Tris-HCl buffer solution and incubated at 37 °C for 20 min with continuous shaking (100 rpm). The reaction was stopped by adding 0.5 mL of 20% TCA and 0.5 mL of 0.67% of TBA, and the solution was boiled at 95 °C for 15 min. Then, the solution was centrifuged at 5000× *g* for 10 min and the absorbance of the supernatant was measured at 532 nm using a spectrophotometer (Hewlett Packard 8452A, Palo Alto, CA, USA). TBARS levels were calculated using a standard curve of malonaldehyde (MDA).

2.8. Immunoglobulin Enzyme-Linked Immunosorbent Assay (ELISA)

The frozen plasma samples were thawed on ice and the IgA and IgG assays were performed in triplicate according to the manufacturer's instructions using a rat IgA and IgG ELISA kit (Genway, San Diego, CA, USA). The optical density was measured at 450 nm using a microplate reader (TECAN, Zürich, Switzerland). The concentrations of IgA and IgG in plasma samples were determined using standard curves produced from serial dilutions of the rat IgA and IgG, as provided in the kit.

2.9. Protein Quantitative Assay

Colon tissue (0.15 g) was homogenized in 0.8 mL RIPA buffer containing 0.01 M NaF, 2 mM phenylmethylsulfonyl fluoride, 10 mM sodium pyrophosphate, aprotinin (0.1 mg/mL), pepstatin (1 mg/mL) and leupeptin (1 mg/mL). After protein extraction at 4 °C for 2 h, the homogenate was centrifuged at 15,000× *g* at 4 °C. The supernatant was collected and the protein concentration was analyzed using the Bradford method. Briefly, 50 µL of the protein sample was mixed with 950 µL of the diluted concentrated reagent (1:4) in the disposable cuvette. All protein samples were gently mixed and incubated for 10 min. Absorbance was measured at 595 nm. The standard curve was created using a stock solution of γ-globulin in various concentrations (0, 200, 400, 600, 800, and 1000 µg/mL).

2.10. Western Blotting

An aliquot (40 µL) of protein was mixed with 40 µL of tricine buffer containing 2% 2-mercaptoethanol and denatured at 95 °C for 5 min. Proteins were separated by SDS-polyacrylamide gel electrophoresis and transferred to the PVDF (polyvinylidene difluoride) membrane, where it is blocked with 0.05% Tween-20 (TBST) buffer containing 3% bovine serum albumin (BSA). The membrane was incubated for 10 min and 5 min with TBST solution. Then, the membrane was incubated in primary antibodies and allowed to react overnight at room temperature with gentle agitation. After the primary antibody reaction, the membrane was incubated with TBST solution for 10 min and 5 min for washing. After washing, horseradish peroxidase (HRP)-conjugated goat anti-rabbit immunoglobulin G (Millipore, CA, USA) was added and reacted at room temperature for 1 h. After washing with TBST and TBS solutions, the membranes were reacted with enhanced chemiluminescence (ECL) solution containing luminol and visualized using the Davinch-Chemi™ system (Bio Co., Ltd., Daegu, Republic of Korea). Obtained images were analyzed using Image J software (NIH, Bethesda, MD, USA). All primary antibodies were used at the recommended dilution as given by the manufacturer.

2.11. Analysis of SCFAs in Colon Content Using GC

In rat cecum samples, the gut microbial metabolites, SCFAs, were determined and quantified by gas chromatography (GC) following a previously described protocol with little modification [4]. A Shimadzu GC-17A system (Bellefonte, PA, USA) along with a flame ionization detector (FID) and an automatic liquid sampler was used. A J&W DB-FFAP capillary column (Agilent Technologies Inc., USA; Part # J125-3232, 30 m × 0.53 mm × 1 µm film thickness) was used. One gram of the cecum sample that was previously frozen at −80 °C was thawed and suspended in at least 5 mL of methanol and homogenized for about 3 min, resulting in a 17% (*w*/*w*) cecum suspension. Then, the resulting slurry was adjusted to pH 2–3 by adding HCl and then kept for 10 min at room temperature with occasional shaking every 2 min. The mixture was centrifuged at 5000 rpm for 20 min. The supernatant was then filtered using a filter (0.2 µm pore). The internal standard 2-ethylbutyric acid solution containing 12% formic acid was spiked into the supernatant at a final concentration of 1 mM. The injection volume of the sample was 1.0 µL. The oven temperature was 100 °C for 0.5 min initially then raised to 180 °C at 8 °C/min, held for 1.0 min, then increased to 200 °C at 20 °C/min, and finally held at 200 °C for 5 min. The temperatures of the injection port and FID were 200 °C and 240 °C, respectively. The flow rates of hydrogen, air, and helium, known as makeup gas, were 30, 300, and 25 mL/min, respectively. The concentration of SCFAs was quantified as described previously [15].

2.12. Statistics

All data were expressed as mean ± standard deviation (SEM). A value of $p < 0.05$ was considered to indicate statistical significance using Tukey's multiple-comparison test among more than three mean values for unpaired data.

3. Results

3.1. General Characteristics

The body weight, food, and water intake were measured throughout the experimental period (3 weeks). In Figure 1, all groups treated with DSS display reduced water intake, which is a normal side effect of the DSS-treated model [16]. The DSS groups appeared to have lower food intake as compared to the control group, though a significant difference was not observed. The HO group seemed to show lower food intake and body weight compared to the other groups. However, there are no significant differences among groups.

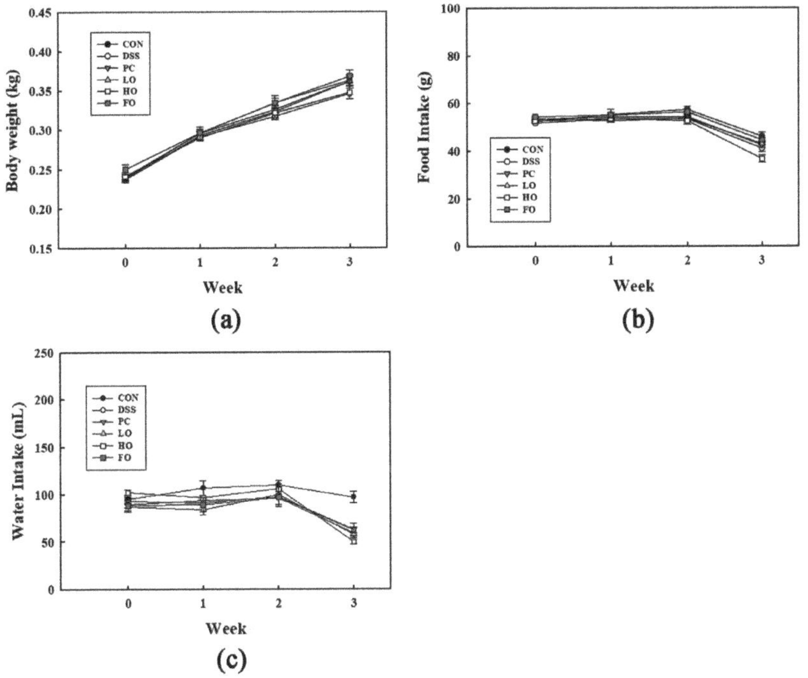

Figure 1. Changes in body weight (**a**), food intake (**b**), and water intake (**c**) of rats administered fermented and fresh onions orally for 3 weeks. CON, control, saline; DSS, dextran sulfate sodium; PC, positive control, probiotic; LO, low dose of fermented onion; HO, high dose of fermented onion; FO, fresh onion. Values are expressed as mean ± SEM (n = 8).

3.2. Colonic Histo-Morphology

Clinical signs of colitis, such as weight loss, diarrhea, and rectal bleeding, were not observed in DSS-induced rats when compared to control rats. H&E-stained sections showed changes in mucus-filled goblet cell numbers. The DSS, PC, LO and FO groups exhibited a depletion in goblet cell count, cellular infiltration in the colonic mucosa, and crypt distortion. DSS and FO groups showed remarkably higher histological damage as compared to the control group. Interestingly, the HO group showed significantly lower damage than the other DSS-treated groups, indicating the preventive effect of the fermented onion. The HO group was even better than PC group, which was treated with a probiotic that was approved for the improvement of gut health by the Ministry of Food and Drug Safety (Figure 2a). Alcian blue-stained colon sections of HO groups show higher goblet cell counts ($p < 0.05$) as compared to that of the DSS group (Figure 2b,c). The FO group, on the other hand, showed a considerable decrease in comparison to the other groups.

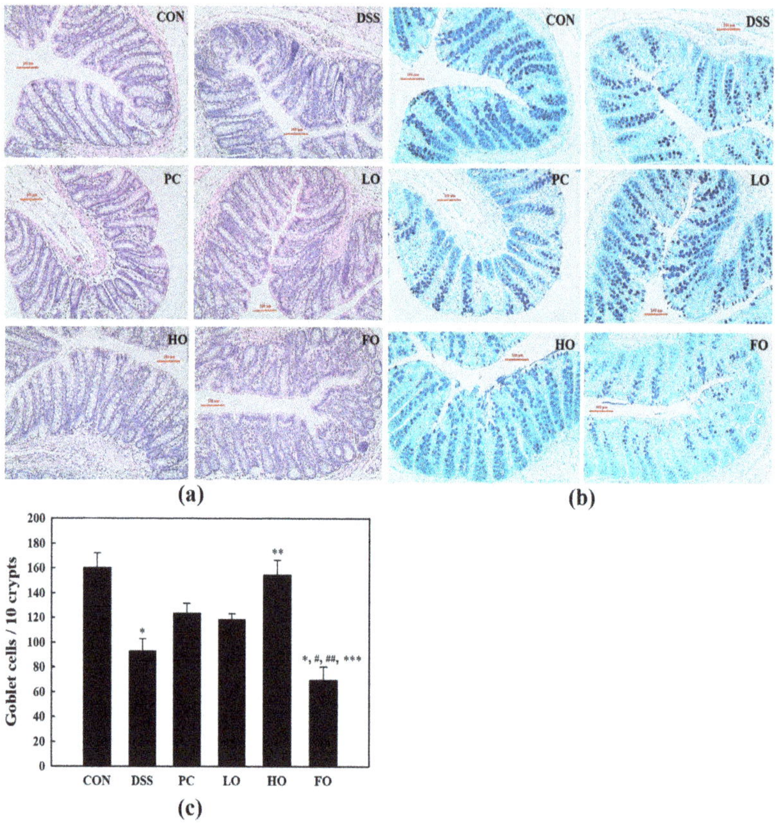

Figure 2. Cross-section of the colonic tissue stained with hematoxylin and eosin (H&E) (**a**), alcian blue and nuclear fast red-stained colonic tissue to show goblet cells (dark blue color) (**b**), and intestinal goblet cells per crypt ratio (**c**). CON, control, saline; DSS, dextran sulfate sodium; PC, positive control, probiotic; LO, low dose of fermented onion; HO, high dose of fermented onion; FO, fresh onion. Values are expressed as mean ± SEM ($n = 8$); different symbols in the bar graph are representative of statistical significance at ($p < 0.05$) measured by one-way ANOVA followed by Tukey's test for multiple comparisons. * $p < 0.05$ vs. CON, ** $p < 0.05$ vs. DSS, $^{\#}$ $p < 0.05$ vs. PC, $^{\#\#}$ $p < 0.05$ vs. LO, *** $p <$ vs. HO.

3.3. Inner Mucus Layer Thickness

The intestinal mucus layer plays a role in intestinal protection from mechanical, chemical, and biological damage, as well as in maintaining intestinal homeostasis by forming a coat that protects the intestinal cells from contact with external toxic substances, digestive enzymes, and bacteria [17]. The mucus layer also helps feces pass from the intestine and reduces the risk of bacterial invasion into the gut epithelium [18]. We collected segments of the colon together with fecal pellets and investigated inner mucus layer thickness. Using the PAS staining method, the thickness of adherent mucus was determined. Mucus layer thickness in the DSS group was decreased along with that of the PC-, LO-, and FO-fed groups as compared to that of the control group, whereas the HO group showed a higher mucus layer thickness compared to the other DSS-induced IBD groups, as shown in Figure 3a,b. However, no significant differences were observed among the groups. The mucus layer thickness is related to the number of goblet cells, because mucin-filled goblet cells secrete mucin continuously to replenish the mucus layer [18]. These data are consistent with the previous data on mucus-filled goblet cell count (Figure 2b,c), in which the high dose of fermented onion group presented with a higher number of mucus-filled goblet cells compared to the other DSS-induced groups.

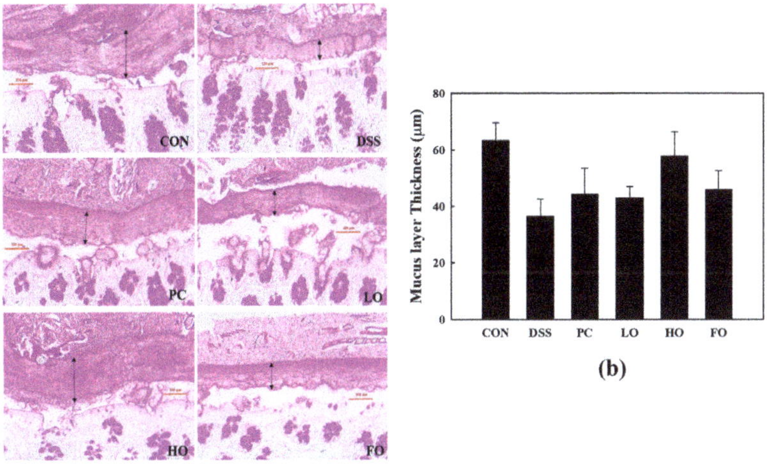

(a) (b)

Figure 3. Cross-section of the colonic tissue stained with periodic acid and Schiff (PAS) reagent showing inner and outer mucus layer, in which arrow indicates the thickness of the intestinal mucus layer. (**a**) and Inner mucus layer thickness, mm (**b**). CON, control, saline; DSS, dextran sulfate sodium; PC, positive control, probiotic; LO, low dose of fermented onion; HO, high dose of fermented onion; FO, fresh onion. Values are expressed as mean ± SEM (n = 8).

3.4. Oxidative Stress Profiling

3.4.1. Protein Carbonyl (PCO) Content

The protein carbonyls (PCOs) assay is frequently used as a biomarker of protein oxidation, measuring protein carbonyl groups, which occur when reactive oxygen species (ROS) attack the amino acid side chains, called protein carbonylation. This process produces a form of reactive ketones or aldehydes that can react with 2,4-dinitrophenylhydrazine (DNPH), and the accumulation of protein carbonyls has been observed in several human diseases [19]. In our result, the DSS-treated group exhibited a significant increase in PCO contents, while the PC, LO, and HO groups that were also treated with DSS displayed profoundly reduced PCO contents (Figure 4a) ($p < 0.05$). However, the PCO content of the FO group was similar to that of the DSS group.

3.4.2. Thiobarbituric Acid Reactive Substances (TBARS) Content

In tissue homogenate, oxidative stress was assessed by estimating the TBARS level. The TBARS assay is one method for detecting lipid oxidation [20] by analyzing malondialdehyde (MDA). The MDA reacts with thiobarbituric acid (TBA), forming a pink chromogen, which is measured at 532–535 nm [21]. In this study (Figure 4b), the DSS group exhibited an increase in TBARS contents, which was followed by the FO group, while the other groups showed similar MDA contents to the control group, while the HO-treated group showed the lowest MDA content as compared to the other groups. However, a significant difference among groups was not observed.

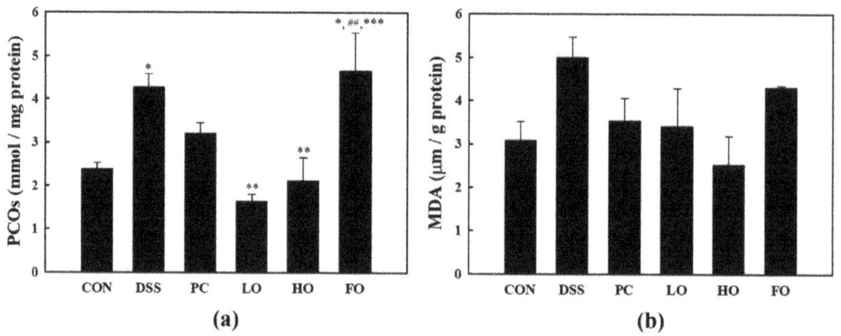

Figure 4. Oxidative stress profilings of colonic protein carbonyl content level (**a**), and colonic MDA level (**b**). CON, control, saline; DSS, dextran sulfate sodium; PC, positive control, probiotic; LO, low dose of fermented on-ion; HO, high dose of fermented onion; FO, fresh onion. Values are expressed as mean ± SEM ($n = 8$); different symbols in the bar graph are representative of statistical significance at ($p < 0.05$) measured by one-way ANOVA followed by Tukey's test for multiple comparisons. * $p < 0.05$ vs. CON, ** $p < 0.05$ vs. DSS, ## $p < 0.05$ vs. LO, *** $p <$ vs. HO.

3.5. Effect on Host Immunity

The innate and adaptive immune system plays a crucial role in the development of IBD. B cells, T cells, and regulatory T/B cells are components of the adaptive immune system. Plasma cells, which are generated from B cells, can release immunoglobulin G, including IgG, IgA, IgD, IgM, and IgE [22]. In Figure 5, IgA and IgG levels in the HO group are shown to be significantly decreased ($p < 0.05$) when compared to those of other DSS-induced groups. These results suggest that a high-dose fermented onion treatment can reduce the distinct clinical characteristics of IBD. Surprisingly, the FO-treated group exhibited a considerable increase in IgG content when compared to other groups.

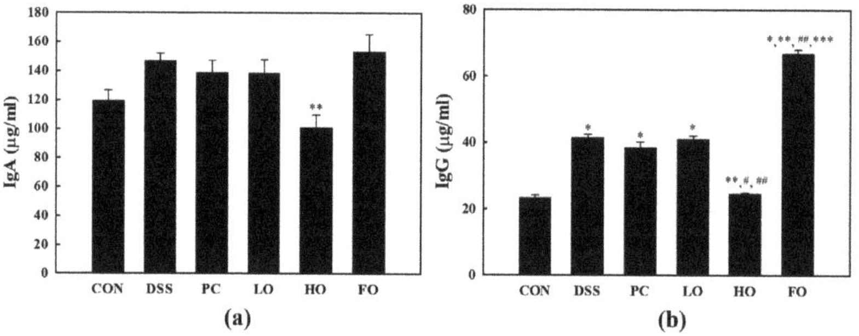

Figure 5. Effects on the plasma immunoglobulins and histology of rats administered fermented and fresh onions orally after 3 weeks of treatment. IgA level in plasma (**a**) and IgG level (**b**). CON, control, saline; DSS, dextran sulfate sodium; PC, positive control, probiotic; LO, low dose of fermented onion; HO, high dose of fermented onion; FO, fresh onion. Values are expressed as mean ± SEM ($n = 8$); different symbols in the bar graph were representative of statistical significance at ($p < 0.05$) measured by one-way ANOVA followed by Tukey's test for multiple comparisons. * $p < 0.05$ vs. CON, ** $p < 0.05$ vs. DSS, # $p < 0.05$ vs. PC, ## $p < 0.05$ vs. LO, *** $p <$ vs. HO.

3.6. Assessment of Inflammation

DSS-induced colonic inflammation is associated with the release of several inflammatory markers [23]. To determine the degree of inflammation in the colon, inflammatory markers including NF-kB, TNF-α, IL-6, and IL-1β were analyzed through Western blotting. The result showed that the NF-kB protein expression level in the colon of the DSS was higher than that in the control, PC, and HO groups (Figure 6a). However, no significant difference was observed. Since NF-kB regulates the expression of numerous inflammatory mediators including interleukins and cytokines, TNF-α, IL-6, and IL-1β were analyzed. Both fermented and fresh onion-treated groups showed slightly decreasing trends in the levels of TNF-α and IL-1β cytokines compared to the DSS-induced group, but no significant differences were observed (Figure 6b,c). For IL-6, the DSS and FO groups exhibited significantly higher levels than the control group (Figure 6d).

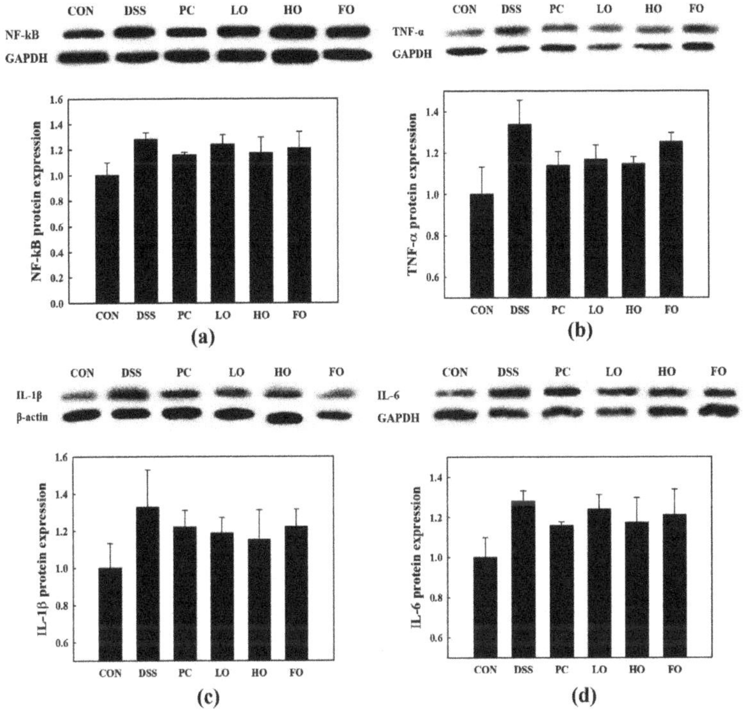

Figure 6. The Western blot analysis of inflammation-related proteins, NF-κB p65 (**a**), TNF-α (**b**), IL-1β (**c**), and IL-6 (**d**). Control, Saline; DSS, dextran sulfate sodium; PC, positive control, probiotic; LO, low dose of fermented onion; HO, high dose of fermented onion; FO, fresh onion. Values are expressed as mean ± SEM ($n = 8$).

3.7. Short-Chain Fatty Acids (SCFAs) or Gut Metabolites

Short-chain fatty acids (SCFAs) are used as a source of energy by the host colonocyte. Therefore, SCFAs are important for gut health. The microbiota in the large intestine produces SCFAs through the anaerobic fermentation of indigestible polysaccharides such as dietary fiber and resistant starch [24]. Among them, acetate, propionate, and butyrate are the most abundant (≥95%) [25]. SCFAs are primarily produced from dietary fiber derived from plant foods which humans lack the enzymes to break down. The SCFAs including acetate, propionate, and butyrate were quantified by GC-FID in fecal samples. In Figure 7, the concentration of acetate in the FO group was significantly lower than that in the other DSS groups, while the HO group had the highest level of acetate, but the difference was not statistically significant. The HO group showed a significantly increased propionate content compared to the control, PC, and LO groups ($p < 0.05$); however, this level was similar to that of the DSS group. Butyrate content was reduced in all DSS-treated groups. Among them, the HO group appeared to show the highest level of butyrate, although the difference was not significant.

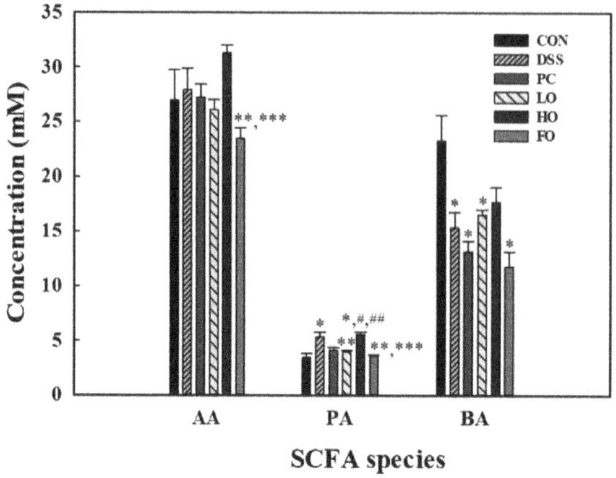

Figure 7. GC analysis of short-chain fatty acid (SCFA) content in the digestive cecum of rats. AA, acetic acid; PA, propionic acid; BA, butyric acid. Control, saline; DSS, dextran sulfate sodium; PC, positive control, probiotic; LO, low dose of fermented onion; HO, high dose of fermented onion; FO, fresh onion. Values are expressed as mean ± SEM ($n = 8$); different symbols in the bar graph were representative of statistical significance at ($p < 0.05$) measured by one-way ANOVA followed by Tukey's test for multiple comparisons. * $p < 0.05$ vs. CON, ** $p < 0.05$ vs. DSS, # $p < 0.05$ vs. PC, ## $p < 0.05$ vs. LO, *** $p <$ vs. HO.

4. Discussion

Onion (*Allium cepa*) is well known for its activities against free radicals and oxidative damage associated with chronic diseases. The phytochemicals in onion include glucosinolates, flavonoids, isoflavones, phenolic acids, phytoestrogens, and carotenoids [26], which provide health benefits such as antioxidant and anti-inflammatory activities. However, onions are considered a high-FODMAP food (10–16% fructans) [27]. FODMAPs include di- and monosaccharides such as fructose and lactose, oligosaccharides originating from fructans and galactans, and polyols [28]. The fructans in onion are poorly absorbed in the human small intestine due to the lack of digestive enzymes. Like many other dietary fibers, they are not digested in the upper gastrointestinal tract and pass through to the large intestine, where they are fermented and utilized by the colonic microbiota, yielding gas [29]. FODMAPs show a negative effect on IBD patients by triggering the symptoms of IBD such as bloating, pain, discomfort, diarrhea, and flatulence. Several studies confirmed that the reduction in FODMAPs in the diet reduces symptoms of irritable bowel syndrome (IBS) [25,28,30–32].

Recent studies have reported that fermentation increases the bioavailability of bioactive compounds including phenolic compounds, flavonoids, and SCFAs, and the consumption of fermented foods increased microbiome diversity and decreased inflammation markers in humans [10,11,33]. Fermentation may improve FODMAP properties and gut health. Fermentation increases the production of cell wall-degrading enzymes such as cellulolytic, ligninolytic, and pectinolytic enzymes that may reduce FODMAP and may increase glycosidases, which are responsible for hydrolyzing the glycosidic bond of phenolic compounds, resulting in the production of an aglycone form (a molecule without a glycoside side chain), which improves absorbability in the intestine [34]. It has been reported that the fermentation of onion raised the content of quercetin (also known as an aglycone) [9], which is one of the most common bioactive compounds found in onions.

In order to determine whether fermented onion also improves DSS-induced IBD in human gut health, we orally administered 10–20 g of fresh onion/kg rat/day to each rat, which is equivalent to about one bulb (100–200 g) of fresh onion/60 kg of person/day.

The crypts in the intestine provide stem cells with a safe environment for the renewal of the intestinal epithelium, which occurs every 3 to 4 days [35]. The high dose of fermented onion group (HO, 20 g/kg rat) showed significantly lower crypt damage compared to DSS and FO groups, in which crypts showed loss of parallelism, and irregularity in size, spacing, and shape (Figure 2). Crypt damage can lead to a reduction in the number of goblet cells that are produced by the stem cells at the bottom of the crypt. In contrast to the HO group, which showed a significantly high number of goblet cells, there is a loss of goblet cells in the DSS and FO groups. The goblet cells help to maintain the integrity of the intestinal epithelium by secreting mucin, which contributes to the ongoing replenishing of the mucus layer [18]. Therefore, the integrity of goblet cells causes increased mucus production, which is followed by an increase in the thickness of the inner mucus layer, as demonstrated in the HO group (Figure 3). To the best of our knowledge, this is the first study that reports the effect of fermented onion on mucus production.

The short-chain fatty acids (SCFAs) contents including acetate, butyrate, and propionate were analyzed (Figure 7). Among the DSS-induced groups, butyrate only increased in the LO and HO groups. In contrast, acetate and propionate were increased in all DSS-induced groups, especially in the HO group. These results are consistent with a previous study, which showed that the DSS significantly increased acetate and propionate concentrations while significantly decreasing butyrate [36]. Further research is required to fully understand this mechanism.

The major immunoglobulins that constitute most humoral and mucosal immunity are immunoglobulin A (IgA) and immunoglobulin G (IgG). IgA is the main antibody generated within the intestines, while IgG is the most prevalent antibody in the peripheral blood [37]. Previous research has shown that patients with IBD had higher levels of serum IgG in the intestinal lumen [38,39]. Our results showed that IgG levels were significantly increased in the DSS-induced groups, especially in the FO group, and dramatically decreased in the HO group. A recent study suggests that IgG might play a pathogenic role in intestinal inflammation via Fc gamma receptor (FcγR) activation by IgG, which leads to the production of IL-1β, type 17 immunity, and the exacerbation of intestinal inflammation [40]. Similar results were shown in IgA, which increased in the DSS-induced groups, except in the HO group, in which IgA decreased. DSS causes impaired mucosal barriers, which result in a leaky gut, allowing substances to easily pass through the gut. These substances have the ability to stimulate IgA secretion. Furthermore, several studies have shown that IBD patients have higher IgA levels than healthy individuals [22,41].

Moreover, we found that the FO groups had significantly higher IL-6 and a slight increase in TNF-α and IL-1β protein expression. In fermented onion, on the other hand, improvement in these inflammation factors was shown. These findings are consistent with all data previously mentioned, suggesting that in DSS-induced IBD rats, fermented onions had better beneficial impacts on gut health than fresh onions, considering that fermentation may reduce FODMAP properties and induce anti-inflammation.

The differences in effects between fermented and fresh onions on gut health might be due to the degradation of FODMAP compounds by fermentation, which results in an increase in butyrate in the fermented onion groups, as butyrate has been shown to control oxidative stress in numerous studies [42]. The differences in the chemical compounds of fresh and fermented onions may be another factor that might play role in the different effects of these two types of onions. In a prior study, chemical analysis revealed that pickled onions had much higher levels of allyl methyl trisulfide, 3,5-diethyl-1,2,4-trithiolane, and 2-hexyl-5-methyl-3(2H)-furanone than fresh onions did [43]. These compounds have also been shown in several studies to have a positive effect on IBD in rats. This may lead to a future study in which we may extensively explore the relationship between the different components of fresh and fermented onion and gut health.

5. Conclusions

In this study, we examined the effects of fermented and fresh onions on gut health in DSS-induced IBD rats. The DSS-induced IBD rats exhibited mucosal inflammation with extensive depletion of goblet cells in the mucosa. We found that fermented onion is associated with repressed oxidative stress and inflammation in the DSS-induced IBD model. Furthermore, fermented onion treatment significantly reduced the mucosal damage by inducing a significant increase in the goblet cell number and the thickness of the mucus layer. Fresh onion, on the other hand, caused a considerable reduction in goblet cells.

Author Contributions: Conceptualization, K.-S.H., N.S., S.C., K.-H.P., Y.-J.P. and K.P.; methodology, N.S., S.C., H.-M.S., D.S. (Doyoung Song) and K.-H.P.; experimental analysis and investigation, N.S., S.C., M.S. and D.S. (Divya Sharma); writing—original draft preparation, N.S.; writing—review and editing, N.S., S.C. and K.-S.H.; supervision, K.-S.H. All authors have read and agreed to the published version of the manuscript.

Funding: This study was supported by a Regional Infrastructure Specialized Support Project from the Ministry of SMEs and Startups (MSS) of Korean government. P0013018.

Institutional Review Board Statement: The animal study protocol was approved by the Institutional Animal Care and Use Committee (IACUC) of Mokpo National University (Muan, Republic of Korea). MNU-IACUC-2020-016.

Informed Consent Statement: Not applicable.

Data Availability Statement: Not applicable.

Conflicts of Interest: Y.-J.P. is a stockholder of a company that specializes in onion processing. The other authors declare no conflicts of interest.

References

1. Limdi, J.K. Dietary practices and inflammatory bowel disease. *Indian J. Gastroenterol.* **2018**, *37*, 284–292. [CrossRef] [PubMed]
2. Bhattacharyya, A.; Chattopadhyay, R.; Mitra, S.; Crowe, S.E. Oxidative stress: An essential factor in the pathogenesis of gastrointestinal mucosal diseases. *Physiol. Rev.* **2014**, *94*, 329–354. [CrossRef] [PubMed]
3. Antoni, L.; Nuding, S.; Wehkamp, J.; Stange, E.F. Intestinal barrier in inflammatory bowel disease. *World J. Gastroenterol.* **2014**, *20*, 1165–1179. [CrossRef] [PubMed]
4. McLean, L.P.; Cross, R.K. Adverse events in IBD: To stop or continue immune suppressant and biologic treatment. *Expert Rev. Gastroenterol. Hepatol.* **2014**, *8*, 223–240. [CrossRef]
5. Andersen, V.; Olsen, A.; Carbonnel, F.; Tjønneland, A.; Vogel, U. Diet and risk of inflammatory bowel disease. *Dig. Liver Dis.* **2012**, *44*, 185–194. [CrossRef]
6. Hong, Z.; Piao, M. Effect of Quercetin Monoglycosides on Oxidative Stress and Gut Microbiota Diversity in Mice with Dextran Sodium Sulphate-Induced Colitis. *Biomed Res. Int.* **2018**, *2018*, 8343052. [CrossRef]
7. Yang, D.; Wang, T.; Long, M.; Li, P. Quercetin: Its Main Pharmacological Activity and Potential Application in Clinical Medicine. *Oxid. Med. Cell. Longev.* **2020**, *2020*, 8825387. [CrossRef]
8. Panche, A.N.; Diwan, A.D.; Chandra, S.R. Flavonoids: An overview. *J. Nutr. Sci.* **2016**, *5*, 9–10. [CrossRef]
9. Yang, E.J.; Kim, S.I.; Park, S.Y.; Bang, H.Y.; Jeong, J.H.; So, J.H.; Rhee, I.K.; Song, K.S. Fermentation enhances the in vitro antioxidative effect of onion (*Allium cepa*) via an increase in quercetin content. *Food Chem. Toxicol.* **2012**, *50*, 2042–2048. [CrossRef]
10. Annunziata, G.; Arnone, A.; Ciampaglia, R.; Tenore, G.C.; Novellino, E. Fermentation of foods and beverages as a tool for increasing availability of bioactive compounds. focus on short-chain fatty acids. *Foods* **2020**, *9*, 999. [CrossRef]
11. Wastyk, H.C.; Fragiadakis, G.K.; Perelman, D.; Dahan, D.; Merrill, B.D.; Yu, F.B.; Topf, M.; Gonzalez, C.G.; Van Treuren, W.; Han, S.; et al. Gut-microbiota-targeted diets modulate human immune status. *Cell* **2021**, *184*, 4137–4153.e14. [CrossRef]
12. Hill, D.; Sugrue, I.; Tobin, C.; Hill, C.; Stanton, C.; Ross, R.P. The *Lactobacillus casei* group: History and health related applications. *Front. Microbiol.* **2018**, *9*, 5–6. [CrossRef]
13. Horvat, I.B.; Gobin, I.; Kresović, A.; Hauser, G. How can probiotic improve irritable bowel syndrome symptoms. *World J. Gastrointest. Surg.* **2021**, *13*, 923. [CrossRef] [PubMed]
14. Reznick, A.Z.; Packer, L. Oxidative damage to proteins: Spectrophotometric method for carbonyl assay. *Methods Enzymol.* **1994**, *233*, 357–363. [CrossRef] [PubMed]
15. Zhao, G.; Nyman, M.; Jönsson, J.Å. Rapid determination of short-chain fatty acids in colonic contents and faeces of humans and rats by acidified water-extraction and direct-injection gas chromatography. *Biomed. Chromatogr.* **2006**, *20*, 674–682. [CrossRef]
16. Chassaing, B.; Aitken, J.D.; Malleshappa, M.; Vijay-Kumar, M. DSS Protocol. *Curr. Protoc. Immunol.* **2015**, *27*, 1–19. [CrossRef]

17. Tarakci, N.G.; Erdem, N.Z.; Dumen, E. Probiotic Foods Are Effective on Weight Loss, Biochemical Parameters, and Intestinal Microbiota in Wistar Albino Rats with Obese Microbiota. *Int. J. Clin. Pract.* **2022**, *2022*, 4569100. [CrossRef] [PubMed]
18. Omotoso, G.O.; Muonagolu, J.N.; Enaibe, B.U. Histological Evaluation of the Jejunum and Ileum of Rats after Administration of High Dose Garlic Aqueous Extract. *Int. J. Health Sci.* **2012**, *6*, 135–140. [CrossRef] [PubMed]
19. Celi, P.; Gabai, G. Oxidant/antioxidant balance in animal nutrition and health: The role of protein oxidation. *Front. Vet. Sci.* **2015**, *2*, 48. [CrossRef] [PubMed]
20. Abeyrathne, E.D.N.S.; Nam, K.; Ahn, D.U. Analytical methods for lipid oxidation and antioxidant capacity in food systems. *Antioxidants* **2021**, *10*, 1587. [CrossRef]
21. Kumar, S.; Krishna Chaitanya, R.; Preedy, V.R. *Assessment of Antioxidant Potential of Dietary Components*; Elsevier Inc.: Amsterdam, The Netherlands, 2017; ISBN 9780128098547.
22. Song, D.J.; Shen, J.; Chen, M.H.; Liu, Z.J.; Cao, Q.; Hu, P.J.; Gao, X.; Qian, J.M.; Wu, K.C.; Lai, L.J.; et al. Association of Serum Immunoglobulins Levels With Specific Disease Phenotypes of Crohn's Disease: A Multicenter Analysis in China. *Front. Med.* **2021**, *8*, 621337. [CrossRef] [PubMed]
23. Marrero, J.A.; Matkowskyj, K.A.; Yung, K.; Hecht, G.; Benya, R.V. Dextran sulfate sodium-induced murine colitis activates NF-κB and increases galanin-1 receptor expression. *Am. J. Physiol. Gastrointest. Liver Physiol.* **2000**, *278*, 797–804. [CrossRef] [PubMed]
24. Silva, Y.P.; Bernardi, A.; Frozza, R.L. The Role of Short-Chain Fatty Acids From Gut Microbiota in Gut-Brain Communication. *Front. Endocrinol.* **2020**, *11*, 25. [CrossRef]
25. Shepherd, S.J.; Lomer, M.C.E.; Gibson, P.R. Short-chain carbohydrates and functional gastrointestinal disorders. *Am. J. Gastroenterol.* **2013**, *108*, 707–717. [CrossRef] [PubMed]
26. Septembre-Malaterre, A.; Remize, F.; Poucheret, P. Fruits and vegetables, as a source of nutritional compounds and phytochemicals: Changes in bioactive compounds during lactic fermentation. *Food Res. Int.* **2018**, *104*, 86–99. [CrossRef] [PubMed]
27. Saeed, F.; Pasha, I.; Arshad, M.U.; Muhammad Anjum, F.; Hussain, S.; Rasheed, R.; Nasir, M.A.; Shafique, B. Physiological and Nutraceutical Perspectives of Fructan. *Int. J. Food Prop.* **2015**, *18*, 1895–1904. [CrossRef]
28. Prince, A.C.; Myers, C.E.; Joyce, T.; Irving, P.; Lomer, M.; Whelan, K. Fermentable carbohydrate restriction (Low FODMAP Diet) in clinical practice improves functional gastrointestinal symptoms in patients with inflammatory bowel disease. *Inflamm. Bowel Dis.* **2016**, *22*, 1129–1136. [CrossRef]
29. Kalantar-Zadeh, K.; Berean, K.J.; Burgell, R.E.; Muir, J.G.; Gibson, P.R. Intestinal gases: Influence on gut disorders and the role of dietary manipulations. *Nat. Rev. Gastroenterol. Hepatol.* **2019**, *16*, 733–747. [CrossRef]
30. Barrett, J.S.; Gearry, R.B.; Muir, J.G.; Irving, P.M.; Rose, R.; Rosella, O.; Haines, M.L.; Shepherd, S.J.; Gibson, P.R. Dietary poorly absorbed, short-chain carbohydrates increase delivery of water and fermentable substrates to the proximal colon. *Aliment. Pharmacol. Ther.* **2010**, *31*, 874–882. [CrossRef]
31. Ong, D.K.; Mitchell, S.B.; Barrett, J.S.; Shepherd, S.J.; Irving, P.M.; Biesiekierski, J.R.; Smith, S.; Gibson, P.R.; Muir, J.G. Manipulation of dietary short chain carbohydrates alters the pattern of gas production and genesis of symptoms in irritable bowel syndrome. *J. Gastroenterol. Hepatol.* **2010**, *25*, 1366–1373. [CrossRef]
32. Murray, K.; Wilkinson-Smith, V.; Hoad, C.; Costigan, C.; Cox, E.; Lam, C.; Marciani, L.; Gowland, P.; Spiller, R.C. Differential effects of FODMAPs (Fermentable Oligo-, Di-, Mono-Saccharides and Polyols) on small and large intestinal contents in healthy subjects shown by MRI. *Am. J. Gastroenterol.* **2014**, *109*, 110–119. [CrossRef] [PubMed]
33. Paramithiotis, S.; Das, G.; Shin, H.S.; Patra, J.K. Fate of Bioactive Compounds during Lactic Acid Fermentation of Fruits and Vegetables. *Foods* **2022**, *11*, 733. [CrossRef] [PubMed]
34. Huynh, N.T.; Van Camp, J.; Smagghe, G.; Raes, K. Improved release and metabolism of flavonoids by steered fermentation processes: A review. *Int. J. Mol. Sci.* **2014**, *15*, 19369–19388. [CrossRef] [PubMed]
35. Car, B.D.; Eng, V.M.; Everds, N.E.; Bounous, D.I. Clinical Pathology of the Rat. In *The Laboratory Rat*; Academic Press: Cambridge, MA, USA, 2006; pp. 127–146. [CrossRef]
36. Borton, M.A.; Sabag-Daigle, A.; Wu, J.; Solden, L.M.; O'Banion, B.S.; Daly, R.A.; Wolfe, R.A.; Gonzalez, J.F.; Wysocki, V.H.; Ahmer, B.M.M.; et al. Chemical and pathogen-induced inflammation disrupt the murine intestinal microbiome. *Microbiome* **2017**, *5*, 47. [CrossRef]
37. Lin, R.; Chen, H.; Shu, W.; Sun, M.; Fang, L.; Shi, Y.; Pang, Z.; Wu, W.; Liu, Z. Clinical significance of soluble immunoglobulins A and G and their coated bacteria in feces of patients with inflammatory bowel disease. *J. Transl. Med.* **2018**, *16*, 1–15. [CrossRef] [PubMed]
38. Lee, H.S.; Lee, K.J. Immunoglobulin G4-related immune responses to common food antigens in patients with ulcerative colitis and Crohn's disease. *Turkish J. Gastroenterol.* **2019**, *30*, 408–414. [CrossRef]
39. Wang, Z.; Zhu, M.; Luo, C.; Zhen, Y.; Mu, J.; Zhang, W.; Ouyang, Q.; Zhang, H. High level of IgG4 as a biomarker for a new subset of inflammatory bowel disease. *Sci. Rep.* **2018**, *8*, 10018. [CrossRef]
40. Castro-Dopico, T.; Dennison, T.W.; Ferdinand, J.R.; Mathews, R.J.; Fleming, A.; Clift, D.; Stewart, B.J.; Jing, C.; Strongili, K.; Labzin, L.I.; et al. Anti-commensal IgG Drives Intestinal Inflammation and Type 17 Immunity in Ulcerative Colitis. *Immunity* **2019**, *50*, 1099–1114.e10. [CrossRef]
41. Kazemi-Shirazi, L.; Gasche, C.H.; Natter, S.; Gangl, A.; Smolen, J.; Spitzauer, S.; Valent, P.; Kraft, D.; Valenta, R. IgA autoreactivity: A feature common to inflammatory bowel and connective tissue diseases. *Clin. Exp. Immunol.* **2002**, *128*, 102–109. [CrossRef]

42. Hamer, H.M.; Jonkers, D.M.A.E.; Bast, A.; Vanhoutvin, S.A.L.W.; Fischer, M.A.J.G.; Kodde, A.; Troost, F.J.; Venema, K.; Brummer, R.J.M. Butyrate modulates oxidative stress in the colonic mucosa of healthy humans. *Clin. Nutr.* **2009**, *28*, 88–93. [CrossRef]
43. Farag, M.A.; Tawfike, A.F.; Donia, M.S.; Ehrlich, A.; Wessjohann, L.A. Influence of pickling process on *Allium cepa* and citrus limon metabolome as determined via mass spectrometry-based metabolomics. *Molecules* **2019**, *24*, 928. [CrossRef] [PubMed]

applied sciences

MDPI

Article

The Influence of Fruit Pomaces on Nutritional, Pro-Health Value and Quality of Extruded Gluten-Free Snacks

Dorota Gumul [1,*], Wiktor Berski [1] and Tomasz Zięba [2]

[1] Department of Carbohydrate Technology and Cereal Processing, Faculty of Food Technology, University of Agriculture in Kraków, Mickiewicza 21, 31-120 Kraków, Poland
[2] Department of Food Storage, The Faculty of Life Sciences and Technology, Wrocław University of Environmental and Life Sciences, Chełmońskiego 37, 51-630 Wrocław, Poland
* Correspondence: rrgumul@cyf-kr.edu.pl

Featured Application: Obtained results give insight into health promoting properties of potential gluten free cornmeal-based snacks.

Abstract: The processing of fruit generates large amounts of different by-products, such as pomace. The extrusion process gives an opportunity for their utilization as a good source of pro-health components. Therefore, this research focused on the utilization of fruit pomaces (cherries, blackcurrants, and chokeberries) as a value-added component of extruded corn snacks. The effect of the level of pomace addition on the content of bioactive polyphenols and nutritional value in cornmeal-based extrudates, as well as antioxidant capacity, was investigated. Additionally, the influence of fruit pomace on the quality of extruded gluten-free snacks was also investigated. It was found that pomace can be a good pro-health addition to corn snacks due to the enrichment of bioactive compounds and dietary fiber in this product. Especially valuable proved to be chokeberry pomace added at a 20% level. Such additions to snacks caused an increase in the content of total phenolic compounds, phenolic acids, flavonoids, flavonols, anthocyanins, and antioxidant activity, respectively, by about 10 times, 2 times, 5 times, 2 times, 10 times, and 5 times, as compared to control snacks. It was observed that the addition of chokeberry pomace did not worsen the physical properties (WBC, hardness, and expansion ratio) of the resulting snacks, which affect the quality of the obtained product. Therefore, such snacks could be recommended for commercial production in order to increase the availability of gluten-free products for people with celiac disease.

Keywords: corn snacks; dietary fiber; extrusion; fruit pomace; nutritional compounds; organic waste utilization; phenolic compounds; quality features

1. Introduction

Food processing leads to inevitable losses. According to a report, "Food wastage footprint: Impacts on natural resources–Summary report" [1], fruit processing, packing, distribution, and consumption respectively generate a huge quantity of fruit waste, approximately 1.81, 6.53, and 32.0 million tones worldwide, and most of this is being disposed of either by composting or dumping in landfills, causing environmental pollution [2]. In fruit and vegetable processing, a significant problem is the management or disposal of waste products—the parts not used during technological processing. This is between 10 and 35% of the weight of the processed raw material and includes peel, seeds, and pomace. By-products of fruit processing are created during the production of juices, pulps, and jams. In the juice production process, the pomace mass depends on the pressing efficiency. During traditional pressing, the proportion of pomace is 20–25% of the initial raw material weight [3]. The application of enzyme preparations that liquefy the pulp and the leaching of the pomace with water allows for a reduction in the pomace share to about 12% of the initial fruit weight. Population growth and dietary changes contribute to greater consumption of fruit and vegetables, which in turn results in an increase in the amount of waste associated with their processing. In Europe, by-products from fruit processing account for 8% of all waste from the food sector. In developing countries, the situation is much worse because the by-products of fruit processing are treated as negligible and redundant compared to processed fruit [3]. Other studies indicate the generation of about 50% of by-products during fruit and vegetable processing in the forms of unripe or damaged fruits and vegetables, cores, peels, and pomaces [4,5].

Poland is an important producer of fruits and products of their processing, such as juices. Such production generates a large quantity of by-products, such as pomaces, that are usually utilized as feed or as raw materials for microbial growth [6,7]. But from a nutritional point of view, fruit pomaces can be treated as an abundant source of health-promoting components, primarily polyphenols and dietary fiber (DF) [8–12]. Although pomaces are not expected to be consumed alone, they can be used as a valuable ingredient to extend the health-promoting properties of food. Such an application will allow for the utilization of a by-product, and at the same time, a value-added product will be created. Extruded snacks seem to fit within this objective, and therefore the research has concentrated on the utilization of fruit pomaces as an additive for extruded snacks with an increased level of health-promoting components (polyphenols and fibers) [13,14].

Extrusion is a high-temperature, short-term physical treatment during which food polymers (usually a cheap one, such as starch or cornmeal) at a relatively low moisture level are subjected to significant structural changes by the combined action of high temperature, pressure, and shear forces. As a consequence, a new product with completely new properties and qualities is created [14], such as breakfast cereals, crisp bread, baby and infant foodstuffs, snack-type foods, meat analogs, textured proteins, noodles, and pasta that could be manufactured in this way [15,16]. However, starch- or cornmeal-based extrudates are characterized by low nutritional value and thus relatively weak health-supporting properties. Moreover, the extrusion process leads to the loss of polyphenols, i.e., components with broad pharmacological effects [17–21]. Therefore, it is suggested to supplement raw cereal material with natural or synthetic phenolic compounds or enrich it with vegetables, fruits, or other ingredients before extrusion [17,22–26]. It should also be taken into consideration that extrusion gives a great opportunity to utilize raw material products with no previously recognized greater economic importance, such as fruit pomace. Moreover, such an approach could be seen as an implementation of the zero-waste strategy [27].

An additional aspect of this research was to pay particular attention to gluten-free products, which are low in many nutrients (particularly protein, ash, and dietary fiber), and by using enriching additives (fruit pomace), we enrich them with health-promoting ingredients, especially polyphenols, which are extremely valuable in the context of the increase in the incidence of celiac disease worldwide [28–30]. In this work, cornmeal snacks enriched with fruit pomace were investigated, creating a new assortment of gluten-free products with improved pro-health and nutritional value. According to Hooper et al. [28], up to 87% of adult celiac disease patients suffer from vitamin (B6, B12, A, and D), mineral (Zn, Cu, and Fe), and dietary fiber (it protects against hypertension and cancer) deficiency. Such a deficiency results in an increased risk of cancer, osteoporosis, and infertility occurrences [31]. Therefore, it is crucial to enrich gluten-free products with components rich in dietary fiber, polyphenols, and minerals, and as a result, fruit pomace can be recognized.

Therefore, the aim of this research was to investigate the effect of the fruit pomaces (cherries, blackcurrants, and chokeberries) at different levels of addition on the content of bioactive compounds and dietary fiber in gluten-free extruded corn snacks. Moreover, the antioxidant activity of such products was evaluated. Moreover, the functional properties of this type of product were analyzed, i.e., color, density, expansion, water absorption, and hardness. Additionally, the nutritional composition of the final product was analyzed.

2. Materials and Methods

2.1. Materials

The study material consisted of fruit pomaces (chokeberry—PCHB; cherry—PCH; blackcurrant—PBC) provided in dried form from a local fruit processing plant (Hortino, Leżajsk, Poland) and cornmeal (Sante, Warsaw, Poland). The fruit pomaces were tested for their bioactive compound content, and the obtained data were collected in Tables 1 and 2. The main research material was extruded corn snacks with the addition of fruit pomace. The control extruded sample was made of pure cornmeal; in regular samples, cornmeal was partially substituted at 5, 10, and 20% levels by fruit pomaces. Snacks produced with the addition of pomaces were denoted as ECHB-05, ECHB-10, and ECHB-20 (chokeberry), ECH-05, ECH-10, and ECH-20 (cherry), and EBC-05, EBC-10, and EBC-20 (blackcurrant pomace) for 5, 10, and 20% replacement of cornmeal, respectively.

Extrusion was performed in a single-screw laboratory extruder 20DN (Brabender, Duisburg, Germany), applying the following parameters: screw speed—190 rpm, dye diameter—4 mm, compression ratio—1:3, and temperature profile—100–120–140 °C. The moisture level of all premixes was equilibrated at 14%.

2.2. Methods

Chemical Analysis

The following analyses of pro-health compounds, bioactive components, and chemical composition were performed on pomaces and resulting extrudates.

2.2.1. Chemical Composition

Protein content (Nx5.7) was determined by the Kjeldahl method (AOAC method No. 920.87) using the Kjeltec 2200 extraction unit (Foss, Hillerød, Denmark), fat content by the Soxhlet method (AOAC method No. 953.38) using the Soxtec Avanti 2055 (Foss, Denmark), and ash (AOAC method: 920.183) and total sugar content (and AOAC method No. 930.05) were determined according to AOAC [32].

2.2.2. Dietary Fiber Content

The dietary fiber (DF) content of non-starch polysaccharides, i.e., soluble (SDF) and insoluble (IDF) dietary fiber, was determined by the enzymatic-gravimetric method [33]. TDF was calculated as the sum of soluble and insoluble fractions. Ground samples were dispersed in water and treated with alpha-amylase, protease, and glucosidase to remove starch and protein. For TDF, enzyme digestate is treated with alcohol to precipitate soluble dietary fiber before filtering, and TDF residue is washed with alcohol and acetone, dried, and weighed. For IDF and SDF, enzyme digestate is filtered, and residue (IDF) is washed with warm water, dried, and weighed. For SDF, the combined filtrate and washes are precipitated with alcohol, filtered, dried, and weighed. TDF, IDF, and SDF residue values are corrected for protein, ash, and blank.

2.2.3. Antioxidants Content and Antioxidant Activity

- Extraction Procedure

Antioxidant content and antioxidant activity were determined in ethanol extracts. 0.6 g of sample was dissolved in 30 mL of 80% ethanol, shaken in the dark for 120 min (electric shaker: type WB22, Memmert, Schwabach, Germany), and separated for 15 min at 1050× *g* using a centrifuge (type MPW-350, MPW MED Instruments, Warsaw, Poland). The supernatant was decanted and stored at −20 °C for further analysis.

- Total Phenolic Content (TPC)

Total phenolic content (TPC) was determined by a spectrophotometric method using the Folin-Ciocalteu reagent, according to Singleton et al. [34]. The ether extract was 10 times diluted to a volume of 50 mL with distilled water in a volumetric flask. Five milliliters of extract was combined with 0.25 mL of Folin–Ciocalteau reagent and 0.5 mL of 7% Na_2CO_3.

The content was vortexed (WF2, Janke, and Kunkel, Staufen, Germany) and stored for 30 min in the dark. The absorbance was measured using Helios Gamma 100–240 (Runcorn, UK), at the wavelength λ = 760 nm. The results were expressed as mg catechin or gallic acid/100 g d.m.

2.2.4. Determination of Total Polyphenols

TPC, phenolic acids, flavonols, and anthocyanins were analyzed by a spectrophotometric method according to Mazza et al. [35] with modification by Oomah et al. [36]. Briefly, 0.1 mL of ethanol extract was taken into a test tube, and 2.4 mL of 2% HCl in 75% ethanol was added. The content of the tube was then mixed on a vortex (type WF2, Janke & Kunkel, Staufen, Germany), and the absorbance was measured in a spectrophotometer (Helios Gamma, 100–240, Runcorn, England) at λ = 280 nm (TPC), λ = 320 nm (phenolic acids), λ = 360 nm (flavonols), and λ = 520 nm (anthocyanins). A blank sample was prepared with 2% HCl in 75% ethanol and 80% ethanol. Results were expressed for TPC: mg catechin/100 g DM, phenolic acids: mg ferulic acid/100 g dm, flavonols: mg quercetin/100 g dm, and anthocyanins: mg glycoside-3-cyanidin/100 g dm.

2.2.5. Determination of Flavonoid Content

Determination of flavonoid content was performed using a method proposed by El Hariri et al. [37]. Specifically, 0.5 mL of ethanol extract was taken into a test tube, and 1.8 mL of distilled water and 0.2 mL of 2-aminoethyl diphenylborate reagent were added. The content of the tube was vortexed (Vortex type WF2, Janke & Kunkel, Staufen, Germany), and the absorbance was measured using a spectrophotometer (Helios Gamma, 100–240, Runcorn, England) at λ = 404 nm. At the same time, a blank test was performed by mixing 0.5 mL of 80% ethanol, 1.8 mL of distilled water, and 0.2 mL of 2-aminoethyl diphenylborate reagent. Flavonoid content was expressed as mg rutin/100 g d.m.

2.2.6. Antioxidant Activity (AA)

Antioxidant activity was assessed using analytical methods, namely ABTS [38]. ABTS was dissolved in the water to a 7 mM concentration. ABTS radical cation (ABTS+•) was produced by reacting ABTS stock solution with 2.45 mM potassium persulfate (final concentration) and allowing the mixture to stand in the dark at room temperature (12–16 h) before use. The bleaching rate of ABTS+• in the presence of the sample was monitored at 734 nm using a Helios Gamma 100–240 (Runcorn, UK) spectrophotometer. The ABTS+• solution was diluted in PBS buffer (pH 7.4) to give an absorption value of 0.700 ± 0.05 for the analysis of ethanol extracts. Volumes of 2.00 mL of ABTS+• and respective ethanol extracts in PBS buffer solution were used. The ABTS+• bleaching was monitored at 37 °C, and the discoloration after 6 min was used as the measure of antioxidant activity. Radical scavenging activity was measured as Trolox equivalent antioxidant capacity (mg Trolox per g of sample dm). Trolox solutions used for the calibration curve were in the concentration range 0–2.5 mM (R^2 = 0.9957). Additionally, AA was expressed as EC_{50} (mg/mL).

2.2.7. Physical Properties of Extruded Corn Snacks

Determination of the Density of Extrudates

The density of extrudates was determined using the displacement method [39] with rapeseed in two cylinders (1 mL graduation) and a RADWAG WPS 600/C balance (accuracy 0.001 g) (Radom, Poland). The commonly used relation mass/volume = density was applied.

Determination of Expansion Ratio of Extrudates

Ten pieces of extrudate were taken from the sample and separated for testing, and the diameter was measured in ten replicates using a caliper (accuracy 0.02 mm) [39]. The expansion ratio (er) was calculated as the ratio of the average diameter of the extrudates to the diameter of the extruder die nozzle using the following formula:

$$er = d/D \tag{1}$$

where:

d—extrudate diameter (mm)
D—extruder die nozzle (mm)

Water Binding Capacity (WBC) of Extrudates

One gram of whole extrudates was weighed, and distilled water at room temperature was added [37]. The whole was rehydrated for 1 h, after which time the extrudates were weighed and the amount of water bound by them was obtained from the difference in masses. The result was then converted to a ratio of 1 g of product. The determination was carried out in at least two replicates, and the result was taken as their average.

Extrudates Texture Profile Analysis (TPA) Using TA-XT Plus Texture Analyzer (Stable Micro Systems, Surrey, UK)

Texture analysis was carried out using a TA-XT Plus texture analyzer equipped with an HDP/KS5 Kramer Shear Cell 2 Blade (Stable Micro Systems Ltd. Manufactured in Surrey, UK). The device was calibrated prior to the test. A test was performed to cut through extrudates with a length of approximately 35 mm. A crushing speed of 2 mm/s and a distance of 20 mm were applied. The test was carried out with fifteen repetitions for each sample. By examining the crumb texture profile of the extrudates, its maximum cutting force was determined and expressed as hardness.

Instrumental Color Measurement in the CIE L* a* b* System

Instrumental measurement of the color components was performed with a Konica Minolta Inc. colorimeter, using the CIE L* a* b* system [38]. Before measuring the test samples, the instrument was calibrated on a white reference plate ($L^* = 98.45$, $a^* = -0.10$, and $b^* = -0.13$). Measurements were carried out with the measuring head set at 8 mm, the observer at 2°, and illumination D65, which is daylight. The parameters analyzed were: brightness L* ($L = 0$ black, $L = 100$ white), red saturation a* (−a share of green, +a share of red), and yellow saturation b* (−b share of blue, +b share of yellow).

Furthermore, the total color difference (ΔE) and whiteness index (WI) were calculated [40]. ΔE was computed as the Euclidean distance between two points in the three-dimensional space defined by L*, a*, and b* using the following equation:

$$\Delta E = ((\Delta a^*)^2 + (\Delta b^*)^2 + (\Delta L^*)^2)^{0.5} \tag{2}$$

where Δa^*, Δb^*, and ΔL^* were the differences in the values of the respective parameters between the investigated samples.

The whiteness index (WI) indicates the whiteness degree, and was calculated as:

$$WI = 100 - ((100 - L)^2 + (\Delta a^*)^2 + (\Delta b^*)^2)^{0.5} \tag{3}$$

Statistical Evaluation of Results

The results were statistically compared based on Duncan's test at a significance level of 0.05 using Statistica 8.0PL software. Moreover, the correlation coefficient was calculated. All measurements were performed at least in duplicate. The results were expressed as the mean ± standard deviation.

3. Results and Discussion

3.1. Characteristics of Fruit Pomaces

The content of bioactive compounds and chemical composition in fruit pomaces is summarized in Tables 1 and 2. The highest protein and fat content was found in blackcurrant pomace because it contained twice as much of these components as compared to chokeberry pomace. The highest amount of total carbohydrates was found in chokeberry pomace, 4 times and 3 times higher than in cherry and currant pomace, respectively. The chokeberry pomace contained the most minerals, 20% more than the other pomace. These results are in agreement with those of Kawecka and Galus [3] and Jurendic and Šcetar [41], who determined protein, fat, and ash in chokeberry pomace at levels of 4.9, 2.9, and 1.4–3.9 g/100 g dm, respectively. In the study of Alba et al. [42], the amount of protein, fat, and ash in currant pomace was 11.1–13.3, 5.9–10.9, and 2.8–3.8 g/100 g dm, respectively.

Table 1. Total phenolic, antioxidant activity, dietary fiber content, and chemical composition of fruit pomaces.

Pomace	Total Phenolic [1] mg Catechin/g dm	ABTS EC_{50} (mg/mL)	ABTS (TEAC) (mgTx/g dm)	Content of Dietary Fiber (g/100 g dm)		Chemical Composition (g/100 g dm)			
				Soluble Fraction	Insoluble Fraction	Protein	Total Sugar	Fat	Ash
PCHB	36.35 ± 0.30 ^{c}	0.47 ± 0.01 ^{a}	19.15 ± 0.09 ^{c}	5.71 ± 0.12 ^{b}	61.18 ± 0.00 ^{b}	6.80 ± 0.20 ^{a}	38.89 ± 0.30 ^{c}	4.70 ± 0.11 ^{b}	3.80 ± 0.00 ^{c}
PCH	18.87 ± 0.53 ^{a}	0.53 ± 0.00 ^{c}	18.15 ± 0.17 ^{a}	4.50 ± 0.23 ^{a}	44.20 ± 0.37 ^{a}	13.01 ± 0.12 ^{b}	10.71 ± 0.45 ^{a}	3.18 ± 0.14 ^{a}	2.8 ± 0.17 ^{a}
PBC	26.73 ± 0.71 ^{b}	0.51 ± 0.00 ^{b}	18.75 ± 0.23 ^{b}	5.36 ± 0.27 ^{b}	60.15 ± 1.11 ^{b}	17.50 ± 0.00 ^{c}	15.23 ± 0.17 ^{b}	10.10 ± 0.00 ^{c}	3.20 ± 0.08 ^{b}

The values in the column marked with the same letter are not significantly different at $p = 0.05$. [1] According to Singleton et al. [34].

Table 2. Selected phenolic compounds content in fruit pomace.

Pomace	Total Phenolic [1] mg Catechin/g dm	Phenolic Acid (mg Ferulic Acid/g dm)	Flavonoids (mg Rutin/g dm)	Flavonols (mg Quercetin/g dm)	Anthocyanins (mg Cyanidin-3-Glucoside/g dm)
PCHB	31.49 ± 0.23 ^{c}	2.163 ± 0.197 ^{b}	11.437 ± 0.053 ^{c}	1.714 ± 0.046 ^{c}	10.841 ± 0.332 ^{c}
PCH	5.89 ± 0.11 ^{a}	0.441 ± 0.020 ^{a}	2.289 ± 0.017 ^{b}	0.298 ± 0.00 ^{a}	0.475 ± 0.023 ^{a}
PBC	6.48 ± 0.34 ^{b}	0.460 ± 0.024 ^{a}	1.321 ± 0.032 ^{a}	0.429 ± 0.021 ^{b}	1.813 ± 0.162 ^{b}

The values in the column marked with the same letter are not significantly different at $p = 0.05$. [1] According to Mazza et al. [35], with modification by Oomah et al. [36].

It was found that the total polyphenol content was highest in PCHB and lowest in PCH. The content of this type of bioactive compound in PBC was lower by 26% as compared to PCHB and higher by 42% in relation to PCH (Table 1). This relationship was confirmed by Nawirska et al. [10]. It was found in their studies that PCHB was characterized by a high content of polyphenols, about 21% higher as compared to PBC. According to Sójka and Król [11], the amount of polyphenols ranged from 18.55–22.85 mg epicatechin/g in PBC, and in PCH, it was at 6.35 mg epicatechin/100 g [8]. According to Mayer-Miebach et al. [9], the amount of total polyphenols in PCHB ranged between 31 and 63 mg catechins/g. So, it could be stated that the total phenolic content identified in this study was consistent with the data provided by other authors.

The content of phenolic acids (PA) was the highest in PCHB, and the amount in PBC was four times lower, and in PCH it was five times lower as compared to PCHB. In the case of flavonoids, their highest content was recorded in the PCHB, analogous to the highest content of PA and total polyphenols in this kind of pomace. PCH and PBC were characterized by, respectively, 5 times and 8.5 times lower contents of flavonoids than PCHB (Table 2). A different trend was observed in the amounts of flavonols and anthocyanins in the studied fruit pomaces as compared to the content of flavonoids. Namely, the amount of flavonols and anthocyanins were in the following order: PCHB > PBC > PCH (Table 2). This difference between the content of flavonoids and the content of their fractions (flavonols and anthocyanins) could be due to other compounds such as isoflavones, chalcones, etc., which were not analyzed. In studies conducted by Vagiri and Jensen [12], the amount of anthocyanins in PCHB was 3.85 mg cyanidin-3-glucoside/g, and in the research of Mayer-Miebach et al. [9], it was 11.9–19.5 mg cyanidin-3-glucoside/g. The results of Jarosławska et al.'s [43] study on anthocyanin content in PBC were 0.4885 mg cyanidin-3-glucoside g. Some differences in the content of anthocyanins in the fruit pomace among the results of this work in comparison with results from other authors could be due to the different methods of extraction and anthocyanin assay.

AA determined by means of a synthetic free radical (ABTS) was in the following order: PCHB > PBC > PCH (Table 1), and it was positively correlated with the total phenol content in the pomace (r^2 = 0.985).

Another health-promoting compound, except for the previously mentioned polyphenols, present in fruit pomace was DF. The amount of this component in the PCHB and PBC was at the same level—approximately 60 g/100 g dm (IDF) and 5.5 g/100 g dm (SDF). Cherry pomace (PCH) was characterized by lower amounts of SDF and IDF, respectively 19% and 27%, in comparison with the remaining pomace (Table 1).

Among the analyzed fruit pomaces, PCHB was the most abundant in bioactive compounds. It contained the highest amounts of total polyphenols, PA, flavonoids, anthocyanins, and flavonols as compared to the remaining pomace. It was consistent with the results of the above-mentioned authors [9,10,12], and the content of DF in PCHB was identical to that in PBC (66 g/100 g dm). In contrast to the results obtained by Kawecka and Galus [3], in which the lowest amount of TDF was determined in chokeberry pomace, a higher one in blackcurrant pomace, and the highest in sour cherry pomace. This study also showed that TDF content in blackcurrant pomace was 59.7 g/100 g dm, and Alba et al. [42] showed a level of 72 g/100 g dm, and in the study by Jurendic and Šcetar [41], the TDF in chokeberry pomace was determined in the range of 63–78 g/100 g dm.

Fruit pomaces are a good source of bioactive compounds with health-promoting properties, and their application as an ingredient in snack production can improve the nutritional and pro-health values of the snacks.

3.2. Characteristics of Extruded Corn Snacks

3.2.1. Chemical Composition and Pro-Health Components in Extruded Corn Snacks with Fruit Pomace

As mentioned, cereal snacks are popular, and nutritionists are interested in their enrichment with all kinds of raw materials, including fruits, mixed fruits, and vegetables. Such snacks focus widespread attention among producers and food consumers. This is due to the fact that the search for a new product is still ongoing, and on the other hand, it relates to the growing consumer awareness of the relationship between the composition of the product and its nutritional and pro-health value. Such value in these snacks would consist of bioactive compounds, particularly polyphenols in the added fruit pomace.

Table 3 shows the chemical composition and contents of total polyphenols in corn extrudates with the addition of blackcurrant, cherry, and chokeberry pomace. It was found that after the application of fruit pomace, the protein content in the snacks decreased from 9% to 21% as compared to the control. This tendency can probably be explained by the fact that the amount of protein in cornmeal/corn grits (8.3%; Sante.pl, 2018) was 1.5 times higher than in chokeberry pomace (6.8/100 g dm). Therefore, the substitution of corn grits with these pomaces reduced the amount of protein in the final product. Definitely less protein reduction occurred in snacks with the participation of cherry and blackcurrant pomace because they contained much more of this compound. However, the amount of protein in extruded snacks decreased with the application of fruit pomace because the extrusion process itself can lead to a reduction in the protein content [15], as protein losses are mainly associated with the combination of simple sugars with amino acids during thermally induced Maillard reactions. Moreover, during the extrusion process, proteins are bound to fats, which can result in some losses of this nutrient [44,45].

Table 3. Chemical composition and total phenolic content, antioxidant activity, and dietary fiber content of corn extrudates with fruit pomace.

Sample	Total Phenolic [1] mg Catechin/g dm	ABTS EC_{50} (mg/mL)	ABTS (TEAC) (mgTx/g dm)	Content of Dietary Fiber (g/100 g dm)		Chemical Composition (g/100 g dm)			
				Soluble Fraction	Insoluble Fraction	Protein	Total Sugar	Fat	Ash
Control	1.19 ± 0.37 [a]	2.34 ± 0.11 [f]	3.74 ± 0.11 [a]	0.12 ± 0.00 [a]	0.83 ± 0.01 [a]	7.32 ± 0.03 [c]	2.32 ± 0.12 [a]	1.88 ± 0.17 [b]	1.19 ± 0.00 [a]
ECHB-05	3.76 ± 0.00 [c]	1.00 ± 0.07 [cd]	9.74 ± 0.23 [f]	0.92 ± 0.11 [b]	2.95 ± 0.05 [b]	6.07 ± 0.15 [a]	3.52 ± 0.07 [c]	1.72 ± 0.08 [b]	1.33 ± 0.12 [ab]
ECHB-10	4.23 ± 0.13 [d]	0.76 ± 0.01 [c]	13.14 ± 0.10 [h]	1.07 ± 0.07 [b]	6.03 ± 0.07 [e]	5.93 ± 0.70 [a]	4.00 ± 0.14 [d]	1.70 ± 0.03 [b]	1.48 ± 0.03 [b]
ECHB-20	13.44 ± 0.43 [g]	0.52 ± 0.03 [a]	18.14 ± 0.09 [i]	1,96 ± 0.17 [c]	8.83 ± 0.00 [h]	5.75 ± 0.23 [a]	4.37 ± 0.00 [e]	1.68 ± 0.12 [a]	1.63 ± 0.07 [c]
ECH-05	2.10 ± 0.07 [b]	1.45 ± 0.17 [e]	4.74 ± 0.34 [b]	0.78 ± 0.11 [b]	2.75 ± 0.17 [b]	6.42 ± 0.17 [a]	2.87 ± 0.09 [b]	1.58 ± 0.13 [a]	1.20 ± 0.00 [a]
ECH-10	3.48 ± 0.28 [c]	1.21 ± 0.00 [e]	7.14 ± 0.25 [d]	0.98 ± 0.05 [b]	5.07 ± 0.23 [d]	6.01 ± 0.00 [a]	3.12 ± 0.10 [bc]	1.71 ± 0.00 [b]	1.28 ± 0.02 [ab]
ECH-20	4.97 ± 0.02 [e]	0.86 ± 0.10 [c]	11.74 ± 0.56 [g]	1.22 ± 0.15 [b]	6.90 ± 0.41 [f]	5.94 ± 0.12 [a]	4,01 ± 0.12 [d]	1.70 ± 0.14 [b]	1.32 ± 0.11 [ab]
EBC-05	2.35 ± 0.17 [b]	1.26 ± 0.07 [e]	5.74 ± 0.24 [c]	0.93 ± 0.23 [b]	3.60 ± 0.00 [c]	6.68 ± 0.10 [b]	3.13 ± 0.14 [bc]	1.90 ± 0.32 [b]	1.23 ± 0.08 [a]
EBC-10	3.38 ± 0.08 [c]	1.17 ± 0.10 [e]	7.94 ± 0.17 [e]	1.03 ± 0.12 [b]	6.28 ± 0.12 [e]	6.47 ± 0.08 [ab]	3.87 ± 0.11 [d]	2.01 ± 0.15 [b]	1.30 ± 0.00 [ab]
EBC-20	5.93 ± 0.03 [f]	0.62 ± 0.00 [b]	17.76 ± 0.47 [i]	1.79 ± 0 [c]	8.53 ± 0.00 [g]	6.07 ± 0.00 [a]	4.12 ± 0.07 [d]	2.37 ± 0.03 [c]	1.42 ± 0.11 [ab]

The values in the column marked with the same letter are not significantly different at $p = 0.05$. [1] According to Singleton et al. [34].

The fat contents in the extrudates with fruit pomace and the control sample were at similar levels, and the highest fat content was a result of the largest addition (20%) of currant pomace (Table 3). A slight reduction in the fat content or its stabilization in the mixture after extrusion can most likely be the result of combining fat with starch or proteins [45,46].

The content of total sugars in the extrudates depended on the share and type of the applied pomace and increased from 24 to 88% in relation to the control (Table 3). This was due to the addition of a fruit component rich in sugar, which is fruit pomace, as well as the degradation of starch chains during the extrusion process [46]. Results similar to those observed in this work were also reported by Gumul et al. [47], where, with the increasing share of defatted blackcurrant seeds from 10% to 50%, the content of total sugars in the extrudates also increased (from 12% to 415%) at 50% addition of blackcurrant seeds).

The ash content in the samples with fruit pomace addition was constant or slightly increased as compared to the control. Changes in the ash content in final products were mainly related to the applied raw materials, i.e., the increasing share of fruit pomace, because, as reported by Henry and Chapman [46], the amount of minerals in the plant material does not increase after extrusion.

It was found that each of the analyzed fruit pomaces increased the content of bioactive compounds in extrudates. For PBC addition to corn snacks, the increase in polyphenol content ranged from 97% to 398%, based on the control sample. PCH contributed to the increase in polyphenols in the range of 76% to 318% in extruded corn snacks as compared to the control. Moreover, extrudates enriched with chokeberry pomace were characterized by a higher content of these ingredients in the range of 216% to 1030% compared to pure corn extrudates (Table 3). A general observation was made: the increase in the content of polyphenols was a function of the amount of pomace introduced into extrudates. PCHB had the greatest impact on polyphenol content because, in snacks with such an addition, it recorded the highest content of the previously mentioned bioactive compounds (Table 3). Total phenolic content was measured by two methods because the Folin–Ciocalteu reagent [34] can react with other compounds such as vitamin C, some alkaloids, amino acids, and proteins. Therefore, another method by Mazza et al. [35] with modification by Oomah et al. [36] was also applied, and the results were collected in Table 4. It was found that PCHB, PCH, and PBC increased the total phenolic content in corn extrudates as measured without applying the Folin–Ciocalteu reagent, respectively, by about 450%, 240%, and 256%, as compared to a reference sample. It was also noted that the increase in total phenolic content in the extrudates was adequate for the amount of supplement added, and the highest content of these bioactive components was determined in ECHB-20, which was similar to the total phenolic content marked with the Folin–Ciocalteu reagent (Tables 3 and 4). It was observed that the introduction of fruit pomace into corn extrudates increased the amount of PA content by about 70% in the case of PCHB addition, 24% for PCH, and 31% for PBC as compared to the control. It was also noted that regardless of the type of pomace, their 5% and 10% additions resulted in an identical increase in phenolic acid content in extrudates, and only 20% supplementation contributed to a significant increase in the low molecular weight bioactive compound content (PA) in the investigated extrudates (Table 4).

Table 4. Selected phenolic compounds and flavonoids content in corn extrudates with addition of fruit pomace.

Sample	Total Phenolic [1] mg Catechin/g dm	Phenolic Acid (mg Ferulic Acid/g dm)	Flavonoids (mg Rutin/g dm)	Flavonols (mg Quercetin/g dm)	Anthocyanins (mg Cyanidin-3-Glucoside/g dm)
Control	0.893 ± 0.032 [a]	0.255 ± 0.000 [a]	0.266 ± 0.021 [a]	0.200 ± 0 [a]	0.068 ± 0.011 [a]
ECHB-05	3.550 ± 0.119 [d]	0.324 ± 0.006 [bc]	0.528 ± 0.014 [e]	0.230 ± 0.002 [b]	0.407 ± 0.00 [d]
ECHB-10	4.197 ± 0.076 [e]	0.379 ± 0.006 [c]	0.891 ± 0.009 [f]	0.258 ± 0.003 [c]	0.457 ± 0.017 [e]
ECHB-20	6.987 ± 0.374 [g]	0.590 ± 0.032 [d]	1.75 ± 0.057 [g]	0.389 ± 0.022 [d]	0.761 ± 0.034 [f]
ECH-05	1.860 ± 0.114 [b]	0.291 ± 0.009 [b]	0.328 ± 0.017 [b]	0.208 ± 0.006 [b]	0.321 ± 0.012 [b]
ECH-10	3.067 ± 0.139 [c]	0.295 ± 0.012 [b]	0.384 ± 0.004 [c]	0.216 ± 0.008 [b]	0.337 ± 0.00 [b]
ECH-20	4.138 ± 0.145 [e]	0.365 ± 0.015 [c]	0.511 ± 0 [e]	0.248 ± 0.008 [bc]	0.368 ± 0.013 [c]
EBC-05	2.002 ± 0.114 [b]	0.298 ± 0.014 [b]	0.447 ± 0.028 [d]	0.231 ± 0.009 [b]	0.364 ± 0.007 [c]
EBC-10	3.061 ± 0.410 [cd]	0.327 ± 0.026 [bc]	0.503 ± 0.017 [e]	0.261 ± 0.017 [c]	0.380 ± 0.008 [c]
EBC-20	4.500 ± 0.062 [f]	0.375 ± 0.017 [c]	0.907 ± 0.042 [f]	0.287 ± 0.009 [c]	0.430 ± 0.012 [e]

The values in the column marked with the same letter are not significantly different at $p = 0.05$. [1] According to Mazza et al. [35], with modification by Oomah et al. [36].

Taking into account high molecular weight phenolic compounds, i.e., flavonoids, it was found that the lowest content of these compounds was found in the control sample. However, whatever type of fruit pomace was introduced, an increase in the content of high molecular weight phenolic compounds in snacks was observed in the range of 23% to 557%, as compared to the control (Table 4). The highest increase in the amounts of these bioactive compounds reported for PCHB addition is about 98% to 557%. The highest content of flavonoids among all analyzed corn snacks was observed in the extrudate with a 20% addition of PCHB (Table 4).

Moreover, a gradual increase in the amount of high molecular weight polyphenols (flavonoids) was also observed in the analyzed samples, appropriate to an increase in the amount of additive (Table 4). In contrast, an increase in the flavonol content was not adequate for the amount of additive introduced into corn extrudates. Additionally, it was observed that a 5% addition of pomace caused the same increase in flavonols (about 13%), regardless of their type. Only higher levels of these additives guarantee a greater increase in the flavonol content of extruded corn snacks compared to the control. The biggest increase in the flavonol content was recorded in the case of PCHB-supplemented snacks—about 61%—and the lowest was 16% for PCH (Table 4). In the case of the second fraction of flavonoids, i.e., anthocyanins, it was observed that for 5% and 10% PCH and PBC added, the increase in the content of these compounds was, respectively, 3.8-fold and 4.3-fold in extrudates as compared to control. A twenty percent addition of the above-mentioned pomaces caused the greatest increase in the amount of anthocyanins: 4.4-fold for PCH and 5.3-fold for PBC (Table 4). The anthocyanin content increased progressively with the amount of the applied additive in the form of PCHB. While the content of anthocyanins in extrudates with PCHB was low, only this pomace among all those analyzed guaranteed a 5- to 10-fold increase in anthocyanin content in extrudates. Generally, low levels of anthocyanins in the corn extrudates increase with the addition of fruit pomace, resulting from their high thermolability. The anthocyanins' stability during the extrusion depends on their type. In blackcurrants, cherries, chokeberries, blueberries, and cranberries, non-acylated anthocyanins are present, which are less stable and bioavailable than the acylated anthocyanins present in red potatoes [48,49]. Hence their low amount in the extruded snacks (Table 4). Despite this, PCHB ensured an increase of 5 to 10 times in the anthocyanin content of snacks. In the context of the health-promoting properties of anthocyanins, namely their anti-mutagenic, anticarcinogenic, and antihypertensive activity, and reduced risk of chronic diseases and neuronal degeneration [50–52], this seems to be an important achievement of this work. Especially anthocyanins from chokeberries show inhibition of cancer cell proliferation and antimutagenic, hepatoprotective, antidiabetic, and cardioprotective effects [53], which additionally supports the fact that health-promoting extruded corn snacks supplemented with chokeberry pomace can be produced.

Taking into consideration the fact that extrusion could cause a reduction in the polyphenol content in the range of 24% to 46% [17], it should be noted that the additive-enriched corn snacks applied in this work are high in polyphenolic compounds, despite the loss of these components resulting from the partial depolymerization, decarboxylation (in the case of phenolic acids), and the polymerization of phenolic compounds and tannins [18] during extrusion. Among all analyzed fruit pomaces, the largest increase in polyphenols, phenolic acids, flavonols, and anthocyanins in extruded corn snacks guarantees PCHB (Table 4).

The results of the AA analysis of the samples determined by the method with a synthetic free radical ABTS are summarized in Table 3. It was found that each of the additives contributed to an increase in this type of activity of the extrudates, on average by 265% for PCHB added, about 108% for PCH, and 180% in the case of PBC (Table 3). AA of the analyzed snacks incorporating fruit pomace was adequate for the content of total polyphenols, as was evidenced by the strong positive correlation between the total phenolic content and ABTS (r^2 = 0.829). Exceptions were observed for ECHB-20 and EPBC20, which have identical AA despite significant differences in the total polyphenol content. Most likely, this must be attributed to the fact that other compounds, not analyzed in this research (such as tannins as well as new compounds formed by extrusion, such as Maillard reaction products or the conversion of less active antioxidants into more active antioxidants [54]), contributed to AA.

AA was also calculated as EC_{50}, i.e., the concentration of antioxidants necessary to neutralize 50% of the ABTS present in the sample; the lower the EC_{50} value, the higher the antioxidant activity of the polyphenols in the sample. When AA was calculated as Trolox (TEAC) and EC_{50}, an inverse relationship was clear. Among the analyzed samples, snacks with a 20% share of PCHB had the highest AA because their EC50 was the lowest.

In studies conducted by Ainsworth et al. [55], the brewer's spent grain was applied, but with no effect on the content of polyphenols or the antioxidant activity of the investigated samples. Moreover, adding the cauliflower by-product to extrudates did not increase the polyphenol level, but it had a detrimental effect on the antioxidant activity [23]. Similarly, the addition of brewer's spent grain to cereal extrudates resulted in no observed changes in polyphenol levels or antioxidant activity [56]. In Camire et al.'s [22] study on the effect of blueberry, cranberry, concord grape, and raspberry additions on the amount of polyphenols in extrudates, there were no changes in the amount of these compounds, most likely as a result of a too low share—only 1%. However, the addition of red cabbage to cereal snacks resulted in a marked increase in the polyphenol content and antioxidant activity [56]. Other research [57] dealing with cereal snacks supplemented with defatted blackcurrant seed noted a 2- to 10-fold increase in the levels of polyphenols and a 2- to 11-fold increase in antioxidative activity as compared to the control. Moussa-Ayoub et al.'s [58] investigation on the enrichment of cereal-based extrudates with cactus fruits led to an increase in the antioxidative activity of the final product, adequate to the amount of the applied supplement. Similarly, Anton et al. [59] observed a significant increase in total polyphenols by about 40% and the antioxidant activity of 98% of the extruded snack obtained from blends of corn starch and navy red beans. Taking into account the results of the presented research, it can therefore be said that the applied fruit pomace can be a perfect health-promoting addition to corn extrudates because it will enrich the final product with a wide range of bioactive compounds from the polyphenols group, ranging from low molecular phenolic acids to anthocyanins (Table 4). Of particular value proved to be PCHB, as ECHB-20 was characterized by the highest content of polyphenols, phenolic acids, flavonols, flavonoids, and anthocyanins (Table 4).

Another very important pro-health component present in the fruit pomace was DF. IDF content was 67 g/100 g dm and 65 g/100 g dm, respectively, for PBC and PCH pomace, and 48.7 g/100 g dm in PCH. The above-mentioned pomaces were used to enrich the resulting extrudates with DF (Table 3). PCHB contributed to the largest, 10-fold increases in the amount of SDF and 6-fold increases in the amount of IDF in extrudates, and PCH addition resulted in the smallest increase, approximately 7-fold and 5.5-fold, respectively (Table 3). Taking into account the fact that SDF has hypocholesterolemic and hypoglycemic properties [60] and IDF has anticancerogenic characteristics [61], the introduction of fruit pomace, which caused such a high increase in DF content in snacks, appeared to be justified. It should also be stressed that not every supplement helped to increase the amount of DF; for example, the addition of extrudates of cauliflower by-product (5–20%) contributed to the decrease in fiber content, which was the result of pectin degradation into low molecular weight compounds, which were not assayed by the applied analytical method [23]. Results from Gumul et al. [57] indicated that defatted blackcurrant seeds contributed to a 2.5-fold increase in SDF and an 8-fold increase in IDF in corn extrudates. Potter et al. [62] reported that the addition of fruit mixtures to extrudates resulted in an increase in the soluble dietary fiber fraction in the range of 50–350% and that the content of IDF was not changed compared to the control.

3.2.2. Quality of Extruded Corn Snacks with Fruit Pomace

Some of the most important parameters for the qualitative evaluation of extruded snack products are density and expansion, especially in terms of assessing their textural properties [63].

Expansion and density of extrudates are quantities that are most often inversely proportional; that is, as the expansion factor increases, the density decreases [47,62]. Both quantities are largely dependent on the dietary fiber (DF) content, especially its soluble and insoluble fractions (SDF and IDF).

It was observed that adding chokeberry, cherry, and blackcurrant pomace increased the extrudates' density, with the higher the level of additive used, the higher the density relative to the control. With a 20% level of the various additives, the density of the extruded snacks was almost 1.5 to 2 times higher as compared to the control. In contrast to the 5% additive, which did not affect this parameter regardless of the type of additive used (pomace-cherry, chokeberry, or blackcurrant) (Table 5).

Table 5. Physical properties of extrudates of corn extrudates with addition of fruit pomace.

Sample	Density [g/cm^3]	Expansion Ratio [%]	WBC [1] [%]
Control	0.064 ± 0.01 ^{a}	4.24 ± 0.76 ^{a}	4.38 ± 0.12 ^{e}
ECHB-05	0.072 ± 0.01 ^{a}	4.58 ± 0.25 ^{a}	3.21 ± 0.09 ^{c}
ECHB-10	0.118 ± 0.01 ^{c}	4.23 ± 0.13 ^{a}	2.84 ± 0.00 ^{b}
ECHB-20	0.138 ± 0.00 ^{d}	4.18 ± 0.15 ^{a}	2.34 ± 0.07 ^{a}
ECH-05	0.059 ± 0.02 ^{a}	4.11 ± 0.18 ^{a}	3.58 ± 0.00 ^{d}
ECH-10	0.084 ± 0.00 ^{ab}	4.29 ± 0.30 ^{a}	2.99 ± 0.17 ^{b}
ECH-20	0.110 ± 0.02 ^{c}	4.12 ± 0.09 ^{a}	2.57 ± 0.21 ^{a}
EBC-05	0.073 ± 0.02 ^{a}	3.89 ± 0.11 ^{a}	3.37 ± 0.00 ^{c}
EBC-10	0.087 ± 0.02 ^{ab}	4.05 ± 0.25 ^{a}	2.87 ± 0.03 ^{b}
EBC-20	0.127 ± 0.02 ^{d}	4.03 ± 0.23 ^{a}	2.42 ± 0.05 ^{a}

The values in the column marked with the same letter are not significantly different at $p = 0.05$. [1] WBC—water binding capacity.

It was found that the expansion of extruded snacks with fruit pomace did not change in relation to the control, regardless of the proportion of such an additive (Table 5).

The expansion of the extruded snacks is mainly ensured by the starch present in the mixture, which is characterized by viscoelastic properties. Bubbles formed by the evaporation of the water present in the extruded material penetrate into the swelling starch granules and cause them to expand [24]. As a result of replacing cornmeal, which is a source of starch, with fruit pomace, which in turn is a source of dietary fiber, there is, among other things, a reduction in the extensibility of the bubbles formed, which has the effect of reducing the expansion ratio and increasing the density of the obtained extruded products [64].

In addition to the above-described reduction in gas bubbles, the IDF fraction also causes their damage, acting as a disintegrating agent [25]. In addition, the SDF fraction (mainly pectin) present in the pomace, which is characterized by a high water-absorbing capacity, contributes to starch water-absorption reduction, thus inhibiting the expansion process of the product [25]. Although, on the other hand, according to van der Sman and Broeze [64], this fraction can at the same time favor the product expansion process—by increasing the elasticity of the forming "sources" of gas bubbles.

Furthermore, the extrusion process will not only result in pasting but also in the melting of the starch polymer, which, according to Colonna and Mercier [65], partially adheres to the cellulose walls, causing the formation of structural complexes containing cellulose, partially melted starch, and proteins, thus limiting the product's ability to expand and therefore increasing its density [65,66]. An additional component limiting the expansion of these extrudates is the water-soluble saccharides in the extrudate. These saccharides limit the amount of water available for bubble formation and starch pasting, which reduces the extent to which these processes occur and leads to a lower expansion factor [47,67].

In the subject of research on the influence of carbohydrate raw materials used as additives to extruded products on expansion and density, wheat, oat, or rye brans were commonly applied additives. Their addition resulted in an increase in IDF fraction content, a simultaneous increase in the density of the extrudates obtained, and a decrease in the expansion factor. This was due to the high fiber content of the additives used as well as the high fat content of oat bran, whose content also affected expansion [68,69].

The addition of buckwheat flour to corn extrudates showed that the density increased with an increased proportion of this ingredient, while only a 50% addition of buckwheat or higher increased the expansion factor. This was related to the components that buckwheat provides, including protein and dietary fiber [70]. It has been demonstrated that enrichment with dried potato juice and various vegetable additives such as paprika, basil, or parsley causes an increase in the density of extrudates with an increased proportion of the applied additive and a decrease in the coefficient of expansion, with an increase in the content of the additive particularly rich in DF [71].

Among the fruits, the effect of elderberries on the density of the extrudates was investigated; the value increased, but the addition of elderberries could not exceed 15% because this would cause unfavorable quality characteristics in the final product. The addition of apple pomace was also applied, which resulted in an increase in density and a decrease in the expansion factor compared to the control sample, despite a slight increase in the SDF fraction [72].

In the case of the water binding capacity (WBC) of the extruded corn snacks obtained with the addition of fruit pomace, a decrease in this parameter's value was observed with an increase in their proportion compared to the control sample (Table 5). In the case of chokeberry pomace, the decrease in WBC of the extrudates with its share ranged from 27–47% as compared to the control (Table 5). In the case of cherry and currant extrudates, the decrease in WBC of extrudates containing them ranged from 18–41% and 25–45%, respectively, relative to the control (Table 5).

Ačkar et al. [72] also observed a decrease in WBC for corn extrudates with different proportions of apple pomace (from 6% to 23%) as compared to control corn extrudate. Slightly different results were obtained by Gumul et al. [47], where extrudates with 10% and 30% blackcurrant seeds had a higher WBC than the control sample, while a 50% share of blackcurrant seeds resulted in a 46% decrease in WBC. This was probably related to the low starch content and high sugar content of the finished product with 50% blackcurrant seeds (322% less starch and 415% more saccharides in the sample with 50% as compared to the control). Contrary, Drożdż et al. [73] investigated the WBC of corn extrudates with the addition of blackcurrant and chokeberry press pulp and found an unambiguous decrease in WBC as the proportion of both fruits in the extrudates increased. According to Drożdż et al. [73], this can be due to the high fiber content. The results ranged from 6.4–4.6% for extrudates with blackcurrant and 5.9–4.8% for extrudates with chokeberry, while for the control sample, it was 6.5% [73]. The observed decrease in WBC of the corn extrudates analyzed in this study with different proportions of fruit pomace was most likely related to the high content of sugars in them, which significantly reduces the amount of water available for the starch pasting process. This effect can also be attributed to the decreasing starch content (replacing cornmeal as a source of starch with extrudate as a source of fiber) with an increasing proportion of fruit pomace in the extrudate, which contains a huge proportion of fiber. This resulted in competition for water between the fiber present in the fruit pulp and the available starch [74].

In addition, it should be mentioned that during the extrusion process, starch undergoes not only pasting but also thermal and mechanical degradation and complex formation with proteins and fats, as well as dilution in the extrusion mixture, which reduces its water-absorbing capacity [45]. The decrease in WBC of the resulting extrudates may also be partly related to the reduction during extrusion of the hydrophilic protein content due to its thermal denaturation as well as its complexation with starch [75].

Based on an analysis of the color parameters (L^* a^* b^*), it was observed that the color of the extrudates was significantly affected by the applied additive. The addition of chokeberry pomace resulted in a significantly darker color of the finished product as compared to the sample without any additive, with a decrease in brightness and yellow saturation but an increase in red saturation (Table 6). On the other hand, the addition of cherry pomace also caused a decrease in lightness and yellow saturation and an increase in red saturation, but not as much as the addition of chokeberries. These changes were more noticeable with the addition of chokeberry pomace than with cherry pomace (Table 6). In contrast, blackcurrant pomace contributed to intermediate changes in the color parameters of the extruded snacks compared to the other pomace. Many authors who carried out studies on the effect of the additive used on the color parameters of extruded snacks also showed significant discrepancies in the level of brightness or color saturation, this being mainly dependent on the amount and type of additive used. Kowalczewski et al. [71] investigated how dried potato juice and various vegetable additives would affect individual color components. They used paprika, parsley, and basil as additives. All these additives decreased lightness with the increased addition of the enhancement component. The addition of paprika resulted in a significant increase in red color saturation, and in the case of parsley and basil, in green color saturation. Among fruits, the effects of elderberry and chokeberry [76] on individual color components were investigated. These additives caused a decrease in lightness and yellow saturation and a significant increase in red saturation. Drożdż et al. [77] studied the color of corn extrudates with the addition of pomace from apples (10, 15, and 20%) and rosehips (10, 15, and 20%) as compared to the control sample. The increase in pomace content resulted in a decrease in the parameter L^*—the lightness of the investigated extrudates changed from 56.54 to 52.09. These results were much lower than the lightness of the control sample, whose brightness was 64.43 [77]. For the parameter a^*, the addition of both types of pomaces resulted in a higher proportion of red color in the extrudates compared to the control sample, whose value was on the negative side (−4.21). Extrudates with a 20% addition of rosehip pomace contained the reddest color, while extrudates with a 10% addition of apple pomace contained the least [77].

Table 6. Color parameters of corn extrudates with addition of fruit pomace.

Sample	L* [-] (Black to White)	a* [-] (Green to Red)	B [-] (Blue to Yellow)	WI [-]	ΔE [-]
Control	78.01 ± 0.12 [g]	0.78 ± 0.27 [a]	17.46 ± 0.18 [h]	71.91	-
ECHB-05	53.43 ± 1.54 [d]	9.11 ± 0.30 [f]	10.25 ± 0.28 [d]	51.45	26.94
ECHB-10	48.31 ± 0.47 [c]	9.48 ± 0.13 [g]	9.00 ± 0.15 [c]	46.68	32.08
ECHB-20	30.20 ± 0.19 [a]	10.57 ± 0.51 [g]	7.12 ± 0.11 [a]	29.05	49.89
ECH-05	66.16 ± 0.16 [f]	5.19 ± 0.07 [b]	14.46 ± 0.14 [g]	62.84	13.00
ECH-10	60.62 ± 0.22 [e]	6.64 ± 0.06 [b]	13.7 ± 0.18 [e]	57.78	18.73
ECH-20	50.31 ± 1.21 [d]	7.12 ± 0.08 [d]	12.10 ± 0.1 [e]	48.36	28.92
EBC-05	60.73 ± 0.11 [e]	7.39 ± 0.00 [d]	9.90 ± 0.03 [d]	58.83	19.99
EBC-10	53.20 ± 0.76 [d]	8.52 ± 0.05 [e]	8.57 ± 0.12 [c]	51.66	27.47
EBC-20	45.20 ± 0.19 [b]	9.57 ± 0.09 [g]	7.32 ± 0.00 [b]	43.89	35.45

The values in the column marked with the same letter are not significantly different at $p = 0.05$. L*, a*, b*—parameters of color in lab color space; WI—whiteness index; ΔE—color difference between control sample and other samples.

On the other hand, when the parameter b* was investigated, it was noted that samples with pomace showed lower values compared to extrudates made from cornmeal only, which had a value of 38.39. The highest value of the b*parameter was measured in extrudates with a 20% addition of rosehip (27.99) and the lowest in extrudates with a 20% apple pomace addition [77].

The introduction of fruit pomaces into snacks resulted in a decrease in WI values, indicating a shift from white color (Table 6), which was increasing with the share of added pomaces. The greatest decrease was observed for extrudates prepared with the addition of chokeberry (ECHB) and the smallest for those with cherry pomace (ECH) addition.

Moreover, the color difference (ΔE) between the control and other investigated samples was evaluated. Similarly, for WI, the same trend was observed—an increasing share of pomace resulted in increased ΔE values. In addition, the greatest differences between control and experimental samples were observed for ECHB samples, whereas the smallest were observed for ECH.

It can be assumed that a ΔE value up to 2.3 corresponds to a just noticeable difference (JND) in color between the two samples [78,79]. Such a low difference was only observed between EBC-10 and ECHB-05 (1.8) (Table S1). According to the aforementioned authors for ΔE values < 3:5, an inexperienced observer also noticed the difference, and such a situation occurred between the ECHB-10 and EBC-20 samples.

Data related to the hardness measurement of selected corn extrudates are summarized in Table 7. The addition of fruit pomaces resulted in decreased hardness (cutting force) as compared to the control sample. Such behavior could be explained by the partial replacement of starch polymers, which were subjected to melting during the extrusion process, forming a matrix for other ingredients in the extruded snack. Such replacement resulted in a more brittle structure, and less effort was required to cut the extrudate.

Table 7. Hardness of selected corn extrudates with addition of fruit pomace.

Sample	Hardness [N]
Control	36.37 ± 8.86 [c]
ECHB-05	24.63 ± 4.8 [b]
ECHB-10	14.32 ± 6.17 [a]
ECH-05	29.24 ± 6.49 [b]
ECH-10	30.58 ± 6.75 [bc]

The values in the column marked with the same letter are not significantly different at $p = 0.05$.

In a study by Gumul et al. [57], a decrease in corn snack hardness was also observed when defatted blackcurrant seeds were added (an average decrease of 53% compared to the control). In contrast, in the study by Stojceska et al. [23], where different proportions of ground dried cauliflower were used to produce corn-wheat extrudates, no increase in hardness was found with increasing proportion of the additive, despite the fact that such an increase was expected based on the study of Yannioti et al. [80], who, as mentioned earlier, proved that the DF addition increases the hardness of the obtained extrudates. It should also be kept in mind, as mentioned in the case of gluten-free extrudates, the partial effect of pectin, which is part of the SDF fraction of the pomace, can reduce the hardness of the obtained expanded products [80].

4. Conclusions

It was found that the applied fruit pomace can be a great addition to pro-health corn extrudates because it enriched the final product with bioactive components from the polyphenols group, as well as nutritional compounds (ash and sugar) and dietary fiber. Particularly valuable proved to be chokeberry pomace, and as an extruded corn snack with a 20% addition of such pomace, they were characterized by the highest content of polyphenols, phenolic acids, flavonoids, flavonols, anthocyanins, SDF, IDF, and ash. It also showed a higher density and no deterioration in expansion as compared to the control extrudate, but was darker than the control as a result of the addition of chokeberry pomace, making it more visually appealing than the regular corn extrudates.

Supplementary Materials: The following supporting information can be downloaded at: https://www.mdpi.com/article/10.3390/app13084818/s1, Table S1: Total color difference ΔE calculated for corn extrudates with addition of fruit pomace.

Author Contributions: Conceptualization, D.G.; methodology, D.G., W.B. and T.Z.; investigation, D.G. and T.Z.; resources, D.G., W.B. and T.Z.; data curation, D.G., W.B. and T.Z.; writing—original draft preparation, D.G. and W.B.; writing—review and editing, D.G. and W.B.; visualization, W.B.; supervision, D.G. All authors have read and agreed to the published version of the manuscript.

Funding: The Research was financed by the Ministry of Science and Higher Education of the Republic of Poland.

Institutional Review Board Statement: Not applicable.

Informed Consent Statement: Not applicable.

Data Availability Statement: Not applicable.

Conflicts of Interest: There are no conflict of interest to declare.

References

1. FAO. *The State of Food and Agriculture 2019. Moving Forward on Food Loss and Waste Reduction*; FAO: Rome, Italy, 2019.
2. Srivastava, N.; Srivastava, M.; Alhazmi, A.; Kausar, T.; Haque, S.; Singh, R.; Ramteke, P.W.; Mishra, P.K.; Tuohy, M.; Leitgeb, M.; et al. Technological advances for improving fungal cellulase production from fruit wastes for bioenergy application: A review. *Environ. Pollut.* **2021**, *287*, 117370. [CrossRef] [PubMed]
3. Kawecka, L.; Galus, S. Wytłoki owocowe—Charakterystyka i możliwości zagospodarowania (Fruit pomace—Characteristics and possibilities of recycling). *Postępy Tech. Przetwórstwa Spożywczego* **2021**, *1*, 156–167. (In Polish)
4. Padayachee, A.; Day, L.; Howell, K.; Gidley, M.J. Complexity and health functionality of plant cell wall fibers from fruits and vegetables. *Crit. Rev. Food Sci. Nutr.* **2017**, *57*, 59–81. [CrossRef]
5. Gómez, M.; Martinez, M.M. Fruit and vegetable by-products as novel ingredients to improve the nutritional quality of baked goods. *Crit. Rev. Food Sci. Nutr.* **2018**, *58*, 2119–2135. [CrossRef]
6. Kruczek, M.; Gumul, D.; Olech, E.; Areczuk, A.; Drygaś, B.; Drygaś, P. Diet and the Context of Fruit Industry. *Econ. Environ. Stud.* **2017**, *17*, 389–398. [CrossRef]
7. Ahlborn, J.; Stephan, A.; Meckel, T.; Maheshwari, G.; Rühl, M.; Zorn, H. Upcycling of food industry side streams by basidiomycetes for production of a vegan protein source. *Int. J. Recycl. Org. Waste Agric.* **2019**, *8*, 447–455. [CrossRef]
8. Kołodziejczyk, K.; Sójka, M.; Abadias, M.; Viñas, I.; Guyot, S.; Baron, A. Polyphenol composition, antioxidant capacity, and antimicrobial activity of the extracts obtained from industrial sour cherry pomace. *Ind. Crop. Prod.* **2013**, *51*, 279–288. [CrossRef]
9. Mayer-Miebach, E.; Adamiuk, M.; Behsnilian, D. Stability of Chokeberry Bioactive Polyphenols during Juice Processing and Stabilization of a Polyphenol-Rich Material from the By-Product. *Agriculture* **2012**, *2*, 244–258. [CrossRef]

10. Nawirska, A.; Sokół-Łętowska, A.; Kucharska, A.Z. Właściwości przeciwutleniające wytłoków z wybranych owoców kolorowych (Antioxidant characteristics of pomace from different fruits). *Żywność Nauka Technol. Jakość* **2007**, *53*, 120–125. (In Polish)
11. Sójka, M.; Król, B. Composition of industrial seedless black currant pomace. *Eur. Food Res. Technol.* **2009**, *228*, 597–605. [CrossRef]
12. Vagiri, M.; Jensen, M. Influence of juice processing factors on quality of black chokeberry pomace as a future resource for colour extraction. *Food Chem.* **2017**, *217*, 409–417. [CrossRef] [PubMed]
13. Liu, Y.; Hsieh, F.; Heymann, H.; Huff, H. Effect of Process Conditions on the Physical and Sensory Properties of Extruded Oat-Corn Puff. *J. Food Sci.* **2000**, *65*, 1253–1259. [CrossRef]
14. Repo-Carrasco-Valencia, R.; de La Cruz, A.A.; Alvarez, J.C.I.; Kallio, H. Chemical and Functional Characterization of Kañiwa (Chenopodium pallidicaule) Grain, Extrudate and Bran. *Plant Foods Hum. Nutr.* **2009**, *64*, 94–101. [CrossRef] [PubMed]
15. Dust, J.M.; Gajda, A.M.; Flickinger, E.A.; Burkhalter, T.M.; Merchen, N.R.; Fahey, G.C. Extrusion Conditions Affect Chemical Composition and in Vitro Digestion of Select Food Ingredients. *J. Agric. Food Chem.* **2004**, *52*, 2989–2996. [CrossRef]
16. Riaz, M.N. *Extruders in Food Applications*; CRC Press LLC: Boca Raton, FL, USA; London, UK; New York, NY, USA; Washington, DC, USA, 2002.
17. Viscidi, K.A.; Dougherty, M.P.; Briggs, J.; Camire, M.E. Complex phenolic compounds reduce lipid oxidation in extruded oat cereals. *LWT Food Sci. Technol.* **2004**, *37*, 789–796. [CrossRef]
18. Dlamini, N.R.; Taylor, J.R.N.; Rooney, L.W. The effect of sorghum type and processing on the antioxidant properties of African sorghum-based foods. *Food Chem.* **2007**, *105*, 1412–1419. [CrossRef]
19. Mink, P.J.; Scrafford, C.G.; Barraj, L.M.; Harnack, L.; Hong, C.-P.; Nettleton, J.A.; Jacobs, D.R. Flavonoid intake and cardiovascular disease mortality: A prospective study in postmenopausal women. *Am. J. Clin. Nutr.* **2007**, *85*, 895–909. [CrossRef]
20. Yokohira, M.; Yamakawa, K.; Saoo, K.; Matsuda, Y.; Hosokawa, K.; Hashimoto, N.; Kuno, T.; Imaida, K. Antioxidant Effects of Flavonoids Used as Food Additives (Purple Corn Color, Enzymatically Modified Isoquercitrin, and Isoquercitrin) on Liver Carcinogenesis in a Rat Medium-Term Bioassay. *J. Food Sci.* **2008**, *73*, C561–C568. [CrossRef]
21. Repo-Carrasco-Valencia, R.; Peña, J.; Kallio, H.; Salminen, S. Dietary fiber and other functional components in two varieties of crude and extruded kiwicha (Amaranthus caudatus). *J. Cereal Sci.* **2009**, *49*, 219–224. [CrossRef]
22. Camire, M.E.; Dougherty, M.P.; Briggs, J.L. Functionality of fruit powders in extruded corn breakfast cereals. *Food Chem.* **2007**, *101*, 765–770. [CrossRef]
23. Stojceska, V.; Ainsworth, P.; Plunkett, A.; İbanoğlu, E.; İbanoğlu, Ş. Cauliflower by-products as a new source of dietary fibre, antioxidants and proteins in cereal based ready-to-eat expanded snacks. *J. Food Eng.* **2008**, *87*, 554–563. [CrossRef]
24. Karkle, E.L.; Alavi, S.; Dogan, H. Cellular architecture and its relationship with mechanical properties in expanded extrudates containing apple pomace. *Food Res. Int.* **2012**, *46*, 10–21. [CrossRef]
25. O'Shea, N.; Arendt, E.; Gallagher, E. Enhancing an Extruded Puffed Snack by Optimising Die Head Temperature, Screw Speed and Apple Pomace Inclusion. *Food Bioprocess Technol.* **2014**, *7*, 1767–1782. [CrossRef]
26. Masli, M.D.P.; Gu, B.-J.; Rasco, B.A.; Ganjyal, G.M. Fiber-Rich Food Processing Byproducts Enhance the Expansion of Cornstarch Extrudates. *J. Food Sci.* **2018**, *83*, 2500–2510. [CrossRef] [PubMed]
27. Pietzsch, N.; Ribeiro, J.L.D.; de Medeiros, J.F. Benefits, challenges and critical factors of success for Zero Waste: A systematic literature review. *Waste Manag.* **2017**, *67*, 324–353. [CrossRef] [PubMed]
28. Hopper, A.D.; Hadjivassiliou, M.; Butt, S.; Sanders, D.S. Adult coeliac disease. *BMJ* **2007**, *335*, 558–562. [CrossRef] [PubMed]
29. Saturni, L.; Ferretti, G.; Bacchetti, T. The Gluten-Free Diet: Safety and Nutritional Quality. *Nutrients* **2010**, *2*, 16–34. [CrossRef]
30. Tsatsaragkou, K.; Protonotariou, S.; Mandala, I. Structural role of fibre addition to increase knowledge of non-gluten bread. *J. Cereal Sci.* **2016**, *67*, 58–67. [CrossRef]
31. Wierdsma, N.J.; van Bokhorst-de van der Schueren, M.A.E.; Berkenpas, M.; Mulder, C.J.J.; van Bodegraven, A.A. Vitamin and Mineral Deficiencies Are Highly Prevalent in Newly Diagnosed Celiac Disease Patients. *Nutrients* **2013**, *5*, 3975–3992. [CrossRef]
32. AOAC. *Official Methods of Analysis of AOAC International*, 18th ed.; Association of Official Analytical Chemists International: Gaithersburg, MD, USA, 2006.
33. AACC. *Approved Methods of the AACC*, 10th ed.; AACC: St. Paul, MN, USA, 2000.
34. Singleton, V.L.; Orthofer, R.; Lamuela-Raventós, R.M. Analysis of total phenols and other oxidation substrates and antioxidants by means of folin-ciocalteu reagent. In *Oxidants and Antioxidants Part A*; Academic Press: Cambridge, MA, USA, 1999; Volume 299, pp. 152–178. ISBN 0076-6879.
35. Mazza, G.; Fukumoto, L.; Delaquis, P.; Girard, B.; Ewert, B. Anthocyanins, Phenolics, and Color of Cabernet Franc, Merlot, and Pinot Noir Wines from British Columbia. *J. Agric. Food Chem.* **1999**, *47*, 4009–4017. [CrossRef]
36. Oomah, B.D.; Cardador-Martínez, A.; Loarca-Piña, G. Phenolics and antioxidative activities in common beans (*Phaseolus vulgaris* L.). *J. Sci. Food Agric.* **2005**, *85*, 935–942. [CrossRef]
37. El Hariri, B.; Sallé, G.; Andary, C. Involvement of flavonoids in the resistance of two poplar cultivars to mistletoe (*Viscum album* L.). *Protoplasma* **1991**, *162*, 20–26. [CrossRef]
38. Re, R.; Pellegrini, N.; Proteggente, A.; Pannala, A.; Yang, M.; Rice-Evans, C. Antioxidant activity applying an improved ABTS radical cation decolorization assay. *Free Radic. Biol. Med.* **1999**, *26*, 1231–1237. [CrossRef] [PubMed]
39. Sokhey, A.S.; Kollengode, A.N.; Hanna, M.A. Screw Configuration Effects on Corn Starch Expansion During Extrusion. *J. Food Sci.* **1994**, *59*, 895–896. [CrossRef]

40. Liszka-Skoczylas, M.; Berski, W.; Witczak, M.; Skoczylas, Ł.; Kowalska, I.; Smoleń, S.; Szlachcic, P.; Kozieł, M. The Influence of Hydroponic Potato Plant Cultivation on Selected Properties of Starch Isolated from Its Tubers. *Molecules* **2022**, *27*, 856. [CrossRef]
41. Jurendić, T.; Ščetar, M. Aronia melanocarpa Products and By-Products for Health and Nutrition: A Review. *Antioxidants* **2021**, *10*, 1052. [CrossRef]
42. Alba, K.; MacNaughtan, W.; Laws, A.P.; Foster, T.J.; Campbell, G.M.; Kontogiorgos, V. Fractionation and characterisation of dietary fibre from blackcurrant pomace. *Food Hydrocoll.* **2018**, *81*, 398–408. [CrossRef]
43. Jarosławska, J.; Juszkiewicz, J.; Zduńczyk, Z.; Matusevicius, P. Polyphenol-rich Blackcurrant Pomace Counteracts Impaired Antioxidant Status and Serum Lipid Profile in Rabbits Fed a Diet High in Unsaturated Fat. *J. Nutr. Food Sci.* **2016**, *6*, 1000531. [CrossRef]
44. Sotillo, E.; HETTIARACHCHY, N.S. Corn Meal-Sunflower Meal Extrudates and Their Physicochemical Properties. *J. Food Sci.* **1994**, *59*, 432–435. [CrossRef]
45. Singh, B.; Sekhon, K.S.; Singh, N. Effects of moisture, temperature and level of pea grits on extrusion behaviour and product characteristics of rice. *Food Chem.* **2007**, *100*, 198–202. [CrossRef]
46. Henry, C.; Chapman, C. *The Nutrition Handbook for Food Processors*; Elsevier: Amsterdam, The Netherlands, 2002.
47. Gumul, D.; Ziobro, R.; Zięba, T.; Rój, E. Physicochemical Characteristics of Cereal Extrudates with Different Levels of Defatted Blackcurrant Seeds. *J. Food Qual.* **2013**, *36*, 385–393. [CrossRef]
48. Brennan, C.; Brennan, M.; Derbyshire, E.; Tiwari, B.K. Effects of extrusion on the polyphenols, vitamins and antioxidant activity of foods. *Trends Food Sci. Technol.* **2011**, *22*, 570–575. [CrossRef]
49. Lachman, J.; Hamouz, K.; Šulc, M.; Orsák, M.; Pivec, V.; Hejtmánková, A.; Dvořák, P.; Čepl, J. Cultivar differences of total anthocyanins and anthocyanidins in red and purple-fleshed potatoes and their relation to antioxidant activity. *Food Chem.* **2009**, *114*, 836–843. [CrossRef]
50. Ahmed, F.A.; Ali, R.F.M. Bioactive Compounds and Antioxidant Activity of Fresh and Processed White Cauliflower. *BioMed Res. Int.* **2013**, *2013*, e367819. [CrossRef] [PubMed]
51. Konczak, I.; Zhang, W. Anthocyanins—More Than Nature's Colours. *BioMed Res. Int.* **2004**, *2004*, 239–240. [CrossRef]
52. Teow, C.C.; Truong, V.-D.; McFeeters, R.F.; Thompson, R.L.; Pecota, K.V.; Yencho, G.C. Antioxidant activities, phenolic and β-carotene contents of sweet potato genotypes with varying flesh colours. *Food Chem.* **2007**, *103*, 829–838. [CrossRef]
53. Kulling, S.E.; Rawel, H.M. Chokeberry (Aronia melanocarpa)—A review on the characteristic components and potential health effects. *Planta Med.* **2008**, *74*, 1625–1634. [CrossRef]
54. Nicoli, M.C.; Anese, M.; Parpinel, M. Influence of processing on the antioxidant properties of fruit and vegetables. *Trends Food Sci. Technol.* **1999**, *10*, 94–100. [CrossRef]
55. Ainsworth, P.; İbanoğlu, Ş.; Plunkett, A.; İbanoğlu, E.; Stojceska, V. Effect of brewers spent grain addition and screw speed on the selected physical and nutritional properties of an extruded snack. *J. Food Eng.* **2007**, *81*, 702–709. [CrossRef]
56. Stojceska, V.; Ainsworth, P.; Plunkett, A.; İbanoğlu, Ş. The effect of extrusion cooking using different water feed rates on the quality of ready-to-eat snacks made from food by-products. *Food Chem.* **2009**, *114*, 226–232. [CrossRef]
57. Gumul, D.; Ziobro, R.; Zięba, T.; Rój, E. The Influence of Addition of Defatted Blackcurrant Seeds on Pro-Health Constituents and Texture of Cereal Extrudates. *J. Food Qual.* **2011**, *34*, 395–402. [CrossRef]
58. Moussa-Ayoub, T.E.; Youssef, K.; El-Samahy, S.K.; Kroh, L.W.; Rohn, S. Flavonol profile of cactus fruits (Opuntia ficus-indica) enriched cereal-based extrudates: Authenticity and impact of extrusion. *Food Res. Int.* **2015**, *78*, 442–447. [CrossRef]
59. Anton, A.A.; Fulcher, R.G.; Arntfield, S.D. Physical and nutritional impact of fortification of corn starch-based extruded snacks with common bean (*Phaseolus vulgaris* L.) flour: Effects of bean addition and extrusion cooking. *Food Chem.* **2009**, *113*, 989–996. [CrossRef]
60. Ragaee, S.; Abdel-Aal, E.-S.M.; Noaman, M. Antioxidant activity and nutrient composition of selected cereals for food use. *Food Chem.* **2006**, *98*, 32–38. [CrossRef]
61. Robin, F.; Schuchmann, H.P.; Palzer, S. Dietary fiber in extruded cereals: Limitations and opportunities. *Trends Food Sci. Technol.* **2012**, *28*, 23–32. [CrossRef]
62. Potter, R.; Stojceska, V.; Plunkett, A. The use of fruit powders in extruded snacks suitable for Children's diets. *LWT Food Sci. Technol.* **2013**, *51*, 537–544. [CrossRef]
63. Chessari, C.J.; Sellahewa, J.N. *Extrusion Cooking Technologies and Applications*; Woodhead: Cambridge, UK, 2000; pp. 83–107.
64. van der Sman, R.G.M.; Broeze, J. Structuring of indirectly expanded snacks based on potato ingredients: A review. *J. Food Eng.* **2013**, *114*, 413–425. [CrossRef]
65. Colonna, P.; Mercier, C. Macromolecular modifications of manioc starch components by extrusion-cooking with and without lipids. *Carbohydr. Polym.* **1983**, *3*, 87–108. [CrossRef]
66. Chang, Y.K.; Silva, M.R.; Gutkoski, L.C.; Sebio, L.; Da Silva, M.A.A.P. Development of extruded snacks using jatobá (*Hymenaea stigonocarpa*Mart) flour and cassava starch blends. *J. Sci. Food Agric.* **1998**, *78*, 59–66. [CrossRef]
67. Altan, A.; McCarthy, K.L.; Maskan, M. Evaluation of snack foods from barley–tomato pomace blends by extrusion processing. *J. Food Eng.* **2008**, *84*, 231–242. [CrossRef]
68. Makowska, A.; Józefacki, A. Możliwość wykorzystania wybranych produktów odpadowych w produkcji wysokobłonnikowych wyrobów ekstrudowanych (Possibility of using of some waste by-products for producon of high fiber content extrudates). *Apar. Badaw. I Dydakt.* **2010**, *15*, 39–44. (In Polish)

69. Makowska, A.; Polcyn, A.; Chudy, S.; Michniewicz, J. Application of oat, wheat and rye bran to modify nutritional properties, physical and sensory characteristics of extruded corn snacks. *Acta Sci. Pol. Technol. Aliment.* **2015**, *14*, 375–386. [CrossRef]
70. Wójtowicz, A.; Kolasa, A.; Mościcki, L. Influence of Buckwheat Addition on Physical Properties, Texture and Sensory Characteristics of Extruded Corn Snacks. *Pol. J. Food Nutr. Sci.* **2013**, *63*, 239–244. [CrossRef]
71. Kowalczewski, P.; Lewandowicz, G.; Makowska, A.; Olejnik, A.; Obuchowski, W. Charakterystyka ekstrudowanych przekąsek zbożowych zawierających sok z ziemniaka (Characterisation of the extruded cereal snacks containing potato juice) I Polona. *Biul. IHAR* **2012**, *266*, 319–329. (In Polish)
72. Ačkar, Đ.; Jozinović, A.; Babić, J.; Miličević, B.; Panak Balentić, J.; Šubarić, D. Resolving the problem of poor expansion in corn extrudates enriched with food industry by-products. *Innov. Food Sci. Emerg. Technol.* **2018**, *47*, 517–524. [CrossRef]
73. Drożdż, W.; Boruczkowska, H.; Boruczkowski, T.; Tomaszewska-Ciosk, E.; Zdybel, E. Use of blackcurrant and chokeberry press residue in snack products. *Pol. J. Chem. Technol.* **2019**, *21*, 13–19. [CrossRef]
74. Artz, W.E.; Warren, C.; Villota, R. Twin-screw extrusion modification of a corn fiber and corn starch extruded blend. *J. Food Sci. Off. Publ. Inst. Food Technol.* **1990**, *55*, 746–754. [CrossRef]
75. Fernández-Gutiérrez, J.A.; San Martín-Martínez, E.; Martínez-Bustos, F.; Cruz-Orea, A. Physicochemical Properties of Casein-Starch Interaction Obtained by Extrusion Process. *Starch Stärke* **2004**, *56*, 190–198. [CrossRef]
76. Nowak, G.; Wójtowicz, A.; Golian, M. *Aronia Czarnoowocowa Jako Innowacyjny Dodatek Funkcjonalny do Ekstrudowanych Chrupek Kukurydzianych;* Innowacje w Praktyce (Black Chokeberry as Innovative Functional Addition to Extruded Corn Snacks, Practical Innovation); Centrum Innowacji Naukowo-Edukacyjnych: Lublin, Poland, 2017; pp. 113–114. (In Polish)
77. Drożdż, W.; Tomaszewska-Ciosk, E.; Zdybel, E.; Boruczkowska, H.; Boruczkowski, T.; Regiec, P. Effect of Apple and Rosehip Pomaces on Colour, Total Phenolics and Antioxidant Activity of Corn Extruded Snacks. *Pol. J. Chem. Technol.* **2014**, *16*, 7–11. [CrossRef]
78. Mokrzycki, W.S.; Tatol, M. Color difference ΔE: A survey. *Mach. Graph. Vis.* **2011**, *20*, 383–411.
79. Karma, I.G.M. Determination and Measurement of Color Dissimilarity. *Int. J. Eng. Emerg. Technol.* **2020**, *5*, 67–71. [CrossRef]
80. Yanniotis, S.; Petraki, A.; Soumpasi, E. Effect of pectin and wheat fibers on quality attributes of extruded cornstarch. *J. Food Eng.* **2007**, *80*, 594–599. [CrossRef]

applied sciences

MDPI

Article

Bioavailability of Citrulline in Watermelon Flesh, Rind, and Skin Using a Human Intestinal Epithelial Caco-2 Cell Model

Inah Gu [1], Olugbenga Balogun [2], Cindi Brownmiller [1], Hye Won Kang [3] and Sun-Ok Lee [1,*]

1 Department of Food Science, University of Arkansas, Fayetteville, AR 72704, USA
2 Applied Science and Technology, North Carolina Agricultural and Technical State University, Greensboro, NC 27411, USA
3 Food and Nutritional Sciences, Department of Family and Consumer Sciences, North Carolina Agricultural and Technical State University, Greensboro, NC 27411, USA
* Correspondence: sunok@uark.edu; Tel.: +1-479-575-6921

Abstract: Watermelon produces many byproducts (watermelon rind and skin) even though those components contain various bioactive compounds, including citrulline. This study evaluated the citrulline concentration, total phenolic content, and antioxidant activity (DPPH and FRAP assays) of different parts of watermelon and investigated the bioavailability of citrulline from different parts of watermelon using an in vitro human intestinal epithelial Caco-2 cell monolayer model. Solid-phase extracted watermelon flesh, rind, and skin samples were treated on a Caco-2 cell monolayer for 1, 2, and 4 h. The collected basolateral solution at each time point was analyzed for the percentage of citrulline transport. Watermelon flesh had the highest citrulline content, but the watermelon skin had the highest total phenolic content and antioxidant activity compared to other watermelon parts. The citrulline bioavailability showed greater % transport in watermelon skin than in watermelon flesh, rind, and L-citrulline standard. It may be due to the different food matrices of watermelon parts. This suggests that the utilization of watermelon by-products such as skin would help develop value-added products with better bioavailability of citrulline. However, since this study was conducted with an in vitro cell model, more extensive research with in vivo studies will be needed.

Keywords: watermelon; citrulline; bioavailability; watermelon flesh; watermelon rind; watermelon skin; Caco-2 cells

1. Introduction

Watermelon (*Citrullus lanatus*) belongs to the Cucurbitaceae family and is one of the most cultivated fruits, with approximately 102 million tons of world production in 2021 [1,2]. Watermelon is composed of flesh, rind, and skin. The flesh is the red pulp; the rind is the white or light green part between flesh and skin; and the skin is the green peel surface. The flesh part accounts for 40–50% of a watermelon, and the other 50–60% of the total mass of a watermelon is the rind and skin part [3,4]. The flesh is generally considered an edible part, and the rind and skin are discarded as by-products or used as animal feed [5]. The rind and skin produce an enormous amount of food waste from watermelon production to watermelon consumption, causing many environmental challenges [6].

Watermelon contains many nutrients and bioactive compounds, including vitamins, lycopene, citrulline, and phenolic compounds [7]. Due to these bioactive compounds, watermelon has shown many health benefits such as antioxidant, anti-diabetic, and anticancer effects [8–11]. Although watermelon rind and skin are normally discarded as by-products, watermelon rind and skin contain similar or higher total phenolic and citrulline content compared to watermelon flesh, showing strong antioxidant activities [12,13]. Thus, there has been increased attention towards the watermelon by-products and their utilization [14–16]. Converting watermelon by-products with high bioactive value to value-added products would benefit functional food industries and the human diet and have a better environmental impact with less food waste [3].

Citrulline is a non-essential non-proteinogenic amino acid, and watermelon is one of the rich sources of citrulline [17]. Watermelon rind has especially shown to have greater citrulline content than watermelon flesh [12,18]. Citrulline is a product of the nitric oxide (NO) cycle and a precursor of arginine [19]. Since arginine is a conditionally essential amino acid related to the NO system, citrulline has a potential role in vasodilation and cardiovascular functions [20–22]. Citrulline also showed hydroxyl radical scavenging activities [23]. Thus, many studies have been on the therapeutic activities of citrulline in watermelon [24–26]. However, the bioavailability of citrulline needs to be considered since it depends on its intestinal absorption and different food matrix [24,27–29]. There have been some studies on the bioavailability of citrulline in watermelon [27,30,31], but little is known about the bioavailability of citrulline from different watermelon parts. Therefore, this study evaluated the chemical composition of different watermelon parts (flesh, rind, and skin) and investigated the intestinal uptake of citrulline from watermelon flesh, rind, and skin using the differentiated monolayer human intestinal Caco-2 cell model.

2. Materials and Methods

2.1. Materials

Pre-extracted, dried samples of watermelon flesh, rind, and skin were prepared as described previously [5]. All media components and reagents were obtained from Gibco® through Thermo Fisher Scientific (Waltham, MA, USA).

2.2. Citrulline Concentration of Watermelon Flesh, Rind, and Skin

2.2.1. Solid Phase Extraction (SPE)

Solid phase extraction (SPE) was performed on pre-extracted watermelon flesh, skin, and rind samples before the analysis of citrulline. The pre-extracted watermelon samples (1 g) were reconstituted in 10 mL of deionized water. Sep-Pak® Vac 20 cc (5 g) cartridges (Waters Corp., Milford, MA, USA) were activated by washing with 20 mL of 100% methanol, followed by a 20 mL deionized water wash. The reconstituted aqueous extracts were bound to the columns, followed by a 20 mL deionized water wash. The remaining was eluted with 10 mL of 100% methanol. The methanol fraction was dried under nitrogen and reconstituted with 10 mL of deionized water.

2.2.2. Citrulline

Citrulline in watermelon samples was determined using the citrulline assay (Abcam, Cambridge, UK). Briefly, 50 µL of the sample was mixed with 5 µL of SDS solution and proteinase K solution in a tube. After 2 h incubation at 37 °C, assay reagents were added and incubated for 30 min at 95 °C. After cooling down at 4 °C for 5 min, the tube was centrifuged at 13,000× g for 10 min. The supernatant (200 µL) was transferred to a 96-well plate, and the absorbance was read on a microplate reader (Synergy HT Multi-Mode Microplate Reader, BioTek Instruments, Inc., Winooski, VT, USA) at 540 nm.

2.3. Total Phenolic Content and Antioxidant Activity

The total phenolic content in watermelon flesh, skin, and rind samples after SPE was determined using the modified Folin-Ciocalteu assay in Slinkard and Singleton [32], using gallic acid as standard. Briefly, samples were diluted 50 times in distilled water. After putting gallic acid standards and diluted samples (20 µL) in a 96-well plate, 100 µL of 0.2 N Folin–Ciocalteu reagent and 80 µL of 0.7 M sodium carbonate. The plate was incubated for 2 h at room temperature. The absorbance was read with a microplate reader (Synergy HT multimode microplate reader, BioTek, Winooski, VT, USA) at 760 nm. The results were expressed as milligrams of gallic acid equivalents (GAE) per gram of sample.

The free radical-scavenging activity of watermelon flesh, skin, and rind samples after SPE was evaluated using an adapted method of Akkari et al. [33] with 2,2-diphenyl-1-picrylhydrazyl (DPPH). Trolox was used as a reference standard. Samples were diluted 50 times in methanol. Trolox standards and diluted samples (10 μL) were added with 140 μL of 0.1 mM DPPH in methanol and incubated for 30 min in the dark at room temperature. The results were measured at 517 nm. The free radical-scavenging activity was calculated with the following equation: scavenging effect = $[(A0 - A1)/A0] \times 100$, where A0 was the absorbance of the control, and A1 was the absorbance in the presence of the sample. Data were expressed as micromole of Trolox equivalent (TE) per gram of sample, using the scavenging effect of the Trolox standard curve.

Ferric reduction antioxidant potential (FRAP) of watermelon flesh, skin, and rind samples after SPE was determined using QuantiChrom™ FRAP Assay Kit (BioAssay Systems, Hayward, CA, USA). Samples diluted 100 times were mixed with working reagent and incubated at room temperature for 40 min. The absorbance was measured at 590 nm. The results were expressed as micromole of ferrous (Fe^{2+}) equivalent per gram of sample.

2.4. Cell Culture and Measurement of Transepithelial Electrical Resistance (TEER)

Caco-2 (ATCC® HTB-37™) human intestinal epithelial cells were purchased from the American Type Culture Collection (ATCC, Manassas, VA, USA). Caco-2 cells were cultured in Dulbecco's modified Eagle's medium (DMEM) supplemented with 1% non-essential amino acids (NEAA), 10% fetal bovine serum (FBS), and 100 U/mL penicillin and 100 μg/mL streptomycin, named working media (WMEM) in this experiment and incubated at 37 °C in a humidified, 5% CO_2 incubator (VWR® Water Jacketed CO_2 incubator, VWR International, Radnor, PA, USA). Cells between passages 24 and 30 were used in the experiments. Caco-2 cells were seeded onto transwell inserts (polycarbonate membrane, 12 mm i.d., 0.4 μm pore size, Corning Inc., Kennebunk, ME, USA) in 12-well plates at a density of 1.0×10^5 cells/well. The media was changed every 2 days until the cells were used for the experiment 21 days after seeding. The integrity of the cell monolayer was measured by TEER using a Millicell-ERS Volt-Ohm Meter (MD Millipore Corporation, Billerica, MA, USA) at 21 days post-seeding. When TEER value reached 400 or greater ohms, the cells were treated with watermelon flesh, rind, and skin samples as well as L-citrulline standard to measure the rate of transport across the cell monolayer to determine the bioavailability of citrulline in watermelon from the three parts.

2.5. Cell Viability Assay

The concentration of watermelon flesh, rind, and skin samples for cell treatment was determined by cell viability assay using the CellTiter 96® AQueous One Solution Cell Proliferation Assay (MTS assay) (Promega Co., Madison, WI, USA) according to the manufacturer's instruction. Caco-2 cells were seeded at a density of 1.0×10^4 cells/well in 96-well plates and incubated in a 37 °C, 5% CO_2 incubator for 24 h. The cells were treated with 3 different citrulline concentrations (100, 200, and 300 μM citrulline) of watermelon flesh, rind, and skin and L-citrulline standard for 4 h. After 4 h, cell supernatant was replaced with fresh cell culture medium, and 20 μL of MTS reagent was added into each well. After 1 h incubation, the absorbance was read at 490 nm with a microplate reader. Cell viability was calculated as the percentage of living cells over control cells.

2.6. In Vitro Bioavailability Assay

Prior to each assay, WMEM in both apical and basolateral chambers of the transwell were aspirated. To measure the uptake of citrulline from the apical side of the Caco-2 cells, the cells were first deprived of amino acids (AA) in incubation for 15 min at 37 °C in phosphate-buffered saline (PBS). PBS was aspirated, and the cells were rinsed three times with ice-cold PBS. The treatment (500 μL) was added to the apical chamber, and 1.5 mL of WMEM was added to the basolateral chamber. The solution from the basolateral chamber was collected at 1, 2, and 4 h. The collected samples were mixed with 50 μL stop solution (25 μL of 100% trifluoroacetic acid and 25 μL of 70% ethanol) and frozen immediately. The concentration of citrulline in the watermelon flesh, rind, and skin treatments and collected basolateral solution at 1, 2, and 4 h was measured using a citrulline assay kit (Abcam, Cambridge, UK), as described in Section 2.2.2.

2.7. Statistical Analysis

All statistical analyses were performed using SAS 9.4 (SAS Institute Inc., Cary, NC, USA) with a one-way analysis of variance test (ANOVA) and Tukey's test. Data were expressed as means ± standard deviation (SD) for citrulline, TPC, and DPPH assay and as means ± standard error of the mean (SEM) for cell culture experiments. A difference of $p < 0.05$ was considered significant.

3. Results

3.1. Citrulline Concentration of Watermelon Flesh, Rind, and Skin

Citrulline concentration of watermelon flesh, rind, and skin was measured using citrulline assay (Figure 1). The watermelon flesh contained the highest citrulline concentration (1505 ± 66.9 μM), followed by skin (1113.5 ± 92.8 μM) and rind (780.5 ± 232 μM). Citrulline concentration in watermelon flesh was significantly higher than in the skin ($p <$ 0.05) and in the rind ($p < 0.01$).

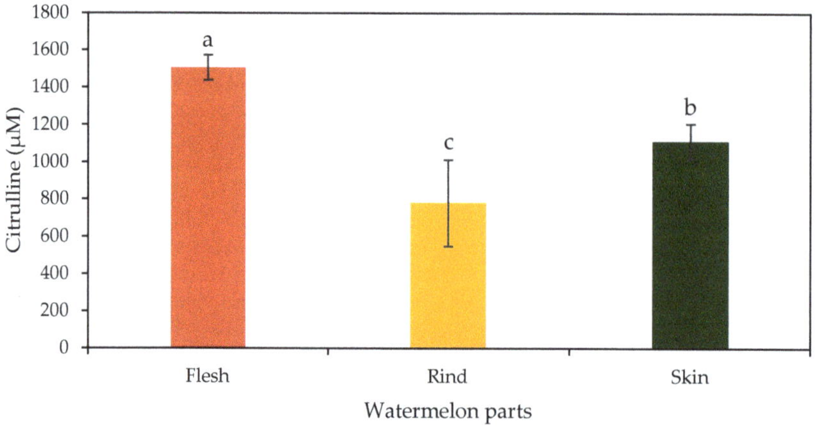

Figure 1. Citrulline concentration of watermelon flesh, rind, and skin. Data are expressed as the mean ± SD (n = 5). Means followed by a different letter are significantly different ($p < 0.05$).

3.2. Total Phenolic Content and Antioxidant Activity

The total phenolic content of watermelon flesh, rind, and skin was 12.7 ± 0.5, 2.4 ± 0.4, and 15.2 ± 0.6 mg GAE/g, respectively (Table 1). Watermelon skin contained the highest total phenolic content, followed by the flesh and rind. The total phenolic content of the rind was significantly lower than other parts ($p < 0.05$). Antioxidant activity of watermelon flesh, rind, and skin tested with DPPH and FRAP assays showed a similar pattern as the total phenolic content. As the result of DPPH assay, watermelon skin (31.1 ± 6.2 μmol TE/g) showed the highest antioxidant activity, followed by watermelon flesh (28.7 ± 3.5 μmol TE/g) and rind (7.0 ± 2.7 μmol TE/g). The free radical-scavenging activities of watermelon skin and flesh were significantly higher than the rind ($p < 0.05$). In FRAP assay, ferric reduction antioxidant potential of watermelon flesh, rind, and skin was 181.9 ± 7.2, 40.1 ± 9.0, and 240.0 ± 8.4 μmol Fe^{2+}/g, respectively. Watermelon skin showed significantly higher FRAP compared to the flesh ($p < 0.05$) and skin ($p < 0.01$).

Table 1. Total phenolic content and antioxidant activity of watermelon parts.

Sample	Total Phenolic Content (mg GAE/g)	DPPH (μmol TE/g)	FRAP (μmol Fe^{2+}/g)
Flesh	12.7 ± 0.5 [b]	28.7 ± 3.5 [a]	181.9 ± 7.2 [b]
Rind	2.4 ± 0.4 [c]	7.0 ± 2.7 [b]	40.1 ± 9.0 [c]
Skin	15.2 ± 0.6 [a]	31.1 ± 6.2 [a]	240.0 ± 8.4 [a]

Values represent mean ± SD (n = 20 for total phenolic content and DPPH; n = 15 for FRAP). Means within columns with different letters are significantly different ($p < 0.05$). DPPH = 2,2-diphenyl-1-picrylhydrazyl; FRAP = ferric reduction antioxidant potential; GAE = gallic acid equivalent; TE = trolox equivalent.

3.3. Cell Viability of Watermelon Flesh, Rind, and Skin

The cell cytotoxicity of watermelon flesh, rind, and skin on Caco-2 cells was determined by using MTS assay (Figure 2). Watermelon parts (flesh, rind, and skin) containing 100–300 μM citrulline and L-citrulline were treated onto Caco-2 cells for 4 h. As a result, all treatments containing 100 μM citrulline did not affect the cell cytotoxicity of Caco-2 cells. However, the rind with 200 and 300 μM citrulline and skin with 300 μM citrulline showed significant cytotoxicity compared to the control ($p < 0.05$). Therefore, watermelon parts with 100 μM citrulline were used as the treatment concentration for in vitro bioavailability assay.

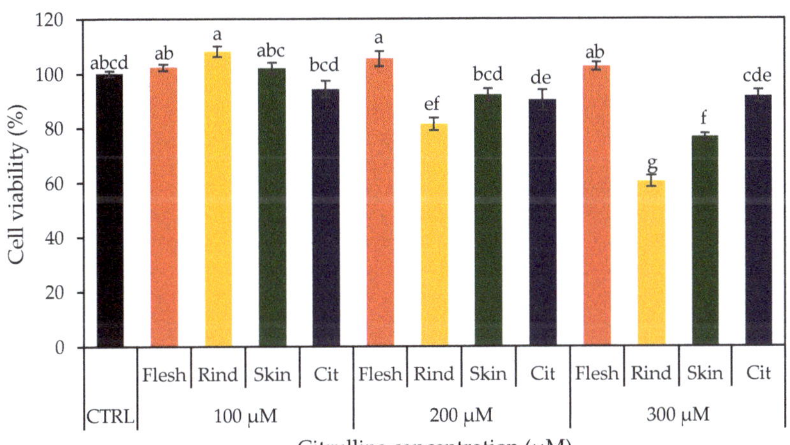

Figure 2. Cell viability of Caco-2 cells treated with different parts of watermelon samples and L-citrulline (100–300 μM citrulline). Data are expressed as the mean ± SEM (n = 12). Means followed by a different letter are significantly different at the 5% level. CTRL = control; Cit = L-citrulline.

3.4. In Vitro Bioavailability of Citrulline in Watermelon Flesh, Rind, and Skin

The concentration of citrulline in watermelon flesh, rind, and skin treatments and in basolateral solution collected at 1, 2, and 4 h was measured using citrulline assay (Figure 3). There were no significant differences in the % transport of each treatment among the different treatment times. This showed that the absorption of citrulline in different watermelon parts and L-citrulline were mostly processed in 1 h. The watermelon skin showed higher % transport at all time points than L-citrulline. Especially at 1 h treatment, % transport was significantly higher in the watermelon skin treatment compared to L-citrulline treatment. The highest % transport in each treatment occurred at 1 h for watermelon skin, at 2 h for watermelon flesh and L-citrulline, and at 4 h for watermelon rind. Overall, it showed transport of citrulline in all watermelon parts and the standard ranged from 30.3 to 36.6%.

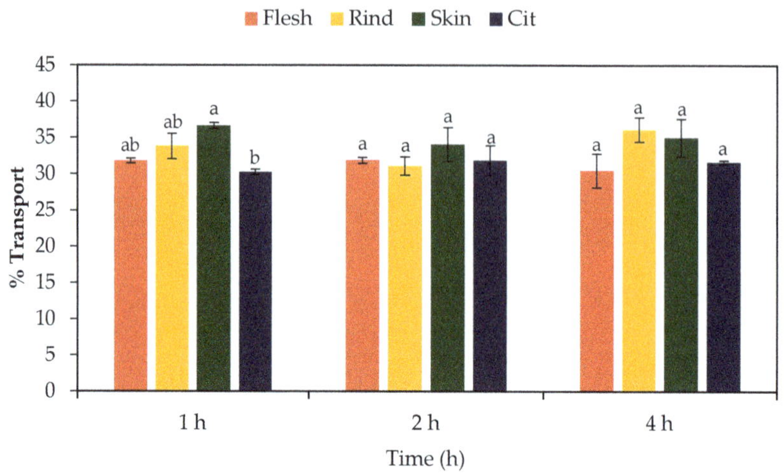

Figure 3. Transport of citrulline in watermelon samples using Caco-2 cell assays after 1, 2, and 4 h incubation. Data are presented as mean ± SEM (n = 2). % Transport was calculated as the percentage transported across the cell monolayer from the apical to the basolateral chamber. Different letters in the same treatment time indicate significant differences between results at the 5% level.

4. Discussion

Watermelon (*Citrullus lanatus*) has many nutritional and beneficial properties. A total of 71 phenolic and other polar compounds have been characterized in watermelon [34]. One of those compounds, citrulline, is a non-essential amino acid. Citrulline was first identified in the juice of watermelon [35,36]. It has been reported that watermelon juice contains ~2.33 g of citrulline per L of unpasteurized watermelon juice [37]. Citrulline has potent antioxidant activity and plays an important role in the nitric oxide (NO) system because it is a precursor of arginine. Citrulline may offer therapeutic strategies for controlling NO metabolism disorders and improving cardiovascular ability [4,24]. Citrulline has also been shown to reduce the recovery heart rate and muscle soreness after 24 h [31].

As the attention to food waste and byproducts for being potential biosources increased, many studies have identified and quantified citrulline content in watermelon rind and skin as well as watermelon flesh. Jayaprakasha et al. [18] measured the concentration of L-citrulline in three different varieties of watermelon juice and rinds. In all varieties, rinds contained higher L-citrulline (13.95–28.46 mg/g dry weight (dw)) compared to watermelon juice (11.25–16.73 mg/g dw). Ridwan et al. [38] investigated L-citrulline content in watermelons (flesh and rind) grown and consumed in Malaysia and showed that L-citrulline content was higher in the rind (45.02 mg/g) than in the flesh (43.81 mg/g) of red watermelon juice extract and showed similar trends in yellow crimson watermelon juice extract, with 16.61 mg/g in rind and 15.77 mg/g in flesh detected. Casacchia et al. [39] determined bioactive compounds extracted from watermelon pulp and rind with nine different watermelon cultivars from different origins. The concentration of L-citrulline in fresh rind was significantly higher than in fresh pulp except in watermelons from Latina, Italia, and Santana, Romania, showing up to 2.6 mg/g L-citrulline in fresh watermelon rind. Pp et al. [40] quantified citrulline from the 'Sugar Baby' variety of watermelon rinds, and the concentration of citrulline was 13.36 mg/g dw in the rind and 9.78 mg/g dw in the skin. Tarazona-Díaz et al. [12] determined the citrulline content in the rind and flesh of five different watermelon cultivars. Citrulline level in the rind (2.0–7.2 g/kg fresh weight (fw)) was higher than that in flesh (1.1–4.7 g/kg fw) in all five watermelon cultivars. Akashi et al. [41] investigated the accumulation pattern of citrulline in mature watermelons. Citrulline showed a bipolar accumulation pattern in mature watermelon, as the concentration of citrulline was highest in the outer peels (4.4 ± 0.8 g/kg fw), decreased in inner rinds (2.1 ± 0.94 g/kg fw) and peripheral flesh fractions (0.83 ± 0.36 g/kg fw), but increased toward the center of flesh. Rimando and Perkins-Veazie [4] measured the citrulline level in the flesh and rind of watermelons with different flesh colors (red, yellow, and orange). A higher level of citrulline was detected in the rind (15.6–29.4 mg/g dw) than in the flesh (7.9–28.5 mg/g dw), but it was the opposite on a fresh weight basis, showing higher citrulline content in fresh flesh (1.0–3.5 mg/g fw) than in fresh rind (0.8–1.5 mg/g fw). This can be explained by the higher moisture level of rind (95%) than flesh (90%). Fan et al. [30] compared the bioactive components of watermelon flesh, rind, and seeds and showed that the highest citrulline content was detected in watermelon rind (19.3 ± 1.1 mg/100 kcal) followed by watermelon flesh (10.0 ± 1.0 mg/100 kcal) and watermelon seeds (1.4 ± 0.2 mg/100 kcal). These previous studies reported that watermelon rind has higher L-citrulline than watermelon flesh. The present study appears to have the opposite results showing the highest citrulline content in watermelon flesh, followed by the skin and rind (Figure 1). However, the opposite results can be explained by the word 'rind' used in some previous studies [12,18,30,38,39], because 'rind' in these previous studies indicated the white rind and green skin, which are all byproduct parts other than flesh.

In addition to the citrulline concentration, total phenolic content and antioxidant activity of watermelon parts were investigated in the present study. Watermelon skin showed the highest total phenolic content, free-radical scavenging activity, and ferric reduction antioxidant potential followed by watermelon flesh and skin. These results were consistent with previously reported studies [42,43]. Green rind (skin) showed the highest total phenolic content (0.7 mg GAE/g), followed by the flesh (0.3 mg GAE/g) and the white rind (0.2 mg GAE/g), with a similar pattern in DPPH radical scavenging activity [42]. The green rind (skin) extract treated at 250 °C showed the maximum ferric reducing power among the watermelon flesh, white rind, and green rind (skin) parts [42]. Yusoff et al. [43] extracted watermelon rind using ultrasound-assisted extraction, and it contained 15.1 ± 0.6 mg GAE/g, showing similar total phenolic content as the present study. Total phenolic content of red- and yellow-fleshed watermelon rind powders extracted using different solvents (water, methanol, ethanol, and acetone) reported values 1.1–2.2 mg GAE/g, which was lower than our result (rind 2.4 mg GAE/g) [44]. However, there were several studies not consistent with the present study [45,46]. The white part of watermelon peels (rind) contained 63.3 ± 1.5 mg TAE (tannic acid equivalent)/g, showing higher total phenolic content than the red flesh (47.3 ± 0.9 mg TAE/g) [45]. In Neglo et al. [46], watermelon skin showed the highest total phenolic content (0.087 ± 0.002 mg GAE/g) and DPPH (55.8 ± 2.4 %), but watermelon rind was higher than the flesh in both total phenolic content and DPPH.

It is important to investigate intestinal absorption of citrulline because the bioavailability of citrulline depends on intestinal absorption and different food matrices [24,27]. There have been several studies on citrulline uptake [27,30]. Bahri et al. [27] investigated the mechanisms and kinetics of citrulline uptake using a Caco-2 human intestinal epithelial cell model. They found that citrulline uptake was pH-independent and that the uptake rate went down without Na^+. Fan et al. [30] investigated the concentration of L-citrulline in plasma over 24 h after watermelon flesh, rind, and seed intake. The maximum concentration showed between 1 and 2 h after watermelon flesh, rind, and seed intake, especially watermelon flesh and rind showing significant increase compared to the control ($p < 0.05$). However, the transport of citrulline from different watermelon parts in the intestine is poorly understood. In the present study, the bioavailability of citrulline was slightly greater in the watermelon skin than in the watermelon flesh and rind (Figure 3). In addition, the bioavailability of L-citrulline standard was similar or lower than the bioavailability of citrulline in different watermelon parts. This difference may be because the citrulline standard was not contained within the matrix of the watermelon. Tarazona-Díaz et al. [31] performed an in vitro intestinal absorption of L-citrulline by using Caco-2 cells, with unpasteurized watermelon juice, pasteurized watermelon juice, and L-citrulline standard. L-citrulline in unpasteurized watermelon juice showed higher absorption than in the L-citrulline standard, suggesting that the transport and bioavailability of L-citrulline are greater when it was contained in a matrix of watermelon juice than as a pure compound.

5. Conclusions

In this study, we evaluated the composition of different parts of watermelon and investigated the bioavailability of citrulline from different parts of watermelon using an in vitro human intestinal Caco-2 cell monolayer model. Watermelon flesh had the highest citrulline content, but the watermelon skin had the highest total phenolic content with higher radical scavenging activity and ferric reduction antioxidant potential than watermelon flesh and skin. The citrulline bioavailability showed higher % transport in watermelon skin than in watermelon flesh, rind, and even L-citrulline standard. It may be due to the different food matrices of watermelon parts. This suggests that the utilization of watermelon byproducts such as skin would help develop value-added products with higher bioactive compounds and better bioavailability of citrulline. However, since this study was conducted with an in vitro cell model, more extensive research with in vivo studies will be needed.

Author Contributions: Conceptualization, S.-O.L.; validation, I.G., O.B., H.W.K., C.B. and S.-O.L.; investigation, I.G. and C.B; writing—original draft preparation, I.G.; writing—review and editing, S.-O.L., H.W.K., O.B., C.B. and I.G.; supervision, S.-O.L. and H.W.K.; funding acquisition, S.-O.L. and H.W.K. All authors have read and agreed to the published version of the manuscript.

Funding: This work was supported by the National Watermelon Promotion Board.

Institutional Review Board Statement: Not applicable.

Informed Consent Statement: Not applicable.

Data Availability Statement: Not applicable.

Conflicts of Interest: The authors declare no conflict of interest.

References

1. Zamuz, S.; Munekata, P.E.S.; Gullón, B.; Rocchetti, G.; Montesano, D.; Lorenzo, J.M. *Citrullus lanatus* as Source of Bioactive Components: An up-to-Date Review. *Trends Food Sci. Technol.* **2021**, *111*, 208–222. [CrossRef]
2. Feng, Z.; Bi, Z.; Fu, D.; Feng, L.; Min, D.; Bi, C.; Huang, H. A Comparative Study of Morphology, Photosynthetic Physiology, and Proteome between Diploid and Tetraploid Watermelon (*Citrullus lanatus* L.). *Bioengineering* **2022**, *9*, 746. [CrossRef] [PubMed]
3. Zia, S.; Khan, M.R.; Shabbir, M.A.; Aadil, R.M. An Update on Functional, Nutraceutical and Industrial Applications of Watermelon by-Products: A Comprehensive Review. *Trends Food Sci. Technol.* **2021**, *114*, 275–291. [CrossRef]
4. Rimando, A.M.; Perkins-Veazie, P.M. Determination of Citrulline in Watermelon Rind. *J. Chromatogr. A* **2005**, *1078*, 196–200. [CrossRef] [PubMed]
5. Balogun, O.; Otieno, D.; Brownmiller, C.R.; Lee, S.-O.; Kang, H.W. Effect of Watermelon (*Citrullus lanatus*) Extract on Carbohydrates-Hydrolyzing Enzymes In Vitro. *Agriculture* **2022**, *12*, 772. [CrossRef]
6. Esparza, I.; Jiménez-Moreno, N.; Bimbela, F.; Ancín-Azpilicueta, C.; Gandía, L.M. Fruit and Vegetable Waste Management: Conventional and Emerging Approaches. *J. Environ. Manag.* **2020**, *265*, 110510. [CrossRef] [PubMed]
7. Romdhane, M.B.; Haddar, A.; Ghazala, I.; Jeddou, K.B.; Helbert, C.B.; Ellouz-Chaabouni, S. Optimization of Polysaccharides Extraction from Watermelon Rinds: Structure, Functional and Biological Activities. *Food Chem.* **2017**, *216*, 355–364. [CrossRef]
8. Tlili, I.; Hdider, C.; Lenucci, M.S.; Ilahy, R.; Jebari, H.; Dalessandro, G. Bioactive Compounds and Antioxidant Activities during Fruit Ripening of Watermelon Cultivars. *J. Food Compos. Anal.* **2011**, *24*, 923–928. [CrossRef]
9. Ahn, J.; Choi, W.; Kim, S.; Ha, T. Anti-Diabetic Effect of Watermelon (*Citrullus vulgaris* Schrad) on Streptozotocin-Induced Diabetic Mice. *Food Sci. Biotechnol.* **2011**, *20*, 251–254. [CrossRef]
10. Ali, O.M.; Hasanin, M.S.; Suleiman, W.B.; Helal, E.E.-H.; Hashem, A.H. Green Biosynthesis of Titanium Dioxide Quantum Dots Using Watermelon Peel Waste: Antimicrobial, Antioxidant, and Anticancer Activities. *Biomass Conv. Bioref.* **2022**. [CrossRef]
11. Di Sano, C.; Lazzara, V.; Durante, M.; D'Anna, C.; Bonura, A.; Dino, P.; Uasuf, C.G.; Pace, E.; Lenucci, M.S.; Bruno, A. The Protective Anticancer Effect of Natural Lycopene Supercritical CO_2 Watermelon Extracts in Adenocarcinoma Lung Cancer Cells. *Antioxidants* **2022**, *11*, 1150. [CrossRef] [PubMed]
12. Tarazona-Díaz, M.P.; Viegas, J.; Moldao-Martins, M.; Aguayo, E. Bioactive Compounds from Flesh and By-Product of Fresh-Cut Watermelon Cultivars. *J. Sci. Food Agric.* **2011**, *91*, 805–812. [CrossRef]
13. Din, S.N.; Mubarak, A.; Lani, M.N.; Yahaya, M.Z.; Wan Abdullah, W.Z. Development of Pastilles from Flesh and Rind of Watermelon. *Food Res.* **2022**, *6*, 288–297. [CrossRef]

14. Rico, X.; Gullón, B.; Alonso, J.L.; Yáñez, R. Recovery of High Value-Added Compounds from Pineapple, Melon, Watermelon and Pumpkin Processing by-Products: An Overview. *Food Res. Int.* **2020**, *132*, 109086. [CrossRef]
15. Zia, S.; Khan, M.R.; Aadil, R.M.; Shahid, M. Development and Storage Stability of Value-Added Watermelon Fruit Butter by Incorporating Watermelon Rind Byproduct. *J. Food Process. Preserv.* **2022**, *46*, e17031. [CrossRef]
16. Ashoka, S.; Shamshad Begum, S.; Vijayalaxmi, K.G. Byproduct Utilization of Watermelon to Develop Watermelon Rind Flour Based Cookies. *Pharma Innov.* **2021**, *10*, 196–199. [CrossRef]
17. Joshi, V.; Joshi, M.; Silwal, D.; Noonan, K.; Rodriguez, S.; Penalosa, A. Systematized Biosynthesis and Catabolism Regulate Citrulline Accumulation in Watermelon. *Phytochemistry* **2019**, *162*, 129–140. [CrossRef]
18. Jayaprakasha, G.K.; Chidambara Murthy, K.N.; Patil, B.S. Rapid HPLC-UV Method for Quantification of l-Citrulline in Watermelon and Its Potential Role on Smooth Muscle Relaxation Markers. *Food Chem.* **2011**, *127*, 240–248. [CrossRef]
19. Wu, G.; Meininger, C.J. Analysis of Citrulline, Arginine, and Methylarginines Using High-Performance Liquid Chromatography. In *Methods in Enzymology*; Nitric Oxide, Part F; Academic Press: Cambridge, MA, USA, 2008; Volume 440, pp. 177–189.
20. Levine, A.B.; Punihaole, D.; Levine, T.B. Characterization of the Role of Nitric Oxide and Its Clinical Applications. *Cardiology* **2012**, *122*, 55–68. [CrossRef]
21. Volino-Souza, M.; Oliveira, G.V.; Vargas, R.; Tavares, A.C.; Conte-Junior, C.A.; Alvares, T. da S. Effect of Microencapsulated Watermelon (*Citrullus lanatus*) Intake on Plasma Amino Acids and Glycemic Response in Healthy Adults. *Food Biosci.* **2022**, *46*, 101553. [CrossRef]
22. Hong, M.Y.; Hartig, N.; Kaufman, K.; Hooshmand, S.; Figueroa, A.; Kern, M. Watermelon Consumption Improves Inflammation and Antioxidant Capacity in Rats Fed an Atherogenic Diet. *Nutr. Res.* **2015**, *35*, 251–258. [CrossRef] [PubMed]
23. Akashi, K.; Miyake, C.; Yokota, A. Citrulline, a Novel Compatible Solute in Drought-Tolerant Wild Watermelon Leaves, Is an Efficient Hydroxyl Radical Scavenger. *FEBS Lett.* **2001**, *508*, 438–442. [CrossRef] [PubMed]
24. Bahri, S.; Zerrouk, N.; Aussel, C.; Moinard, C.; Crenn, P.; Curis, E.; Chaumeil, J.-C.; Cynober, L.; Sfar, S. Citrulline: From Metabolism to Therapeutic Use. *Nutrition* **2013**, *29*, 479–484. [CrossRef] [PubMed]
25. Romero, M.J.; Platt, D.H.; Caldwell, R.B.; Caldwell, R.W. Therapeutic Use of Citrulline in Cardiovascular Disease. *Cardiovasc. Drug Rev.* **2006**, *24*, 275–290. [CrossRef]
26. Rashid, J.; Kumar, S.S.; Job, K.M.; Liu, X.; Fike, C.D.; Sherwin, C.M. Therapeutic Potential of Citrulline as an Arginine Supplement: A Clinical Pharmacology Review. *Paediatr. Drugs* **2020**, *22*, 279–293. [CrossRef]
27. Bahri, S.; Curis, E.; El Wafi, F.-Z.; Aussel, C.; Chaumeil, J.-C.; Cynober, L.; Zerrouk, N. Mechanisms and Kinetics of Citrulline Uptake in a Model of Human Intestinal Epithelial Cells. *Clin. Nutr.* **2008**, *27*, 872–880. [CrossRef]
28. Xu, Y.; Li, Y.; Xie, J.; Xie, L.; Mo, J.; Chen, W. Bioavailability, Absorption, and Metabolism of Pelargonidin-Based Anthocyanins Using Sprague–Dawley Rats and Caco-2 Cell Monolayers. *J. Agric. Food Chem.* **2021**, *69*, 7841–7850. [CrossRef]
29. Liu, C.-S.; Glahn, R.P.; Liu, R.H. Assessment of Carotenoid Bioavailability of Whole Foods Using a Caco-2 Cell Culture Model Coupled with an in Vitro Digestion. *J. Agric. Food Chem.* **2004**, *52*, 4330–4337. [CrossRef]
30. Fan, J.; Park, E.; Zhang, L.; Edirisinghe, I.; Burton-Freeman, B.; Sandhu, A.K. Pharmacokinetic Parameters of Watermelon (Rind, Flesh, and Seeds) Bioactive Components in Human Plasma: A Pilot Study to Investigate the Relationship to Endothelial Function. *J. Agric. Food Chem.* **2020**, *68*, 7393–7403. [CrossRef]
31. Tarazona-Díaz, M.P.; Alacid, F.; Carrasco, M.; Martínez, I.; Aguayo, E. Watermelon Juice: Potential Functional Drink for Sore Muscle Relief in Athletes. *J. Agric. Food Chem.* **2013**, *61*, 7522–7528. [CrossRef]
32. Slinkard, K.; Singleton, V.L. Total Phenol Analysis: Automation and Comparison with Manual Methods. *Am. J. Enol. Vitic.* **1977**, *28*, 49–55. [CrossRef]
33. Akkari, H.; Hajaji, S.; B'chir, F.; Rekik, M.; Gharbi, M. Correlation of Polyphenolic Content with Radical-Scavenging Capacity and Anthelmintic Effects of Rubus Ulmifolius (*Rosaceae*) against *Haemonchus contortus*. *Vet. Parasitol.* **2016**, *221*, 46–53. [CrossRef] [PubMed]
34. Abu-Reidah, I.M.; Arráez-Román, D.; Segura-Carretero, A.; Fernández-Gutiérrez, A. Profiling of Phenolic and Other Polar Constituents from Hydro-Methanolic Extract of Watermelon (*Citrullus lanatus*) by Means of Accurate-Mass Spectrometry (HPLC–ESI–QTOF–MS). *Food Res. Int.* **2013**, *51*, 354–362. [CrossRef]
35. Koga, Y. Study report on the constituents of squeezed watermelon. *Tokyo Kagaku Kaishi [J. Tokyo Chem. Soc.]* **1914**, *35*, 519–528.
36. Wada, M. On the Occurrence of a New Amino Acid in Watermelon, *Citrullus vulgaris*, Schrad. *Bull. Agric. Chem. Soc. Jpn.* **1930**, *6*, 32–34. [CrossRef]
37. Bailey, S.J.; Blackwell, J.R.; Williams, E.; Vanhatalo, A.; Wylie, L.J.; Winyard, P.G.; Jones, A.M. Two Weeks of Watermelon Juice Supplementation Improves Nitric Oxide Bioavailability but Not Endurance Exercise Performance in Humans. *Nitric Oxide* **2016**, *59*, 10–20. [CrossRef] [PubMed]
38. Ridwan, R.; Abdul Razak, H.R.; Adenan, M.I.; Md Saad, W.M. Development of Isocratic RP-HPLC Method for Separation and Quantification of L-Citrulline and L-Arginine in Watermelons. *Int. J. Anal. Chem.* **2018**, *2018*, 4798530. [CrossRef] [PubMed]
39. Casacchia, T. Nutraceutical properties and health-promoting biological activities of fruits of watermelon cultivars with different origins. *Farmacia* **2020**, *68*, 679–686. [CrossRef]
40. Pp, S.B.; Sagar, V.R.; Kar, A.; Varghese, E.; Singh, S.; Choudhary, H. Identification and Quantification of Physicochemical and Bioactive Components from Sugar Baby Variety of Watermelon (*Citrullus lanatus*). *Agric. Res.* **2022**, *11*, 410–420. [CrossRef]

41. Akashi, K.; Mifune, Y.; Morita, K.; Ishitsuka, S.; Tsujimoto, H.; Ishihara, T. Spatial Accumulation Pattern of Citrulline and Other Nutrients in Immature and Mature Watermelon Fruits. *J. Sci. Food Agric.* **2017**, *97*, 479–487. [CrossRef]
42. Kim, S.-J.; Matsushita, Y.; Fukushima, K.; Aoki, D.; Yagami, S.; Yuk, H.-G.; Lee, S.-C. Antioxidant Activity of a Hydrothermal Extract from Watermelons. *LWT Food Sci. Technol.* **2014**, *59*, 361–368. [CrossRef]
43. Yusoff, I.M.; Chua, L.S.; Taher, Z.M. Valorization of Fruit Waste from Cucurbitaceae Family: Profiling of Phytoconstituent of Benincasa Hispida and *Citrullus lanatus* Rinds Using Ultrasound-Assisted Extraction. *Food Biosci.* **2023**, *51*, 102190. [CrossRef]
44. Ho, L.-H.; Ramli, N.F.; Tan, T.-C.; Muhamad, N.; Haron, M.N. Effect of Extraction Solvents and Drying Conditions on Total Phenolic Content and Antioxidant Properties of Watermelon Rind Powder. *Sains Malays.* **2018**, *47*, 99–107. [CrossRef]
45. Dieng, S.I.M.; Diallo, A.J.; Fall, A.D.; Diatta-Badji, K.; Diatta, W.; Sarr, A.; Bassene, E. Total Polyphenols and Flavonoids Contents of Aqueous Extracts of Watermelon Red Flesh and Peels (*Citrullus lanatus*, Thunb). *J. Pharm. Phytochem.* **2017**, *6*, 801–803.
46. Neglo, D.; Tettey, C.O.; Essuman, E.K.; Kortei, N.K.; Boakye, A.A.; Hunkpe, G.; Amarh, F.; Kwashie, P.; Devi, W.S. Comparative Antioxidant and Antimicrobial Activities of the Peels, Rind, Pulp and Seeds of Watermelon (*Citrullus lanatus*) Fruit. *Sci. Afr.* **2021**, *11*, e00582. [CrossRef]

applied sciences

MDPI

Article

Influence of Cryogenic Grinding on the Nutritional and Antinutritional Components of Rapeseed Cake

Klara Kraljić *, Dubravka Škevin, Nikolina Čukelj Mustač, Maja Benković, Saša Drakula, Sandra Balbino, Ana Mandura Jarić, Karla Mamilović, Ivana Ramljak and Duška Ćurić

Faculty of Food Technology and Biotechnology, University of Zagreb, Pierottijeva 6, 10000 Zagreb, Croatia; dubravka.skevin@pbf.unizg.hr (D.Š.); nikolina.cukelj@pbf.unizg.hr (N.Č.M.); maja.benkovic@pbf.unizg.hr (M.B.); sasa.drakula@pbf.unizg.hr (S.D.); sandra.balbino@pbf.unizg.hr (S.B.); karla.mamilovic@gmail.com (K.M.); iramljak0@gmail.com (I.R.); duska.curic@pbf.unizg.hr (D.Ć.)

* Correspondence: klara.kraljic@pbf.unizg.hr

Abstract: We investigated the influence of cryogenic grinding on the quality of rapeseed cake. Rapeseed cake is a good source of valuable proteins (30%) and oil (14%), with a balanced fatty acid composition and a fair amount of sterols, which may reduce the risk of cardiovascular diseases. However, the presence of antinutritive compounds prevents its use as a food source. Grinding under cryogenic conditions is much more efficient than grinding at room temperature in terms of particle size reduction. The additional cryogenic grinding of the cake had little effect on the nutritional components, as phytosterols and soluble dietary fiber increased slightly. It had no effect on insoluble dietary fiber, polyphenols, and tannins. Prolonged grinding time, both at room and subzero temperatures, reduced the total amount of glucosinolates by 34 and 43%, respectively. However, the reduction in undesirable components is not sufficient to use cryogenic grinding as the sole treatment for the cake, but it could be a good pretreatment for chemical or biological treatments.

Keywords: rapeseed cake; cryogenic grinding; dietary fibre; polyphenols; glucosinolates

1. Introduction

According to the Food and Agriculture Organization of the United Nations (FAO), demand for food will roughly double by 2050 due to the growing world population. One of the biggest challenges will be meeting the growing demand for meat, as it is estimated that meat production will have to increase by over 200 million tons per year [1]. Population pressures aside, environmental considerations and production efficiencies underscore the importance of new high-quality sources of protein for human nutrition, such as plants, for both developed and developing countries. And rapeseed could be a good local source. Rapeseed is used to produce high-quality oil, but the by-products of this production, rapeseed meal and cake, have a high protein value. The balance and bioavailability of amino acids and the digestibility of proteins are comparable to those of soybean meal [2,3]. In fact, rapeseed cake and meal are the second most important feed source after soybean meal [4] and have the chance to be an economical and sustainable source of protein in food [2]. In addition, research has shown that rapeseed meal has good water and oil binding capacity as well as very good emulsifying and foaming properties, which are of great importance for incorporation into novel functional foods [5]. However, despite decades of research, development of various technologies, and large-scale manufacturing, there are still no commercially available rapeseed cake products for human consumption [2]. The use of rapeseed meal and cake in human nutrition is limited by the presence of antinutritional components such as glucosinolates, phenolic compounds, tannins, phytates, and hulls [6].

The main concern when using rapeseed meal and cake as food ingredients are glucosinolates. This is a large group of sulphur-containing secondary plant metabolites found in all economically important varieties of Brassica. There is great diversity among glucosinolates, and more than 120 different compounds have been identified [7]. Although some glucosinolates possess antibacterial and antifungal properties as well as anticancer activity, the antinutritional effects of glucosinolates and their degradation products present in rapeseed meal and cake have limited the use of rapeseed by-products as a food source [8]. They are known to reduce animal growth, cause iodine deficiency, and cause hypertrophy of liver, kidney, and thyroid [7]. In addition, these compounds provide a bitter taste and sulphurous aroma due to their degradation products. Therefore, one of the most important breeding objectives is to reduce the glucosinolate content in the seed to trace amounts [8].

Commercial rapeseed cakes are rich in phenolic compounds, especially derivatives of sinapic acid, which account for about 80–99% of total phenols, mainly in the form of esters and glucosides. Although they are known to be potent free radical scavengers and antioxidants [9], their content in rapeseed meal and cake is up to 30 times higher than in by-products from other seeds. These high concentrations of phenolic compounds may contribute to the dark colour, bitter taste, and astringency of rapeseed meal. In addition, phenols and their oxidized products can form complexes with essential amino acids, enzymes, and other substances, lowering the nutritional value of meal and cake. Another group of phenolic constituents found in canola are condensed tannins. Tannins are complexed phenolic compounds formed by polymerization of flavan-3-ols or flavan-3,4-diols [10]. Similar to phenolic acids, they also have antioxidant activity [11], and they are also of concern due to their ability to bind with proteins. Tannins form complexes with lysine and methionine, making them nutritionally unavailable. Therefore, the phenolic compounds and tannins, as well as their concentrations, are important factors when canola meal is considered a protein source in food formulations [10].

Great efforts have been made to reduce undesirable components in rapeseed meal and cake and to reduce their antinutritional effects, which would allow their use in human nutrition. In recent decades, new varieties with lower glucosinolate and hull content (and thus insoluble dietary fibre—IDF) have been developed. Phytates are removed by enzymatic methods, while various extraction methods are used for phenolic compounds. New techniques are also being explored, such as ultrasound, pulsed electric field, and micronization, but so far none of these techniques are being used commercially [2]. Micronization technology is developing rapidly as it not only improves the functional and sensory properties of the ground material but also increases the bioavailability of nutrients and the rate of biological and chemical reactions. Various methods can be used to reduce particle size [12]. It has been reported that cryogenic grinding increases the production of fine particles and reduces the energy consumption for grinding the materials by increasing their brittleness. This could be very important for oily materials such as cake and meal residues after oil extraction, as the extremely low temperatures in the grinder solidify the oils, causing the materials to crumble easily [13]. In our research group, experiments were performed on millet bran [14] and pumpkin seed cake [15] to investigate the effects of cryogenic grinding on the composition, functional properties, and extractability of nutritive compounds. The aim of this study was to determine the effects of cryogenic grinding on the content and composition of the nutritional and antinutritional components of rapeseed cake and its fitness as a treatment or pre-treatment for further use of rapeseed by-products in food formulations.

2. Materials and Methods

2.1. Rapeseed Cake

Rapeseed cake was obtained as a by-product in the laboratory production of virgin rapeseed oil. Rapeseed was grown in the experimental field of the Faculty of Agriculture, University of Zagreb. Prior to oil extraction, the seeds were ground, heated at 80 °C for 30 min, and the oil was extracted by double pressing with a Komet screw press (model CA/53, Monforts and Reiners, Rheydt, Germany). The obtained rapeseed cake was ground using a laboratory disc mill (Buehler—Miag, Helvoirt, The Netherlands) and stored at −20 °C until further analysis.

2.2. Grinding of the Cake

Additional grinding of the cake was performed using the ball mill "CryoMill" (Retsch, Haan, Germany). The rapeseed cake (8 g) together with 12 metal balls (10 mm in diameter) were placed in a 50 mL metal jar in which grinding was performed. Grinding in this type of mill is performed by the radial oscillations of the jar, which cause high friction and impact. A constant flow of liquid nitrogen flows through a hollow double wall of the grinding jar before and during the grinding process, maintaining the temperature at −196 °C. Samples were ground for 2, 4, 8, and 12 min without liquid nitrogen cooling (at room temperature—RT) and with liquid nitrogen cooling (cryogenic cooling—CC). The experiments with cryogenic cooling were performed with automatic pre-cooling. In this setting, grinding starts after a countdown of 3 min, which is initiated when the temperature sensor indicates that the surface of the jar has cooled down to −190 °C.

2.3. Particle Size Distribution

The particle size distribution of the ground rapeseed cake was determined using a laser diffraction particle size method as previously described by Benković et al. [16] with some modifications. Briefly, the analysis was performed using the Mastersizer 2000 particle size analyzer connected to the Scirocco 2000 dry disperser (Malvern Instruments, Worchester shire, UK) operating at a feed pressure of 1 bar and a laser obscuration of 2–6%. The feed rate was adjusted between 50 and 100% during the measurement to ensure that the laser obscuration was within the defined range. Six different particle size parameters were determined: d (0.1), d (0.5), d (0.9), Sauter diameter (D_{32}, span, and specific surface area.

2.4. Basic Quality Parameters of the Cake

Rapeseed cake quality parameters were determined using standard methods. The moisture and volatile matter content were determined by drying to a constant mass according to the standard ISO 665:2000 method [17]. The oil content was determined by the Soxhlet reference method according to ISO 659:2009 [18]. Crude protein content and total ash were determined according to the grain analysis methods ISO 20483:2013 [19] and ISO 2171:2007 [20], respectively.

2.5. Extraction of the Non-Polar Components

The extraction of the nonpolar components of rapeseed cake was carried out according to the method described by Li et al. [21] with some minor modifications. The sample (2 g) was placed in a 50-mL plastic tube along with 20 mL of hexane. The test tube was placed horizontally on the orbital shaker (IKA MS 3 basic, IKA-Works, Staufen im Breisgau, Germany) and shaken for 30 min. After extraction, the phases were separated by centrifugation for 10 min at 5000 rpm (Rotina 380, Hettich GmbH, Kirchlengern, Germany). The extraction process was repeated 2 more times. The extracts were combined and then filtered through a sintered glass filter (pore size 10–16 µm), and the solvent was evaporated under reduced pressure at 40 °C using a rotary evaporator (Heidolph, Schwabach, Germany). Nonpolar extracts were used for the determination of fatty acid composition and sterol content and composition, while defatted rapeseed cakes were used for the extraction of phenolic components, tannins, dietary fibre, and glucosinolates.

2.6. Fatty Acid Composition

The fatty acid composition was determined by gas chromatography after transmethylation of the nonpolar dry extracts with methanolic potassium hydroxide [22]. The analysis of the prepared methyl esters was performed according to the method described in our previous research [23]. Briefly, 1 µL of methyl esters were injected into an Agilent Technologies 6890N Network GC system (Santa Clara, CA, USA) with flame ionization detector. Fatty acid methyl esters were separated on a DB-23 capillary column (60 m × 0.25 mm × 0.25 μm; Agilent Technologies). The injector and detector temperatures were set at 250 °C and 280 °C, respectively. The Owen temperature was programmed to increase by 7 °C/min from 60 °C to 220 °C, where it was held for 17 min. The split ratio was 30:1, and helium was used as the carrier gas with a constant flow of 1.5 mL/min. The fatty acid methyl esters were identified by comparing their retention times with those of commercial standards. The content of each fatty acid is expressed as a percentage of total fatty acids.

2.7. Content and Composition of Sterols

The sterol composition was determined according to the standard ISO 12228-1:2014 method for animal and vegetable fats and oils [24] with α-cholestanol as the internal standard. After saponification of the oily extract of rapeseed cakes, followed by extraction and purification of the sterol fraction, the sterols were silylated with 100 μL of pyridine-hexamethyl-disilazane-chlorotrimethylsilane solution (5/2/1, *v*/*v*/*v*). The prepared silylated sterol fraction (1 μL) was injected into an Agilent Technologies 6890N Network GC system (Santa Clara, CA, USA) equipped with an Agilent Technologies 5973 Inert Mass Selective Detector and an Agilent DB-17MS capillary column (30 m × 0.32 mm × 0.25 μm). The split ratio was 13.3:1. Helium was used as the carrier gas at a constant flow rate of 1.5 mL/min. The temperature of the injector was set at 290 °C. The temperature of the oven was programmed to increase by 6 °C/min from 180 to 270 °C and remained at the maximum temperature for 30 min. The transfer line temperature was set at 280 °C. Sterols were identified from an internal mass spectral library created using the β-sitosterol, campesterol, and stigmasterol standards and the NIST 2 and NIST 5 online libraries. All sterols were quantified using an internal standard method. The concentration of total sterols was calculated based on the mass of the cake, taking into account the amount of extracted oil from each sample.

2.8. Determination of Dietary Fibre

Insoluble and soluble dietary fibre and total dietary fibre were determined according to the official AOAC method [25] using the Megazyme assay kit. After extraction of insoluble dietary fibre (IDF) and dietary fibre soluble in water but precipitated in 78% aqueous ethanol (SDFP), dietary fibre soluble in water and not precipitated in 78% aqueous ethanol (SDFS) was analysed using a Shimadzu HPLC system with RID detector. Prior to injection, samples were deionized with a freshly prepared mixture of Amberlite FPA 53 and Ambersep 200 resins (1/1, *w*/*w*). The deionized sample (20 μL) was injected onto the METACARB 67C column (300 mm × 6.5 mm; Agilent Technologies, Santa Clara, CA, USA) heated to 80 °C. The components were separated on a Shimadzu HPLC system by isocratic chromatography using calcium disodium EDTA hydrate solution (50 mg/L) as the mobile phase at a flow rate of 0.5 mL/min for 30 min. Quantification of low molecular weight soluble dietary fibres was performed by comparing their peak with the peak of the internal standard (D-sorbitol).

2.9. Extraction of the Phenolic Components

The extraction of the free and bound phenolic components from the rapeseed cakes was performed according to the modified method described by Martini et al. [26].

For the extraction of free phenolic compounds, 100 μL of an internal standard solution (3,5-dichloro-4-hydroxibenzoic acid, 5 mg/mL) was added to 0.5 g of the defatted rapeseed cake, homogenized for 15 s (IKA MS 3 basic, IKA-Works, Staufen im Breisgau, Germany) with 3 mL of an 80% ethanolic solution (*v*/*v*), and then placed in an ultrasonic bath (Sonorex, Berlin, Germany) for 10 min. After centrifugation at 4000 rpm for 15 min (Rotina 380, Hettich GmbH, Kirchlengern, Germany), the supernatants of the three successive extractions were combined in a 10-mL flask and made up to the mark with ethanol solution

(80%, v/v). The crude rapeseed cake residue after extraction of the free phenolic compounds was used to extract the bound phenolic compounds.

After discarding the ethanol extract, 100 μL of the internal standard solution was again added to the rapeseed cake, and the samples were hydrolyzed with 4 mL of 2 M sodium hydroxide for 4 h. After 4 h, the pH of the mixture was adjusted to pH 2 ± 0.1 (Jenway pH meter, London, UK) to deionize the released phenolic compounds. The extraction was performed three times, the first time with 4 mL of ethyl acetate, and the second and third times with 2 mL of ethyl acetate, as described for the extraction of free phenolics. The supernatants were combined, the solvent was evaporated under nitrogen, and the dry extract was dissolved in 10 mL of methanol.

The content and composition of free and bound phenolic compounds were determined by HPLC using a diode array detector and a Kinetex C18 column (150 mm × 4.6 mm × 2.6 μm, Phenomenex, Torrance, CA, USA) under chromatography conditions published by Panić et al. [27]. All extracts were filtered through a PVDF filter with a pore size of 0.20 μm before injection (5 μL) into the system. Chromatograms of the phenolic compounds were recorded at a wavelength of 280 nm. Retention times and spectral data of the external standards were compared to identify the individual components. Detected peaks whose retention times did not match any of the external standards used, but whose UV spectra matched the UV spectra of the phenolic compounds (hydroxycinnamic acids or hydroxybenzoic acids), were classified as non-identified phenolic compounds. The internal standard mentioned above was used for the quantification of the identified and non-identified phenolic compounds.

2.10. Extraction and Determination of the Tannins

Tannins were determined according to the method published by Price et al. [28]. Defatted rapeseed cake (1 g) was weighed into a 50-mL plastic test tube, and 15 mL of acetone (70%, v/v) was added. Extraction was performed using a laboratory homogenizer (IKA T18 Ultraturrax, IKA-Works, Staufen im Breisgau, Germany) for 1 min at 10,000 rpm. The supernatant was separated using a centrifuge (10 min, 5000 rpm) (Rotina 380, Hettich GmbH, Kirchlengern, Germany), and the acetone solution was evaporated using a rotary evaporator (Heidolph, Schwabach, Germany) at 40 °C under reduced pressure. The dry extract was dissolved in 10 mL of methanol and filtered through a 0.20 μm PVDF filter.

For spectrophotometric determination of tannin concentration, the vanillin reagent was prepared by mixing equal amounts of a 1% vanillin solution in methanol and an 8% solution of HCl in methanol and stabilized overnight. 0.5 mL of the tannin extract was transferred into the spectrophotometric cuvette, and 2.5 mL of vanillin reagent was added. After incubation at 30 °C for 20 min, the absorbance was measured at 500 nm against the blank (0.5 mL extract + 2.5 mL 4% HCl solution in methanol) (Specord 50 plus, Analytik Jena, Germany). Tannin concentration was calculated using the calibration curve of catechin (10–1000 μg/mL).

2.11. Determination of Glucosinolates

The content of glucosinolates was determined according to the standard ISO method [29] using HPLC with sinigrin monohydrate (potassium allyl-glucosinolate monohydrate) as an internal standard. After extraction and desulfation of glucosinolates, an Agilent Technology HPLC 1200 system with DAD detector and Phenomenex C18 column (Kinetex 150 mm × 4.6 mm, 2.6 μm, 100 Å) was used for separation and identification of each component. The prepared sample (20 μL) was injected into the system. The mobile phases for the separation of glucosinolates were water (solvent A) and 20% (v/v) acetonitrile solution in water (solvent B), and the gradient used was as follows: 0–1 min 100% A; 1–21 min 100–0% A; 21–26 min 0–100% A; 26–31 min 100% A. The flow rate was 1 mL/min throughout the run. The column temperature was 30 °C. Chromatograms were recorded at a wavelength of 229 nm, and UV spectra were also recorded throughout the analysis time. The identification of each component was performed by comparing the relative retention times with those in the aforementioned ISO method, and for their quantification the internal standard was used, taking into the account response factors for each component given in the same ISO standard method.

2.12. Statistical Analysis

Rapeseed cake was ground in three replicates. All analyses were performed in triplicate on all ground rapeseed cake samples and on the original cake as a control, except for the analysis of basic quality parameters, which were determined only on the original rapeseed cake. Results are given as mean ± SD. To estimate the effects of time and grinding conditions and their interactions on the nutritional and antinutritional components of rapeseed cake, the two-way ANOVA was used in combination with Tukey's multiple comparison tests. Statistical analyses were performed using XLSTAT 2023 software solution (Lumivero: Denver, CO, USA) with the statistical significance threshold set at $p = 0.05$.

3. Results and Discussion

3.1. Quality of the Rapeseed Cake

The quality of rapeseed cake is determined by the composition of macro- and micronutrients and the presence of desirable and undesirable components. Rapeseed cake produced in our laboratory contained 9.5% water, which was due to the addition of water during the production process (Table 1). The oil content of the cake produced was 14%, a result comparable to cakes produced in industry [30]. Since rapeseed cake contains a relatively high amount of oil, the composition of these nonpolar components is important for the incorporation of the by-product in food formulations. The fatty acid composition of the cake affects not only the nutritional value but also the oxidative stability of the cake and the products in which it is used. The fatty acid composition of the residual oil in rapeseed cake is shown in Table 1. According to the Codex Standard for Named Vegetable Oils [31], it can be classified as low erucic acid rapeseed, with oleic acid being the predominant fatty acid and a significant amount of linoleic and linolenic acid in the ratio 2.8:1. Due to its favourable fatty acid composition, rapeseed oil is beneficial to health from various aspects: it reduces the risk of cardiovascular disease, improves lipid and glucose metabolism, and has a positive effect on atherosclerosis and inflammation [32].

Table 1. Quality parameters, composition of the non-polar fraction and dietary fiber of laboratory produced rapeseed cake (RC).

Parameter		RC
Quality parameter (%)	Water	9.5 ± 0.1
	Oil	14.1 ± 0.3
	Protein	29.5 ± 0.4
	Ash	6.4 ± 0.3
Fatty acid (% of total)	C14:0	0.1 ± 0.0
	C16:0	5.1 ± 0.0
	C16:1	0.4 ± 0.0
	C17:1	0.1 ± 0.0
	C18:0	1.9 ± 0.1
	C18:1	64.4 ± 0.3
	C18:2	19.1 ± 0.0
	C18:3	6.8 ± 0.2
	C20:0	0.6 ± 0.0
	C20:1	1.1 ± 0.1
	C22:0	0.3 ± 0.0

Table 1. *Cont.*

Parameter		RC
Sterol (% of total)	Brassicasterol	11.6 ± 0.5
	Campesterol	29.4 ± 1.0
	Campestanol	1.1 ± 0.0
	Stigmasterol	0.7 ± 0.1
	β-sitosterol	51.5 ± 0.5
	Δ5-avenasterol	3.5 ± 2.3
	Δ7-avenasterol	0.8 ± 0.4
	Δ7-stigmasterol	0.1 ± 0.1
	Δ-5,24-stigmastadienol	1.3 ± 0.1
	Total sterols (mg/100 g)	148 ± 6
Dietary fiber (%)	IDF	25.7 ± 0.6
	SDFP	3.0 ± 0.1
	SDFS	2.0 ± 0.0

IDF—insoluble; SDFP—soluble in water but precipitated in 78% ethanol; SDFS soluble in water and not precipitated in 78% ethanol.

In addition, the sterols of rapeseed oil are known as assistants in lowering the risk of cardiovascular disease. Based on the mass of the cake, the rapeseed cake produced in this experiment contained 148 mg/100 g of sterols (Table 1). The composition of sterols is typical of low erucic rapeseed [31] with β-sitosterol as the dominant sterol (51.5%), followed by campesterol (29.4%) and brassicasterol (11.6%).

The analyzed rapeseed cake contained 6.4% ash. It has already been reported that rapeseed cake and meal are good sources of available calcium, iron, manganese, and selenium. Although they are high in phytate, they are also one of the richest sources of non-phytate phosphorus [33]. However, the most valuable components of rapeseed cake and the reason for numerous research studies to incorporate it in human nutrition are its proteins, which have a balanced amount of essential amino acids, are rich in sulfur-containing amino acids, and have good functional properties. Their digestibility is reported to be as high as 84%, while egg and milk proteins have a digestibility of 94% and 95%, respectively. Rapeseed cake proteins have good solubility, emulsification, foaming, and stabilization properties. To date, there are several rapeseed protein products that have been approved by the U.S. Food and Drug Administration (FDA) as safe (GRAS) or by the European Food Safety Authority (EFSA) as novel foods and can be used in a variety of bakery products, fruit and vegetable juices, egg substitute products, and processed meats [2]. Rapeseed cake produced and used in this study contained nearly 30% crude protein (Table 1).

Regardless of the enumerated advantages, the use of rapeseed cake as a food source is limited by the presence of antinutritional components. One of these components are dietary fibres, which accounts for the largest portion of rapeseed cake, 30.7%. However, dietary fibre can also have some beneficial properties that depend mainly on their nature. Khajali and Slominski, in their review [33], reported the average values of several different studies, and the amounts of total dietary fibre were comparable to the results of this study. The vast majority of the dietary fibre fraction, more than 83%, was insoluble dietary fibre (IDF) (Table 1). This is the undesirable fraction, consisting mainly of non-starch polysaccharides (cellulose and a variety of non-cellulose polysaccharides) and polyphenolic lignin, components resistant to hydrolysis by enzymes in the digestive tract. Their antinutritional properties are reflected in their inhibitory effect on the digestibility of nutrients [34,35]. However, soluble dietary fibre (SDF), which represent a desirable but small proportion of rapeseed dietary fibre (about 17% in the present study), could have a beneficial effect on total and LDL cholesterol levels [36].

Polyphenolic compounds are also among the components with ambiguous characteristics. They are known for their antioxidant and free radical scavenging activities and potential health benefits [37]. However, high concentrations of phenolic compounds, as found in our study, where phenolic compounds accounted for 2.1% of the produced cake (Table 2), have negative effects on the nutritional value and organoleptic properties of rapeseed cake. Phenolic compounds are known to be responsible for the dark colour, bitter taste, and astringency of rapeseed meal. As mentioned earlier, phenolic compounds and their oxidised products can also interact with amino acids, enzymes, and other food components [6]. Free phenols accounted for nearly 70% of the total phenols in this study (Table 2). Of these, free phenolic acids accounted for less than 15%, and the remainder were esterified phenolic compounds. The dominant phenolic compound was sinapine, choline ester of sinapic acid (607 mg/100 g). In contrast to previously reported results [10], sinapic acid was not the predominant free phenolic acid in the produced cake. The reason for this could be due to the production of oil and cake, i.e., the conditioning of the seeds, which leads to the decarboxylation of sinapic acid and the formation of canolol, decreasing the concentration of sinapic acid in the cake. The absence of a carboxyl group gives canolol a more lipophilic character than its precursor, which is why it accumulates in the oil [38] and its concentration in the cake was low (9 mg/100 g). The bound phenolic compounds represent a smaller fraction of phenolics and accounted for slightly more than 30% of the total phenolic compounds. Among them, sinapic acid was the dominant compound. It accounted for 80% of the bound phenolic compounds. Minor phenolic compounds were ferulic acid, p-coumaric acid, and syringaldehyde. Tannins are another group of phenolic compounds. Their concentration in rapeseed cake was slightly lower than 0.5%. This amount is lower than those reported in the literature [39], but the concentration of tannins should be further reduced because they can form complexes with proteins and proteolytic enzymes in the gastrointestinal tract, which negatively affects the nutritional value of the cake [35].

Table 2. Composition of phenolic compounds and glucosinolates of laboratory produced rapeseed cake (RC).

Parameter		RC
Free phenolic compound (mg/100 g)	Gallic acid	27 ± 0
	Chlorogenic acid	75 ± 9
	Ferulic acid	65 ± 2
	Sinapic acid	85 ± 2
	p-coumaric acid	97 ± 1
	Syringaldehyde	14 ± 6
	Canolol	9 ± 1
	Sinapine	607 ± 188
	Non identified	825 ± 22
	Total	1927 ± 51
Bound phenolic compound (mg/100 g)	Ferulic acid	32 ± 0
	Sinapic acid	688 ± 4
	p-coumaric acid	10 ± 0
	Syringaldehyde	4 ± 0
	Non identified	132 ± 1
	Total	865 ± 4
Tannins (mg/100 g)		489 ± 36
Glucosinolate (µmol/g) *	Glucoiberin	0.1 ± 0.1
	Progoitrin	6.9 ± 0.7
	Glucoraphinin	0.5 ± 0.0
	Gluconapoleiferin	0.4 ± 0.0
	Gluconapin	4.0 ± 0.0
	4-hidroxyglucobrassicin	0.3 ± 0.1
	Glucobrassicanapin	1.1 ± 0.0
	Glucobrassicin	0.5 ± 0.0
	4-metoxygucobrassicin	0.1 ± 0.0
	Total	14.0 ± 0.8

* Expressed on the mass of defatted and dry rapeseed cake.

The concentration of glucosinolates in the produced cake was low, only 14 µmol per g of defatted and dry cake. For a seed to be classified as a canola or 00-rapeseed variety, the concentration of glucosinolates should be less than 30 µmol/g [40]. The major glucosinolates detected were progoitrin (6.9 µmol/g), gluconapin (4.0 µmol/g), and glucobrassicanapin (1.1 µmol/g) (Table 2). All other compounds detected (gucoiberin, epi-progoitrin, glucoraphanin, gluconapoleiferin, glucobrasicin, 4-hidroxyglucobrassicin, 4-metoxyglucobrassicin) are present at concentrations less than 1 µmol/g. This composition is typical for Brassica napus cultivars [35]. Although the produced cake contains low concentrations of glucosinolates, and most of which are aliphatic glucosinolates, which mainly have a negative effect on the organoleptic properties of the cake, their concentration should be reduced due to degradation products such as goitrin, a degradation product of progoitrin, which has antithyroid activity [35].

3.2. *Influence of the Grinding on the Particle Size Distribution*

Particle size is one of the most important properties affecting the flow properties of powders used as raw materials, as well as the textural and sensory properties of the products in which powders are used. In general, during the grinding process, feed particles are broken down into smaller particles that become more reactive due to a larger specific surface area, which is beneficial in processes involving heat and mass transfer [41,42]. The effects of grinding time as well as cooling with liquid nitrogen on particle size are shown in Table 3.

Table 3. Particle size distribution parameters determined for the rapeseed cake (RC = sample before additional grinding; RT = room temperature; CC = cryogenic cooling).

Sample	d(0.1) (µm) $^{\$,£,§}$	d(0.5) (µm) $^{\$,£,§}$	d(0.9) (µm) $^{\$,£,§}$	D_{32} (µm) $^{\$,£,§}$	Span $^{\$,£,§}$	Specific Surface Area $^{\$,£,§}$ (m^2/g)
RC	128.2 ± 3.8 a	393.6 ± 9.0 a	881.6 ± 19.7 a	238.8 ± 6.0 a	1.9 ± 0.0 g	0.010 ± 0.0006 h
2RT	60.1 ± 0.3 b	273.4 ± 2.0 b	770.9 ± 6.2 b	130.6 ± 0.9 b	2.6 ± 0.0 d	0.019 ± 0.0000 g
4RT	48.5 ± 1.0 c	229.4 ± 7.1 c	698.4 ± 20.3 c	108.6 ± 2.1 c	2.8 ± 0.0 c	0.022 ± 0.0006 f
8RT	41.3 ± 0.5 d	202.4 ± 5.0 d	669.7 ± 13.6 c	95.1 ± 1.1 d	3.1 ± 0.0 b	0.026 ± 0.0006 e
12RT	33.5 ± 0.4 f	174.9 ± 5.0 e	615.2 ± 10.6 d	79.7 ± 1.1 ef	3.3 ± 0.0 a	0.031 ± 0.0006 d
2CC	39.6 ± 2.9 de	164.2 ± 2.8 e	360.4 ± 1.5 e	83.1 ± 4.2 e	1.9 ± 0.0 fg	0.029 ± 0.0015 d
4CC	35.5 ± 0.1 ef	137.4 ± 0.1 f	288.2 ± 1.1 f	75.1 ± 0.2 f	1.8 ± 0.0 h	0.033 ± 0.0000 c
8CC	20.7 ± 0.4 g	92.3 ± 0.2 g	204.9 ± 0.7 g	49.8 ± 0.5 g	2.0 ± 0.0 f	0.049 ± 0.0006 b
12CC	13.8 ± 0.2 h	61.8 ± 1.2 h	148.8 ± 1.0 h	34.3 ± 0.5 h	2.2 ± 0.0 e	0.071 ± 0.0012 a

$ Grinding time had significant influence ($p \leq 0.05$); £ grinding temperature had significant influence ($p \leq 0.05$); § interaction of the grinding time and temperature had significant influence ($p \leq 0.05$); values with different letters in each column are statistically different ($p \leq 0.05$) according to Tukey's multiple comparison tests.

The rapeseed cake was ground for 2, 4, 8, and 12 min with cryocooling (samples CC) or without cooling at room temperature (samples RT). For RT samples, particle size decreased with grinding time, ending after 12 min with a 44.4% decrease from the original median particle size. The grinding duration affects the particle size of different crops [43], but it is important to emphasize that the effect of grinding duration also depends on the properties of the ground material [41]. For samples CC, the reduction in particle size was significantly higher ($p < 0.05$) than for samples RT. For example, the median particle size decreased to 164 µm after only 2 min of grinding, which according to the Tukey test is comparable to the particle size measured for the sample RT after 12 min of grinding (175 µm). The need for shorter grinding times is very important when considering the economics of particle size reduction, as it also means that less energy is required and only a small amount of liquid nitrogen is needed to speed up the process. After 12 min of grinding under liquid nitrogen, a significant reduction in particle size was achieved, with the resulting particles being only 15% of the original size.

Span and specific surface area of powders affect the flow and caking properties of powders [44]. The samples without cooling (RT) had a larger distribution span, ranging from 2.6 to 3.3 and increasing significantly with increasing grinding time. This was an indication that while longer grinding time resulted in a smaller median particle size, it also resulted in particles with different particle diameters and a larger spread from the median diameter. Powders with larger spans often exhibit poor flow properties that are difficult to control [41]. The effect of increasing the span was not as pronounced at CC. The span of the initial sample before grinding (1.9) was comparable to the values obtained after grinding for two minutes with cryogenic cooling, with a slight increase in span for sample 12CC (2.2). The specific surface area increased from 0.010 m^2/g to 0.031 m^2/g without cooling and to 0.071 m^2/g with cooling. Again, the positive effect of cooling is particularly pronounced, as the specific surface area increases sevenfold, whereas without cooling it increases only threefold. As with the median particle size, the same increase in specific surface area was obtained with 12 min of RT grinding as with only 2 min of CC grinding. The smaller particle size of rapeseed cake and distribution span of particle size, as well as the increase in specific surface area of samples grinded under cryogenic conditions, are in accordance with the results of previous research [13–15]. Pre-cooling rapeseed cake with liquid nitrogen solidifies the oil since the melting point of rapeseed oil ranges from −20 to −10 °C [40]. In addition, cooling the system during the grinding absorbs the heat generated during the grinding operation and maintains the desired low temperature [13]. This embrittles the material, which allows finer grinding and a more uniform particle size.

3.3. *Influence of the Grinding on the Rapeseed Cake Quality*

As expected, grinding had no effect on the fatty acid composition of the residual oil in the produced rapeseed cake (Table 4). The grinding process was much too short, and the final temperatures of the cake in the experiments conducted at room temperature were far too low for oxidation processes to occur. Sterol content and sterol composition, on the other hand, were influenced by the grinding process (Table 5). Stigmasterol was under direct influence of grinding temperature ($p < 0.05$), with generally higher amounts in CC samples according to Tukey test. The levels of brassicasterol and Δ5-avenasterol were also affected by grinding temperature, but also by the duration of the process and the interaction of the two factors. Time and the interaction of time and temperature had significant effects on the content of campesterol, Δ-5,24-stigmastadienol, and concentration of total sterols, while β-sitosterol, the dominant sterol in rapeseed cake, was affected only by the interaction of the factors studied. According to the Tukey test, the highest amounts of total sterols were found in the sample CC ground for twelve minutes. These values were slightly higher but comparable to those of the RT samples ground for 8 and 12 min, respectively. Longer mechanical treatment by the ball mill probably resulted in greater disruption of the cell structure and thus better availability and extractability of the sterols. Using wheat bran as an example, it has been shown that cryogenic grinding causes destruction of the cell wall, resulting in the release of the cell contents and thus higher bioavailability of certain compounds. It also facilitates the interaction between the compound and the solvent and the extraction of the desired components [45].

Table 4. Fatty acid composition of rapeseed cakes additionally grinded at room temperature (RT) and with cryogenic cooling (CC).

Fatty Acid (% of Total)	Sample							
	2RT	**4RT**	**8RT**	**12RT**	**2CC**	**4CC**	**8CC**	**12CC**
C14:0	nd *	0.1 ± 0.0	nd *	nd *	nd *	nd *	nd *	nd *
C16:0	5.1 ± 0.0	5.1 ± 0.0	5.1 ± 0.1	5.1 ± 0.0	5.1 ± 0.0	5.1 ± 0.2	5.1 ± 0.0	5.2 ± 0.0
C16:1	0.4 ± 0.0	0.4 ± 0.0	0.4 ± 0.0	0.4 ± 0.0	0.5 ± 0.0	0.5 ± 0.0	0.5 ± 0.0	0.5 ± 0.0
C17:1	0.1 ± 0.0	0.1 ± 0.0	0.1 ± 0.0	0.1 ± 0.0	0.1 ± 0.0	0.1 ± 0.0	0.1 ± 0.0	0.1 ± 0.0
C18:0	1.9 ± 0.0	2.0 ± 0.0	2.0 ± 0.2	2.0 ± 0.1	1.9 ± 0.0	1.9 ± 0.0	1.9 ± 0.1	1.9 ± 0.0

Table 4. *Cont.*

Fatty Acid (% of Total)	Sample							
	2RT	4RT	8RT	12RT	2CC	4CC	8CC	12CC
C18:1	64.3 ± 0.6	64.3 ± 0.4	64.2 ± 0.1	64.2 ± 0.2	64.2 ± 0.6	64.2 ± 0.3	64.2 ± 0.2	63.9 ± 0.4
C18:2	19.1 ± 0.3	19.1 ± 0.0	19.2 ± 0.2	19.2 ± 0.0	19.2 ± 0.2	19.1 ± 0.1	19.1 ± 0.0	19.4 ± 0.4
C18:3	6.8 ± 0.0	6.8 ± 0.2	6.8 ± 0.1	6.8 ± 0.1	6.8 ± 0.0	6.9 ± 0.0	6.9 ± 0.2	6.8 ± 0.1
C20:0	0.6 ± 0.0	0.6 ± 0.0	0.6 ± 0.0	0.6 ± 0.0	0.6 ± 0.0	0.6 ± 0.0	0.6 ± 0.0	0.6 ± 0.0
C20:1	1.0 ± 0.0	1.0 ± 0.0	1.0 ± 0.0	1.0 ± 0.0	1.0 ± 0.1	1.0 ± 0.0	1.0 ± 0.2	1.0 ± 0.0
C22:0	0.3 ± 0.0	0.3 ± 0.0	0.3 ± 0.1	0.3 ± 0.0	0.3 ± 0.0	0.3 ± 0.0	0.4 ± 0.0	0.3 ± 0.1

* not detected.

Table 5. Content and composition of sterols determined in rapeseed cakes additionally grinded at room temperature (RT) and with cryogenic cooling (CC).

Sterol (% of Total)	Sample							
	2RT	4RT	8RT	12RT	2CC	4CC	8CC	12CC
Brasicasterol $^{\$,£,§}$	11.0 ± 0.0 c	11.4 ± 0.5 bc	11.8 ± 0.2 ab	11.7 ± 0.0 ab	11.5 ± 0.3 bc	12.1 ± 0.2 a	11.4 ± 0.2 bc	11.6 ± 0.4 ab
Campesterol $^{\$,§}$	27.2 ± 0.4 c	28.9 ± 1.4 ab	30.0 ± 0.2 a	29.1 ± 0.0 ab	28.7 ± 0.5 abc	28.6 ± 0.7 abc	28.8 ± 1.1 bc	28.2 ± 0.9 bc
Campestanol	1.1 ± 0.0	1.1 ± 0.1	1.1 ± 0.0	1.2 ± 0.0	1.1 ± 0.0	1.1 ± 0.0	1.1 ± 0.1	1.1 ± 0.0
Stigmasterol $^{£}$	0.7 ± 0.0 ab	0.7 ± 0.0 ab	0.5 ± 0.4 b	0.7 ± 0.0 ab	0.7 ± 0.0 ab	0.7 ± 0.0 ab	0.8 ± 0.1 a	0.9 ± 0.0 a
β-sitosterol §	51.8 ± 0.6 b	51.2 ± 0.9 ab	52.3 ± 0.4 a	51.3 ± 0.4 ab	51.8 ± 0.7 ab	52.8 ± 1.8 a	51.1 ± 1.0 ab	51.7 ± 0.6 ab
Δ5-avenasterol $^{\$,£,§}$	5.8 ± 1.1 a	4.7 ± 2.2 ab	2.2 ± 1.0 bc	3.2 ± 0.1 bc	3.5 ± 1.0 bc	1.1 ± 0.2 c	4.7 ± 2.3 ab	3.6 ± 1.8 bc
Δ7-avenasterol	0.6 ± 0.1	0.7 ± 0.6	0.8 ± 0.2	1.0 ± 0.0	0.9 ± 0.4	1.2 ± 1.5	0.8 ± 0.4	1.1 ± 0.1
Δ7-stigmasterol	0.3 ± 0.1	nd *	nd *	nd *	0.1 ± 0.1	0.7 ± 0.2	0.1 ± 0.1	0.2 ± 0.1
Δ-5,24-stigmastadienol $^{\$,§}$	1.4 ± 0.1 ab	1.2 ± 0.0 b	1.2 ± 0.1 ab	1.4 ± 0.0 ab	1.3 ± 0.1 ab	1.2 ± 0.0 ab	1.4 ± 0.2 a	1.3 ± 0.1 ab
Total sterols (mg/100 g) $^{\$,§}$	138 ± 6 b	143 ± 3 b	149 ± 8 ab	147 ± 2 ab	149 ± 3 ab	140 ± 7 c	145 ± 8 b	157 ± 5 a

* Not detected; $^{\$}$ grinding time had significant influence ($p \leq 0.05$); $^{£}$ grinding temperature had significant influence ($p \leq 0.05$); § interaction of the grinding time and temperature had significant influence ($p \leq 0.05$); values with different letters in each row are statistically different ($p \leq 0.05$) according to Tukey's multiple comparison tests.

The content of IDF was not affected by the grinding process and particle size reduction, while it significantly affected soluble dietary fiber (SDF) (Figure 1). The amount of SDFP was significantly affected by the grinding time and the interaction of grinding time and temperature. According to the Tukey test, their amount increased with increasing grinding time until 8th minute, both with and without cryogenic cooling, and then decreased after 12 min. Samples ground for 12 min contained significantly lower amounts of SDFP than samples prepared in shorter procedures. The increase was also observed by Liu et al. [46] on an orange peel. The reason could be the release of soluble fibres from stable complexes with a longer grinding time, while further grinding leads to mechanical destruction of the molecules and consequently to a decrease in their amount. This hypothesis is confirmed by the fact that the number of SDFS increased slightly (by about 8%) after 12 min of grinding with cryogenic cooling. Both SDFP and SDFS were significantly affected by the duration of the grinding process, but SDFS were also affected by the temperature of the process, with cryogenic grinding yielding a slightly higher value. Čukelj Mustač et al. [14] also reported an increase in SDFS of millet bran with decreasing particle size.

Grinding process had a significant effect on the concentration of free and bound phenolic compounds, but not as much as expected. Several authors reported no significant effect of cryogenic grinding on phenolic compounds [14,45,47], but the results of our previous studies on pumpkin seed cake showed a significant effect of grinding time and grinding conditions on total phenolic content. Grinding the pumpkin cake for 12 min under cryogenic conditions in the aforementioned study resulted in 36% more extracted phenolic compounds [15]. The results of this study were somewhat different. In the present study, the duration of grinding treatment had a significant effect on the concentrations of gallic, ferulic, sinapic, and *p*-coumaric acids, as well as on sinapine and consequently on total free phenolics (Table 6). A significant effect of grinding temperature was found for the concentration of all free phenolic compounds except for canolol. According to the Tukey test, the concentration of sinapine, the dominant phenolic compound in rapeseed cake, and the total amount of free phenolics were significantly higher in the RT samples than in the CC samples. The total amount of bound phenols was also significantly affected by the grinding conditions applied, but an opposite trend to the free phenolics was observed (Table 7). The Tukey test showed that the concentration of bound phenolics was significantly higher in the cryogenically ground samples. Interestingly, a very small scatter is observed between the results of total phenolics, and free and bound phenols combined (2783 ± 127 mg/100 g). This suggests that the increase in free phenolics is due to the damage to the cell walls caused by grinding and the release of phenolic compounds bound to the cell wall material, and that the heat generated during grinding at room temperature favours this process.

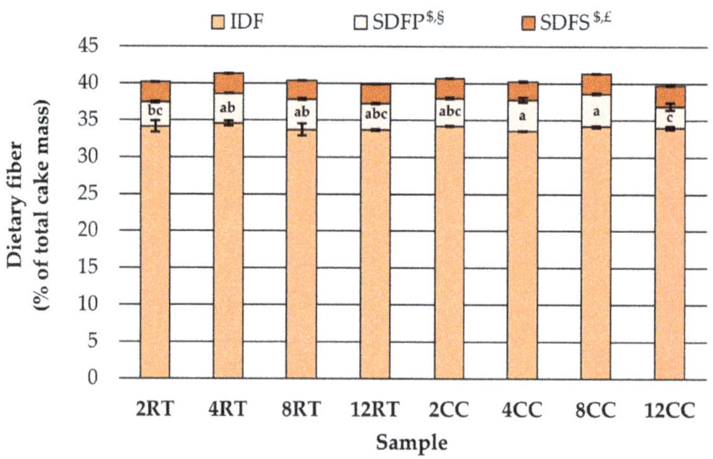

Figure 1. Dietary fiber content and composition of rapeseed cake ground at room temperature (RT) and with cryogenic cooling (CC) expressed as % of defatted and dried cake (IDF—insoluble dietary fiber; SDFP—soluble in water but precipitated in 78% ethanol; SDFS soluble in water and not precipitated in 78% ethanol; [$] grinding time had significant influence ($p \leq 0.05$); [£] grinding temperature had significant influence ($p \leq 0.05$); [§] interaction of the grinding time and temperature had significant influence ($p \leq 0.05$); values with different letters are statistically different ($p \leq 0.05$) according to Tukey's multiple comparison tests.

Table 6. Content and composition of free phenolics determined in defatted and dry rapeseed cake (RC) and cakes ground at room temperature (RT) and with cryogenic cooling (CC).

Phenolic Compound (mg/100 g)	Sample							
	2RT	4RT	8RT	12RT	2CC	4CC	8CC	12CC
Gallic acid [$,£]	58 ± 11 ab	47 ± 15 abc	68 ± 11 a	59 ± 1 ab	43 ± 6 bc	38 ± 13 bc	62 ± 11 ab	32 ± 8 c
Chlorogenic acid [£]	84 ± 4 a	84 ± 4 a	84 ± 4 a	77 ± 3 ab	75 ± 5 ab	76 ± 1 ab	79 ± 6 ab	72 ± 2 b
Ferulic acid [$,£]	76 ± 3 a	69 ± 4 ab	74 ± 3 a	69 ± 3 ab	70 ± 5 ab	67 ± 2 ab	71 ± 5 ab	62 ± 5 b
Sinapic acid [$,£]	97 ± 3 a	93 ± 7 a	98 ± 7 a	91 ± 3 a	90 ± 5 ab	87± 0 ab	92 ± 6 a	78 ± 8 b
p-coumaric acid [$,£]	112 ± 6 a	106 ± 4 ab	108 ± 6 ab	100 ± 4 ab	100 ± 8 ab	102 ± 5 ab	105 ± 8 ab	95 ± 8 b
Syringaldehyde [£]	18 ± 1	19 ± 2	19 ± 2	18 ± 1	15 ± 4	16 ± 3	19 ± 1	16 ± 1
Canolol	9 ± 2	9 ± 0	8 ± 1	8 ± 1	8 ± 2	7 ± 1	8 ± 1	9 ± 1
Sinapine [$,£]	913 ± 49 a	809 ± 44 ab	882 ± 25 a	798 ± 55 ab	826 ± 52 ab	816 ± 31 ab	830 ± 73 ab	710 ± 78 b
Non identified [$,£]	954 ± 23 a	791 ± 65 bcd	974 ± 2 a	827 ± 30 bc	876 ± 70 ab	816 ± 12 bc	690 ± 26 d	731 ± 56 cd
Total [$,£]	2322 ± 70 a	2027 ±145 bc	2225 ± 53 ab	2047 ± 92 abc	2102 ±145 ab	2025 ± 35 bc	2115 ± 142 ab	1804 ± 167 c

[$] grinding time had significant influence ($p \leq 0.05$); [£] grinding temperature had significant influence ($p \leq 0.05$); values with different letters in each row are statistically different ($p \leq 0.05$) according to Tukey's multiple comparison tests.

Table 7. Content and composition of bound phenolic compounds determined in defatted and dry rapeseed cake (RC) and cakes ground at room temperature (RT) and with cryogenic cooling (CC).

Phenolic Compound (mg/100 g)	Sample							
	2RT	4RT	8RT	12RT	2CC	4CC	8CC	12CC
Ferulic acid £	25 ± 1	28 ± 2	30 ± 1	26 ± 3	28 ± 1	29 ± 4	28 ± 1	33 ± 2
Sinapic acid $,£,§	413 ± 14 c	502 ± 21 c	533 ± 6 bc	427 ± 37 c	535 ± 48 bc	646 ± 33 ab	515 ± 10 bc	694 ± 48 a
p-coumaric acid $,£	7± 0 c	8 ± 1 b	8 ± 0 bc	7 ± 0 c	9 ± 1 ab	10 ± 0 a	9 ± 1 ab	9 ± 1 ab
Syringaldehyde	4 ± 1	3 ± 1	5 ± 0	3 ± 0	4 ± 0	5 ± 0	4 ± 0	4 ± 0
Non identified $,£,§	93 ± 7 c	114 ± 3 c	110 ± 1 c	101 ± 8 c	182 ± 21 a	137 ± 12 bc	92 ± 14 c	168 ± 24 ab
Total $,£,§	541± 8 e	656 ± 22 cde	685 ± 4 cd	565 ± 48 de	758 ± 10 bc	826 ± 25 ab	647 ± 23 cde	907 ± 26 a

$ grinding time had significant influence ($p \leq 0.05$); £ grinding temperature had significant influence ($p \leq 0.05$); § interaction of the grinding time and temperature had significant influence ($p \leq 0.05$); values with different letters in each row are statistically different ($p \leq 0.05$) according to Tukey's multiple comparison tests.

The changes in tannin concentration in ground rapeseed cakes are shown in Figure 2. According to the results, the initial rapeseed cake contained lower amounts of tannins (Table 2) than all other samples. The results show that a longer grinding time improves the extractability of tannins ($p \leq 0.05$). It is observed that better extraction is obtained at higher temperatures, but the grinding temperature was not found to be significant. Tukey test also showed that there was no significant difference in tannin concentration between samples. Similar to phenolic compounds, tannins are also released from their complexes by cell disruption.

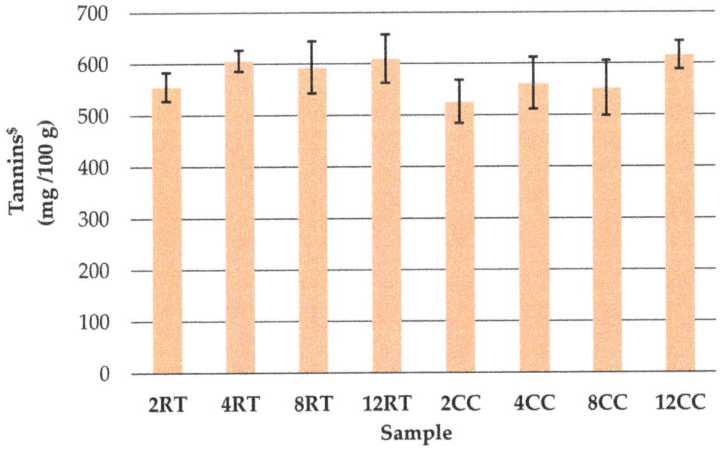

Figure 2. Tannin content cakes grinded at room temperature (RT) and with cryogenic cooling (CC); $ grinding time had significant influence on the concentration of tannins ($p \leq 0.05$).

As mentioned above, glucosinolates are considered the major antinutritive constituents of rapeseed cake. Their degradation products, including goitrin, inhibit iodine uptake and the biosynthesis of thyroid hormones T3 and T4, leading to hypothyroidism and subsequent enlargement of the thyroid gland [35]. The concentration and composition of glucosinolates in rapeseed cake are shown in Table 8. The concentration of major glucosinolates (progoitrin, gluconapin, glucobrasicanapin) and glucoraphanin decreased with the duration of the grinding process, according to Tukey test ($p \leq 0.05$). This resulted in a 34% decrease in total glucosinolates after 12 min of grinding at room temperature and 43% after 12 min of cryogenic grinding. Wang et al. [48] reported that a larger surface area of ground horseradish correlated positively with the concentration of isothyocianates, which may be attributed to more extensive hydrolysis of glucosinolates by endogenous myrosinase. Although we achieved a significant reduction in glucosinolate concentration, the remaining concentrations are still too high, requiring further processing of the cake to use it as a food source.

Table 8. Content and composition of glucosinolates determined in defatted cakes ground at room temperature (RT) and with cryogenic cooling (CC).

Glucosinolate (μmol/g)	Sample							
	2RT	4RT	8RT	12RT	2CC	4CC	8CC	12CC
Glucoiberin	0.2 ± 0.1	0.1 ± 0.0	0.1 ± 0.1	0.1 ± 0.0	0.1 ± 0.1	tr *	tr	tr
Progoitrin $^{\$,\S}$	5.0 ± 0.4 ab	5.2 ± 0.4 ab	5.2 ± 0.6 ab	4.8 ± 0.7 ab	5.9 ± 1.3 a	5.1 ± 0.2 ab	5.0 ± 1.4 ab	4.4 ± 0.6 b
Epi-progoitrin	nd **	0.1 ± 0.1	0.1 ± 0.1	nd **	0.1 ± 0.1	0.1 ± 0.1	0.1 ± 0.0	nd **
Glucoraphinin $^{\$}$	0.4 ± 0.2	0.2 ± 0.1	0.3 ± 0.0	0.3 ± 0.0	0.3 ± 0.1	0.3 ± 0.0	0.2 ± 0.1	0.2 ± 0.0
Gluconapoleiferin	0.2 ± 0.1	0.4 ± 0.0	0.4 ± 0.0	0.3 ± 0.1	0.4 ± 0.1	0.4 ± 0.0	0.3 ± 0.1	0.3 ± 0.0
Gluconapin $^{\$,\S}$	2.9 ± 0.5 a	2.5 ± 0.3 ab	2.4 ± 0.3 ab	2.5 ± 0.2 ab	2.9 ± 0.7 a	2.4 ± 0.3 ab	2.6 ± 0.4 ab	2.1 ± 0.4 b
4-hidroxyglucobrassicin	0.3 ± 0.2	0.2 ± 0.1	0.2 ± 0.1	0.2 ± 0.0	0.2 ± 0.0	0.3 ± 0.1	0.2 ± 0.1	0.1 ± 0.0
Glucobrassicanapin $^{\$}$	0.8 ± 0.2 a	0.7 ± 0.1 ab	0.7 ± 0.1 ab	0.7 ± 0.0 bc	0.8 ± 0.2 a	0.7 ± 0.1 ab	0.8 ± 0.2 bc	0.6 ± 0.1 c
Glucobrassicin $^{\$,\S}$	0.6 ± 0.2 a	0.3 ± 0.1 ab	0.2 ± 0.0 ab	0.3 ± 0.0 ab	0.3 ± 0.0 ab	0.3 ± 0.0 ab	0.1 ± 0.0 b	0.2 ± 0.1 ab
4-metoxygucobrassicin $^{\$,\S}$	0.1 ± 0.0	0.1 ± 0.0	0.1 ± 0.0	0.1 ± 0.1	0.1 ± 0.0	0.1 ± 0.0	0.1 ± 0.0	tr
Total $^{\$,\S}$	10.6 ± 0.7 a	9.6 ± 0.7 ab	9.5 ± 1.2 ab	9.3 ± 0.9 ab	10.8 ± 2.3 a	9.5 ± 0.4 ab	9.4 ± 1.3 ab	8.0 ± 1.3 b

* traces (<0.05) ** not detected; $^{\$}$ grinding time had significant influence ($p \leq 0.05$); § interaction of the grinding time and temperature had significant influence ($p \leq 0.05$); values with different letters in each row are statistically different ($p \leq 0.05$) according to Tukey's multiple comparison tests.

4. Conclusions

Cryogenic grinding is a much more efficient method of reducing particle size than room temperature grinding. Grinding the rapeseed cake with a ball mill for only 2 min at −196 °C resulted in the same particle size as grinding for 12 min at room temperature. Twelve minutes of cryogenic grinding resulted in particles about 6.5 times smaller and a specific surface area 7 times larger compared to the starting material.

Micronization of the rapeseed cake had some effect on the nutritional and antinutritional components, but not to the extent we expected. On the positive side, grinding the cake had no effect on the fatty acid composition of the cake, but the cake ground for 12 min under cryogenic conditions had an increased content of phytosterols. Soluble dietary fibre content also slightly increased with decreasing particle size. However, the content of insoluble dietary fibre was not affected. The mechanical destruction of the cell walls combined with the heat generated during grinding at room temperature resulted in the release of the bound phenolics and thus an increase in the total free phenolic compounds in these samples. The same occurred with the tannins, making them more extractable and consequently resulting in higher tannin concentrations after longer grinding times. However, the best results were obtained in reducing the components that raise the most concerns, glucosinolates. Room temperature grinding reduced glucosinolate content by up to 34%, and even better results were obtained under cryogenic conditions: 43% less after 12 min.

Although these results are promising, they are not sufficient to recommend micronization as the sole treatment for rapeseed cake. It could be used as a pre-treatment for chemical or biological treatments to reduce antinutritive components, since cell disruption and increase in specific surface area increases the extractability and availability of undesirable components. These studies should also investigate whether the high cost of the cryogenic process is justified, as it did not significantly affect the antinutritional components of rapeseed cake.

Author Contributions: Conceptualization, K.K., D.Š., S.B. and D.Ć.; Data curation, K.K. and N.Č.M.; Formal analysis, K.K. and S.B.; Funding acquisition, D.Ć.; Investigation, M.B., S.D., A.M.J., K.M. and I.R.; Methodology, K.K., N.Č.M. and M.B.; Project administration, N.Č.M.; Resources, D.Š. and D.Ć.; Supervision, D.Ć.; Validation, K.K., N.Č.M. and S.D.; Visualization, K.K.; Writing—original draft, K.K.; Writing—review and editing, K.K., D.Š. and D.Ć. All authors have read and agreed to the published version of the manuscript.

Funding: This research was founded by the Croatian Science Foundation (IP-2016-06-3789 project "From grain by-products to functional food through innovative processing").

Institutional Review Board Statement: Not applicable.

Informed Consent Statement: Not applicable.

Data Availability Statement: The data that support the findings of this study are available from the corresponding author upon reasonable request.

Acknowledgments: Authors thank Milan Pospišil, Faculty of Agriculture, University of Zagreb for the rapeseed.

Conflicts of Interest: The authors declare no conflict of interest.

References

1. Wiebe, K. FAO's Director-General on How to Feed the World in 2050. *Popul. Dev. Rev.* **2009**, *35*, 837–839. [CrossRef]
2. Campbell, L.; Rempel, C.; Wanasundara, J. Canola/Rapeseed Protein: Future Opportunities and Directions—Workshop Proceedings of IRC 2015. *Plants* **2016**, *5*, 17. [CrossRef] [PubMed]
3. Hald, C.; Dawid, C.; Tressel, R.; Hofmann, T. Kaempferol 3-O-(2‴-O-Sinapoyl-β-Sophoroside) Causes the Undesired Bitter Taste of Canola/Rapeseed Protein Isolates. *J. Agric. Food Chem.* **2019**, *67*, 372–378. [CrossRef] [PubMed]
4. USDA ERS Oil Crops Yearbook. Available online: https://www.ers.usda.gov/topics/crops/soybeans-oil-crops/oil-crops-sector-at-a-glance/ (accessed on 7 February 2023).
5. Khattab, R.Y.; Arntfield, S.D. Functional Properties of Raw and Processed Canola Meal. *LWT Food Sci. Technol.* **2009**, *42*, 1119–1124. [CrossRef]
6. Shahidi, F.; Naczk, M. An Overview of the Phenolics of Canola and Rapeseed: Chemical, Sensory and Nutritional Significance. *J. Am. Oil Chem. Soc.* **1992**, *69*, 917–924. [CrossRef]
7. Tripathi, M.K.; Mishra, A.S. Glucosinolates in Animal Nutrition: A Review. *Anim. Feed Sci. Technol.* **2007**, *132*, 1–27. [CrossRef]
8. Szydłowska-Czerniak, A.; Bartkowiak-Broda, I.; Karlović, I.; Karlovits, G.; Szłyk, E. Antioxidant Capacity, Total Phenolics, Glucosinolates and Colour Parameters of Rapeseed Cultivars. *Food Chem.* **2011**, *127*, 556–563. [CrossRef]
9. Thiyam, U.; Stöckmann, H.; Schwarz, K. Antioxidant Activity of Rapeseed Phenolics and Their Interactions with Tocopherols during Lipid Oxidation. *J. Am. Oil Chem. Soc.* **2006**, *83*, 523–528. [CrossRef]
10. Kozlowska, H.; Naczk, M.; Shahidi, F.; Zadernowski, R. Phenolic Acids and Tannins in Rapeseed and Canola. In *Canola and Rapeseed*; Springer: Boston, MA, USA, 1990; pp. 193–210.
11. Amarowicz, R.; Naczk, M.; Shahidi, F. Antioxidant Activity of Crude Tannins of Canola and Rapeseed Hulls. *J. Am. Oil Chem. Soc.* **2000**, *77*, 957. [CrossRef]
12. Chen, T.; Zhang, M.; Bhandari, B.; Yang, Z. Micronization and Nanosizing of Particles for an Enhanced Quality of Food: A Review. *Crit. Rev. Food Sci. Nutr.* **2018**, *58*, 993–1001. [CrossRef]
13. Singh, K.K.; Goswami, T.K. Cryogenic grinding of cloves. *J. Food Process. Preserv.* **2007**, *24*, 57–71. [CrossRef]
14. Čukelj Mustač, N.; Novotni, D.; Habuš, M.; Drakula, S.; Nanjara, L.; Voučko, B.; Benković, M.; Ćurić, D. Storage Stability, Micronisation, and Application of Nutrient-Dense Fraction of Proso Millet Bran in Gluten-Free Bread. *J. Cereal Sci.* **2020**, *91*, 102864. [CrossRef]
15. Balbino, S.; Dorić, M.; Vidaković, S.; Kraljić, K.; Škevin, D.; Drakula, S.; Voučko, B.; Čukelj, N.; Obranović, M.; Ćurić, D. Application of Cryogenic Grinding Pretreatment to Enhance Extractability of Bioactive Molecules from Pumpkin Seed Cake. *J. Food Process. Eng.* **2019**, *42*, e13300. [CrossRef]
16. Benković, M.; Srečec, S.; Špoljarić, I.; Mršić, G.; Bauman, I. Flow Properties of Commonly Used Food Powders and Their Mixtures. *Food Bioprocess Technol.* **2013**, *6*, 2525–2537. [CrossRef]
17. *ISO 665:2000*; Oilseeds—Determination of Moisture and Volatile Matter Content. International Organization for Standardization: Geneva, Switzerland, 2000.
18. *ISO 659:2009*; Oilseeds—Determination of Oil Content (Reference Method). International Organization for Standardization: Geneva, Switzerland, 2009.
19. *ISO 20483:2013*; Cereals and Pulse—Determination of the Nitrogen Content and Calculation of the Crude Protein Content—Kjeldahl Method. International Organization for Standardization: Geneva, Switzerland, 2013.
20. *ISO 2171:2007*; Cereals, Pulses and by-Products—Determination of Ash Yield by Incineration. International Organization for Standardization: Geneva, Switzerland, 2007.
21. Li, T.; Beveridge, T.; Drover, J. Phytosterol Content of Sea Buckthorn (*Hippophae rhamnoides* L.) Seed Oil: Extraction and Identification. *Food Chem.* **2007**, *101*, 1633–1639. [CrossRef]
22. *ISO 12966-2:2017*; Animal and Vegetable Fats and Oils—Gas Chromatography of Fatty Acid Methyl Esters—Part 2: Preparation of Methyl Esters of Fatty Acids. International Organization for Standardization: Geneva, Switzerland, 2017.
23. Kraljić, K.; Stjepanović, T.; Obranović, M.; Pospišil, M.; Balbino, S.; Škevin, D. Influence of Conditioning Temperature on the Quality, Nutritional Properties and Volatile Profile of Virgin Rapeseed Oil. *Food Technol. Biotechnol.* **2018**, *56*, 562–572. [CrossRef]

24. *ISO 12228-1:2014*; Determination of Individual and Total Sterols Contents—Gas Chromatographic Method—Part 1: Animal and Vegetable Fats and Oils. International Organization for Standardization: Geneva, Switzerland, 2014.
25. AOAC. *AOAC Official Method 2011.25 Insoluble, Soluble, and Total Dietary Fiber in Foods*; AOAC: Rockville, MD, USA, 2012.
26. Martini, D.; D'Egidio, M.G.; Nicoletti, I.; Corradini, D.; Taddei, F. Effects of Durum Wheat Debranning on Total Antioxidant Capacity and on Content and Profile of Phenolic Acids. *J. Funct. Foods* **2015**, *17*, 83–92. [CrossRef]
27. Panić, M.; Stojković, M.R.; Kraljić, K.; Škevin, D.; Redovniković, I.R.; Srček, V.G.; Radošević, K. Ready-to-Use Green Polyphenolic Extracts from Food by-Products. *Food Chem.* **2019**, *283*, 628–636. [CrossRef]
28. Price, M.L.; Van Scoyoc, S.; Butler, L.G. A Critical Evaluation of the Vanillin Reaction as an Assay for Tannin in Sorghum Grain. *J. Agric. Food Chem.* **1978**, *26*, 1214–1218. [CrossRef]
29. *ISO 9167-1:1992*; Rapeseed—Determination of Glucosinolates Content—Part. 1: Method Using High-Performance Liquid Chromatography. International Organization for Standardization: Geneva, Switzerland, 1992.
30. Leming, R.; Lember, A. Chemical Composition of Expeller-Extracted and Cold-Pressed Rapeseed Cake. *Agraarteadus* **2005**, *16*, 103–109.
31. WHO; FAO. *Codex Standard for Named Vegetable Oils*; Food and Agriculture Organization of the United Nations/World Health Organization: Rome, Italy, 1999.
32. Hammad, S.; Pu, S.; Jones, P.J. Current Evidence Supporting the Link Between Dietary Fatty Acids and Cardiovascular Disease. *Lipids* **2016**, *51*, 507–517. [CrossRef] [PubMed]
33. Khajali, F.; Slominski, B.A. Factors That Affect the Nutritive Value of Canola Meal for Poultry. *Poult. Sci.* **2012**, *91*, 2564–2575. [CrossRef] [PubMed]
34. Kaczmarek, P.; Korniewicz, D.; Lipinski, K.; Mazur, M. Chemical Composition of Rapeseed Products and Their Use in Pig Nutrition. *Pol. J. Nat. Sci.* **2016**, *31*, 545–562.
35. Slominski, B.A.; Jia, W.; Rogiewicz, A.; Nyachoti, C.M.; Hickling, D. Low-Fiber Canola. Part 1. Chemical and Nutritive Composition of the Meal. *J. Agric. Food Chem.* **2012**, *60*, 12225–12230. [CrossRef]
36. Brown, L.; Rosner, B.; Willett, W.W.; Sacks, F.M. Cholesterol-Lowering Effects of Dietary Fiber: A Meta-Analysis. *Am. J. Clin. Nutr.* **1999**, *69*, 30–42. [CrossRef]
37. Leopoldini, M.; Marino, T.; Russo, N.; Toscano, M. Antioxidant Properties of Phenolic Compounds: H-Atom versus Electron Transfer Mechanism. *J. Phys. Chem. A* **2004**, *108*, 4916–4922. [CrossRef]
38. Kraljić, K.; Škevin, D.; Pospišil, M.; Obranović, M.; Neđeral, S.; Bosolt, T. Quality of Rapeseed Oil Produced by Conditioning Seeds at Modest Temperatures. *J. Am. Oil Chem. Soc.* **2013**, *90*, 589–599. [CrossRef]
39. Naczk, M.; Amarowicz, R.; Sullivan, A.; Shahidi, F. Current Research Developments on Polyphenolics of Rapeseed/Canola: A Review. *Food Chem.* **1998**, *62*, 489–502. [CrossRef]
40. Przybylski, R.; Mag, T.; Eskin, N.A.M.; McDonald, B.E. Canola Oil. In *Bailey's Industrial Oil and Fat Products*; Shahidi, F., Ed.; John Wiley & Sons, Inc.: Hoboken, NJ, USA, 2005; pp. 61–121. ISBN 9780471678496.
41. Barbosa-Canovas, G.; Ortega-Rivas, E.; Juliano, P.; Yan, H. *Food Powders*; Food Engineering Series; Kluwer Academic/Plenum Publishers: New York, NY, USA, 2005; ISBN 978-0-306-47806-2.
42. Wang, H.; Mustaffar, A.; Phan, A.N.; Zivkovic, V.; Reay, D.; Law, R.; Boodhoo, K. A Review of Process Intensification Applied to Solids Handling. *Chem. Eng. Process. Process Intensif.* **2017**, *118*, 78–107. [CrossRef]
43. Hossen, M.S.; Sotome, I.; Takenaka, M.; Isobe, S.; Nakajima, M.; Okadome, H. Effect of Particle Size of Different Crop Starches and Their Flours on Pasting Properties. *Japan J. Food Eng.* **2011**, *12*, 29–35. [CrossRef]
44. Carpin, M.; Bertelsen, H.; Dalberg, A.; Bech, J.K.; Risbo, J.; Schuck, P.; Jeantet, R. How Does Particle Size Influence Caking in Lactose Powder? *J. Food Eng.* **2017**, *209*, 61–67. [CrossRef]
45. Hemery, Y.; Chaurand, M.; Holopainen, U.; Lampi, A.M.; Lehtinen, P.; Piironen, V.; Sadoudi, A.; Rouau, X. Potential of Dry Fractionation of Wheat Bran for the Development of Food Ingredients, Part I: Influence of Ultra-Fine Grinding. *J. Cereal Sci.* **2011**, *53*, 1–8. [CrossRef]
46. Liu, Y.; Wang, L.; Liu, F.; Pan, S. Effect of Grinding Methods on Structural, Physicochemical, and Functional Properties of Insoluble Dietary Fiber from Orange Peel. *Int. J. Polym. Sci.* **2016**, *2016*, 6269302. [CrossRef]
47. Rosa, N.N.; Barron, C.; Gaiani, C.; Dufour, C.; Micard, V. Ultra-Fine Grinding Increases the Antioxidant Capacity of Wheat Bran. *J. Cereal Sci.* **2013**, *57*, 84–90. [CrossRef]
48. Wang, J.; Zhang, M.; Devahastin, S.; Liu, Y. Influence of Low-Temperature Ball Milling Time on Physicochemical Properties, Flavor, Bioactive Compounds Contents and Antioxidant Activity of Horseradish Powder. *Adv. Powder Technol.* **2020**, *31*, 914–921. [CrossRef]

applied sciences

MDPI

Article

Wheat Bread Enriched with Black Chokeberry (*Aronia melanocarpa* L.) Pomace: Physicochemical Properties and Sensory Evaluation

Grażyna Cacak-Pietrzak [1], Dariusz Dziki [2,*], Urszula Gawlik-Dziki [3], Natalia Parol-Nadłonek [1], Stanisław Kalisz [1,*], Anna Krajewska [2] and Sylwia Stępniewska [4]

[1] Department of Food Technology and Assessment, Institute of Food Sciences, Warsaw University of Life Sciences, 159C Nowoursynowska Street, 02-776 Warsaw, Poland; grazyna_cacak_pietrzak@sggw.edu.pl (G.C.-P.)

[2] Department of Thermal Technology and Food Process Engineering, University of Life Sciences in Lublin, 31 Głęboka Street, 20-612 Lublin, Poland; anna.krajewska@up.lublin.pl

[3] Department of Biochemistry and Food Chemistry, University of Life Sciences in Lublin, 8 Skromna Street, 20-704 Lublin, Poland

[4] Department of Grain Processing and Bakery, Prof. Wacław Dąbrowski Institute of Agricultural and Food Biotechnology—State Research Institute, 36 Rakowiecka Street, 02-532 Warsaw, Poland

* Correspondence: dariusz.dziki@up.lublin.pl (D.D.); stanislaw_kalisz@sggw.edu.pl (S.K.)

Abstract: Fruit pomace is a highly valuable byproduct from a nutritional standpoint. The aim of this study was to determine the physicochemical and sensory properties of wheat bread enriched with freeze-dried and powdered chokeberry pomace in amounts of 1, 2, 3, 4, 5, and 6% relative to the flour weight. The influence of pomace addition on the physical properties of the wheat dough was analyzed, and the physicochemical properties and consumer acceptance of the chokeberry-pomace-enriched bread were determined. Based on the obtained research results, it was shown that the addition of pomace increased the water absorption of the flour but caused a decrease in stability and weakening of the dough, resulting in increased softening. Consequently, the volume of the bread decreased, and the crumb hardness increased. Furthermore, the addition of pomace significantly reduced the brightness and yellowness of the bread crumbs, while making them more red. Importantly, compared to the control bread, the pomace-enriched bread was characterized by higher contents of minerals, dietary fiber, phenolic compounds, and higher antioxidant activity. Sensory evaluation showed that the addition of freeze-dried chokeberry pomace to wheat bread should not exceed 3% in relation to the weight of the flour used. Additionally, a 3% addition of chokeberry pomace increased the dietary fiber content by 80.9%, ash content by 2.6%, fat content by 26.5%, and total phenolics content by 272%. It decreased the protein content by 1.2%, and reduced the carbohydrate content by 4% compared to the control sample.

Keywords: chokeberry pomace; bread; antioxidant activity; consumer acceptance; color; texture; chemical composition

1. Introduction

Bread is one of the basic food products consumed widely in many regions of the world. In Europe, wheat bread is particularly popular, most often produced from low-extraction white flour. The advantages of such bread are a crunchy crust, soft, fluffy, and light-colored crumbs, and easy digestibility [1]. However, white wheat bread is characterized by a relatively low nutritional value, containing a high amount of carbohydrates and little dietary fiber, mineral components, vitamins, or other bioactive compounds, regular consumption of which is an essential element of preventive healthcare [2,3].

In recent years, a popular trend has been to enrich the recipe of food with additional ingredients to obtain a product with functional properties [4]. Bread, due to its frequent consumption, provides a good food matrix for introducing various recipe additives, such as oily plant seeds [5], flour from other types of grains or pseudocereals [6–8], flour from legume seeds [9,10], spices, herbs [11–14], and dried vegetable and fruit extracts [15,16], as well as waste products from the processing of plant raw materials, such as bran, brewer's spent grain, or pomace [17–19]. It should be emphasized that waste products can be particularly valuable recipe ingredients for bread, as they are often characterized by a much higher content of valuable nutritional components than the raw materials from which they were processed [20]. The use of waste products is in line with the EU's policy on waste-free food production and is one of the elements of the circular economy strategy [21].

A plant with exceptional chemical composition of its fruit is the black chokeberry (*Aronia melanocarpa* (Michx.) Elliott), a perennial deciduous shrub from the *Roseaceae* family. This plant comes from North America and is characterized by low soil requirements and high frost and disease resistance, and is currently widely cultivated in the northern and eastern parts of Europe. Aronia fruits are spherical in shape and are gathered in clusters. When ripe, they have a dark pomegranate to black color, are covered with a waxy coating, and are slightly shiny [22]. The dry matter content in aronia fruits is 17.9–26.0% [22,23], including 13.7–15.1% carbohydrates, 4.8–6.3% dietary fiber, 0.7% protein, and 0.2% fat [24]. Aronia fruits are characterized by an exceptionally high content of active substances that have a beneficial effect on the body, which is why they are called a superfruit or superberry [23]. These compounds include polyphenols, including anthocyanins, flavonoids, tannins, and phenolic acids. The total content of polyphenols in aronia fruits is 2000–8000 mg per 100 g of dry matter, including about 500 mg of anthocyanins [22,25]. The most important compounds from the group of anthocyanins are cyanidin glycosides (3-O-galactoside, 3-O-arabinoside, 3-O-xyloside, and 3-O-glucoside) [25]. Aronia fruits are also rich in phenolic acids, mainly chlorogenic and neochlorogenic acid [23]. They also contain flavonoid glycosides, with quercetin 3-O-galactoside and quercetin 3-O-glucoside being the most prevalent. Tannins are also included in the phenolic compounds, with a content of approximately 0.4% [22]. Together with organic acids, they contribute to the tart taste of the fruits. In addition, aronia fruits are a source of vitamins, especially vitamin C, as well as vitamins E, K, B1, B2, B3, B5, B6, and folic acid [24], and macro- and microelements, including magnesium, potassium, calcium, iron, iodine, molybdenum, and manganese [24,26,27].

Ripe aronia berries are characterized by a tart and bitter taste, and therefore are rarely consumed in their unprocessed form as a dessert fruit. They are widely used in processing as a raw material for the production of juices, jams, preserves, beverages, wines, liqueurs, and infusions, as well as fruit teas and jellies, and for obtaining food extracts and dyes [23,28]. During the processing of the berries, waste is produced in the form of pomace. Mayer-Miebach et al. [23] have shown that pomace has a much greater antioxidant activity than the fruit or juice.

The aim of this study was to determine the effect of lyophilized pomace on the properties of dough and the quality of wheat flour bread, as well as to determine the maximum level of this additive acceptable to consumers.

2. Materials and Methods

The scope of the conducted research is schematically presented in Figure 1.

2.1. Materials

The following chemical reagents were used for the research: sodium salicylate, gallic acid, ferrozine (3-(2-pyridyl)-5,6-bis-(4-phenyl-sulfonic acid)-1,2,4-triazone), and ABTS (2,2′-azino-bis-(3-ethylbenzothiazoline-6-sulfonic acid)) in analytical purity. All reagents were purchased from Sigma-Aldrich (Poznań, Poland).

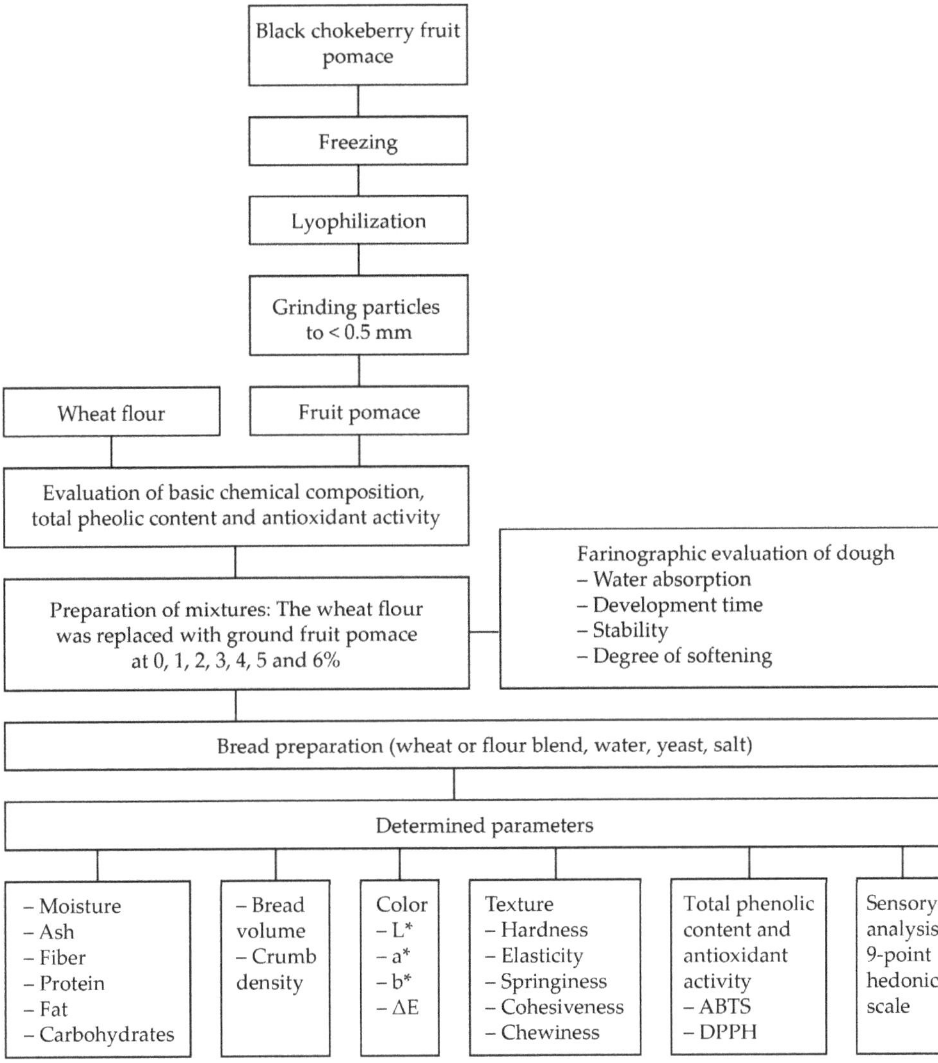

Figure 1. Graphic diagram of conducted research.

For the preparation of the bread dough, the following ingredients were used: commercial bread wheat flour type 750 (Polskie Młyny S.A., Warsaw, Poland), fresh pressed yeast (Lesaffre Polska S.A., Wołczyn, Poland), table salt (Cenos Sp. z o.o., Września, Poland), and freeze-dried pomace from black chokeberries. The pomace used in the study was a waste product obtained during laboratory juice extraction. Immediately after extraction, the pomace was frozen at −25 °C and then subjected to freeze-drying using an Alpha 1–4 LSC Plus freeze-dryer with shelf heating (Martin Christ Gefriertrocknungsanlagen GmbH, Osterode am Harz, Germany). The shelf temperature was set at 20 °C, the condenser temperature at −56 °C, and the pressure at 0.1 kPa. After drying, the pomace was vacuum-sealed in foil pouches. Prior to analysis, the pomace was ground to particles below 0.5 mm using an analytical mill A11 (IKA Works GmbH & Co., Staufen im Breisgau, Germany).

2.2. Basic Composition of Raw Materials and Bread

The basic composition of the raw materials (wheat flour, freeze-dried pomace from black chokeberries), as well as the bread samples, was determined using AACC methods [29]. Moisture content was determined using the gravimetric drying method (Method 44-15.02) using a SUP-65 W drying oven (Wamed, Warsaw, Poland). Total ash content was determined with the incineration method (Method 08-01.01) using an FCF S muffle furnace (Czylok, Jastrzębie Zdrój, Poland). Total dietary fiber content (Method 32-05.01) was determined using a Fibertec 2010 apparatus (Foss, Hillerod, Denmark). Total protein content was determined with the Kjeldahl method (Method 46-11.02) using a Kjeltec 8200 apparatus (Foss, Hillerod, Denmark). Total fat content was determined with the Soxhlet method (Method 30.10.01) using a Soxtec Labtec Line ST 243 apparatus with a Cu 2046 Control Unit Controller (Foss, Hillerod, Denmark). Carbohydrate content was calculated as the difference between 100 and the sum of the other components (water, ash, protein, fat, and fiber).

2.3. Properties of Flour and Dough

In wheat flour, the yield and quality of wet gluten were determined using the mechanical method (Method 38.12) on the Glutomatic 2200 device (Perten Instruments, Stockholm, Sweden), and the falling number was determined using the Hagberg–Perten method (Method 56-81B) on the Falling Number 1400 device (Perten Instrument, Stockholm, Sweden). The water absorption of the flour and the rheological properties of the dough were also evaluated using the Farinograph-E model 810114 (Brabender GmbH & Co. KG, Duisburg, Germany) according to AACC Method 54-21 [29]. Prior to the analysis, mixtures were prepared in which wheat flour was replaced with lyophilized pomace from black chokeberries in amounts of 1, 2, 3, 4, 5, and 6% relative to the flour weight. The control sample consisted of dough made from wheat flour. The water absorption of the flour mixtures was determined based on the amount of water added from a burette needed to achieve a dough consistency of 500 FU, and the rheological properties of the dough (development time, stability time, degree of softening after 12 min of mixing) were determined from the normal curve graph using the Farinograph v.5 computer program.

2.4. Baking Procedure

Bread dough was prepared using the direct method according to the methodology provided by Cacak-Pietrzak et al. [3]. The basic recipe for the bread dough (control sample (CP)) included 500 g of wheat flour, 15.0 g of fresh compressed yeast, 7.5 g of salt, and water in the amount necessary to achieve a dough consistency of 350 FU. This consistency is especially preferred for the process of bread production in molds [3]. The amount of water added was calculated based on the determined farinographic water absorption. Wheat flour was replaced with pomace from black chokeberries in the following amounts: 1, 2, 3, 4, 5, and 6% (CP1, CP2, CP3, CP4, CP5, and CP6, respectively) relative to the flour weight. The dough ingredients were mixed in an SP-800A mixer (Spar Food Machinery, Taichung, Taiwan) for 4 min at speed level 2. The dough was fermented in a fermentation chamber D-32 (Sveba Dahlen, Fristad, Sweden) for 90 min, with punching performed after 60 min for 1 min. After fermentation, the dough was divided into 250 g portions, shaped by hand, placed in molds, and subjected to proofing for 40 min in the fermentation chamber. Baking took place in the D-32 oven (Sveba Dahlen, Fristad, Sweden) at a temperature of 230 °C for 30 min. After baking, bread samples were removed from the molds, cooled, and then packed in polyethylene bags and stored for 24 h at a temperature of 20–22 °C.

2.5. Basic Properties of Bread

After 24 h of baking, the bread loaves were weighed, the bread yield was calculated, and the loaf volume and crumb density were determined. The loaf volume was measured using a 3D scanner (NextEngine, West Los Angeles, CA, USA) with computer software (MeshlLab 2016, ISTI-CNR Research Center, Rome, Italy) following the methodology described by Romankiewicz et al. [5], and then converted to 100 g of bread. The crumb density of the bread was determined according to the methodology provided by Belyaev et al. [30]. The procedure involved cutting cubes of known volume from the middle part of the crumb, weighing them, and calculating the crumb density.

2.6. Bread Texture

Texture profile analysis of the bread crumbs was conducted using a texture analyzer TA.XT2i (Stable Microsystem, Surrey, UK). The measurement was performed according to the methodology described by Armero and Collar [31]. Cylindrical samples (diameter 30 mm) were cut from bread slices with a thickness of 20 mm and subjected to compression using a probe with a diameter of 25 mm. The sample penetration was set at 40%, with a 45-second pause between the first and second compression, and a probe speed of 1 mm s^{-1}. From the obtained curve, the following parameters were determined: hardness, elasticity, springiness, cohesiveness, and chewiness of the crumbs.

2.7. Color Coordinates

The color coordinates of the bread samples were determined using the reflective method in the CIE-Lab* color space, where L* represents lightness, a* indicates redness (+)/greenness (−), and b* indicates yellowness (+)/blueness (−). Measurements were performed using a CR-200 colorimeter (Konica Minolta, Osaka, Japan). From the average values of the Lab* color components, the absolute color difference (ΔE*) between the bread made from wheat flour (control sample) and the bread with the addition of freeze-dried black chokeberry pomace was calculated using the formula provided by Różyło et al. [32].

2.8. The Content of Total Phenolic and Antioxidant Activity

The total polyphenol content (TPC) and antioxidant activity (AA) were determined in the raw materials (wheat flour, freeze-dried black chokeberry pomace) as well as in the bread. The ability to scavenge DPPH free radicals and ABTS•+ cation radicals was assessed.

2.8.1. Extract Preparation for TPC and AA Evaluation

The extracts of the bread samples were prepared according to the methodology described by Gawlik-Dziki et al. [33]. Samples weighing 1 ± 0.0001 g were weighed on an analytical balance and extracted with 25 cm^3 of PBS buffer (pH = 7.4) at room temperature for 60 min. The mixture was then centrifuged, and the supernatant was separated. The extraction process was repeated three times, and the supernatants were combined and stored in a dark place at a temperature of −20 °C.

2.8.2. Total Phenolic Content

The total polyphenol content was determined using the Folin–Ciocalteu spectrophotometric method, following the methodology described by Singleton and Rossi [34]. For the analysis, 0.5 mL of the buffer extract was taken, and 0.5 mL of distilled water, 2 mL of Folin reagent (diluted 1:5 with water), and 2 mL of a 10% Na_2CO_3 solution were added. The mixture was thoroughly mixed and left to stand for 60 min. The absorbance of the solution was measured using a UV Mini 1240 spectrophotometer (Shimadzu, Kyoto, Japan) at a wavelength of 720 nm against a blank sample (PBS buffer). The total polyphenol content was expressed in milligrams of gallic acid equivalents (mg GAE) per gram of dry matter (DM).

2.8.3. The Ability to Scavenge DPPH Free Radicals

The ability to scavenge DPPH free radicals was determined using a spectrophotometric method according to the methodology described by Brand-Williams et al. [35]. For the analysis, 0.01 mL of the buffer extract was taken, and 0.25 mL of an ethanol solution of DPPH was added. After thorough mixing, the mixture was left to stand for 15 min. The absorbance of the solution was measured using a microplate spectrophotometer (BioTek, Santa Clara, CA, USA) at a wavelength of 725 nm. The ability to scavenge DPPH free radicals was expressed as EC_{50}, which represents the concentration at which the DPPH radicals were reduced by 50%.

2.8.4. The Ability to Scavenge ABTS•+ Cation Radicals

The ability to scavenge ABTS•+ cation radicals was determined using a spectrophotometric method according to the methodology described by Re et al. [36]. For the analysis, 0.04 mL of the buffer extract and 1.8 mL of the ABTS•+ radical were taken. After 5 min, the absorbance of the solution was measured using a UV Mini 1240 spectrophotometer (Shimadzu, Kyoto, Japan) at a wavelength of 734 nm. The ability to scavenge ABTS•+ cation radicals was expressed as EC_{50}, which represents the concentration at which the ABTS•+ radicals were reduced by 50%.

2.9. Sensory Evaluation of Bread

The sensory evaluation of the bread was conducted according to the methodology provided by Garcia-Gómez et al. [37], 24 h after baking. The evaluation panel consisted of 46 assessors (employees and students of Warsaw University of Life Sciences) aged 21–56. The bread samples were assessed on a 9-point hedonic scale, where 1 represented very undesirable and 9 represented very desirable. The evaluation took into account the following quality attributes of the bread: loaf appearance, crust appearance, crumb properties, taste and aroma, and overall acceptability.

2.10. Statistical Analysis

All measurements were performed in a minimum of three parallel repetitions. Statistical analysis of the obtained results was conducted using Statistica 13.3 software (TIBCO Software, Palo Alto, CA, USA). Analysis of variance (ANOVA) was performed, and homogeneous groups were determined using Tukey's test. Additionally, the Pearson's correlation coefficients between variables were determined. All statistical tests were performed with a significance level of α set at 0.05.

3. Results and Discussion

3.1. Physical Properties of Dough

The physical properties of dough measured by the farinograph can help to adjust the bread recipe and consequently allow for the optimization of bread production [38]. Addition of CP into wheat dough resulted in a linear increase in water absorption of flour from 59.07% for the control sample to 60.97% for dough with 6% of CP flour (Table 1). This means that the dough can hold more water, which can affect the bread yield and texture. A strong and positive linear relation was found between the level of wheat flour replacement with CP and flour water absorption ($r = 0.982$, $p = 0.001$). Higher water absorption of enriched flour probably results from the high content of fiber in pomace. Similar effects on wheat flour water absorption were observed by others authors after the addition of powders rich in fiber [1,39]. Interestingly, incorporation of AP into wheat flour had no significant influence on the development time but significantly decreased the stability of dough during mixing. These parameters can vary depending on the type of flour and used additives [14]. Typically, higher values of these parameters result in a stronger gluten structure and a firmer dough, while a shorter development time and stability result in a weaker gluten structure and a softer dough [2]. The stability of the dough decreased linearly ($r = -0.951$, $p = 0.001$) from 5.83 (CD) to 2.13 min (CDP6). It indicated that AP considerably weakened the gluten structure. It was also confirmed by the linear increase in degree of dough softening with the increase in percentage of CP in wheat flour. The control dough was characterized by an approximately twofold lower degree of softening compared with the AP6 dough sample. A strong and positive correlation was found between this parameter and the level of CP in dough ($r = 0.965$, $p = 0.001$). Degree of softening is an important indicator of the quality of the flour. A high degree of softening indicates that the flour is weak and has poor gluten-forming ability, which can result in a low volume and dense crumbs. On the other hand, lower values of this parameter indicate that the flour has strong gluten-forming ability, which usually indicates a well-risen bread with a good crumb structure.

Table 1. Physical properties of control dough and dough enriched with chokeberry powder.

Sample	Water Absorption [%]	Development Time [min]	Stability of Dough [min]	Degree of Softening (FU) after 12 min of Mixing
CD	59.07 ± 0.35 [a]	2.00 ± 0.10 [a]	5.83 ± 0.25 [e]	63.33 ± 1.15 [a]
CDP1	59.37 ± 0.78 [ab]	2.23 ± 0.06 [a]	5.00 ± 0.53 [d]	64.67 ± 2.89 [a]
CDP2	59.47 ± 0.21 [ab]	2.17 ± 0.15 [a]	3.63 ± 0.15 [c]	77.00 ± 2.65 [b]
CDP3	59.77 ± 0.49 [abc]	2.17 ± 0.06 [a]	2.93 ± 0.15 [b]	89.33 ± 3.06 [c]
CDP4	60.23 ± 0.15 [abc]	2.03 ± 0.06 [a]	2.67 ± 0.15 [ba]	94.67 ± 2.31 [cd]
CDP5	60.37 ± 0.59 [bc]	2.00 ± 0.00 [a]	2.27 ± 0.12 [ba]	98.67 ± 2.89 [d]
CDP6	60.97 ± 0.06 [c]	2.10 ± 0.10 [a]	2.13 ± 0.12 [a]	99.67 ± 1.53 [d]

CD—control dough, CDP1, CDP2, CDP3, CDP4, CDP5, CDP6—dough enriched with 1, 2, 3, 4, 5, and 6% of chokeberry powder, respectively; the values designated by the different letters [a–e] are significantly different ($p < 0.05$).

3.2. Basic Properties of Bread Samples

The addition of CP into wheat flour resulted in a linear decrease in baking loss ($r = -0.971$, $p = 0.001$). This loss decreased from 13.6% (C) to 10.7% (CP6) (Table 2). Baking loss refers to the reduction in weight that occurs during the baking process, primarily caused by the loss of water but also due to the evaporation of volatile substances such as carbon dioxide, alcohol, and volatile acids. This loss of weight can affect the texture, taste, and appearance of the final product, which is why it is important to understand the factors that contribute to baking loss and take steps to minimize it [40]. The decrease in baking loss with the increased percentage of CP in the bread recipe could have resulted from the higher water holding capacity of the CP. Similar results were obtained by other authors when potato flour was incorporated into wheat bread [41]. Importantly, CP increased the yield of bread from 137.8 to 144.1%. This is because CP is high in dietary fiber, which has the ability to absorb water and hold it in the bread dough during processing, leading to better dough consistency and less water loss during baking. On the other hand, bread volume decreased and, as a result, the density of the bread increased with the higher percentage of CP ($r = -0.989$, $p = 0.001$ and 0.988, $p = 0.001$, respectively). These changes usually have a negative influence on bread texture, leading to a more compact and harder crumbs [9].

Table 2. Basic properties of control and enriched bread samples.

Sample	Baking Loss [%]	Bread Yield [%]	Bread Volume [cm^3 100 g^{-1}]	Crumb Density [kg m^{-3}]
C	13.6 ± 0.2 [e]	137.8 ± 0.6 [a]	349.0 ± 4.2 [f]	0.32 ± 0.01 [a]
CP1	12.9 ± 0.3 [d]	139.4 ± 0.6 [ab]	338.3 ± 2.6 [f]	0.38 ± 0.00 [b]
CP2	12.8 ± 0.2 [d]	139.7 ± 0.8 [b]	318.7 ± 3.8 [e]	0.39 ± 0.01 [b]
CP3	12.4 ± 0.2 [cd]	141.1 ± 0.8 [b]	303.1 ± 3.7 [d]	0.45 ± 0.01 [c]
CP4	12.2 ± 0.2 [b]	142.2 ± 0.8 [cd]	265.3 ± 6.2 [c]	0.52 ± 0.01 [d]
CP5	11.1 ± 0.1 [a]	142.9 ± 0.4 [de]	252.6 ± 4.9 [b]	0.53 ± 0.01 [d]
CP6	10.7 ± 0.1 [a]	144.1 ± 0.3 [e]	237.9 ± 4.7 [a]	0.59 ± 0.01 [e]

C—control bread, CP1, CP2, CP3, CP4, CP5, CP6—bread enriched with 1, 2, 3, 4, 5, and 6% of chokeberry powder, respectively; the values designated by the different letters [a–f] are significantly different ($p < 0.05$).

3.3. Crumb Texture

The increase in the proportion of CP in the bread recipe resulted in a significant increase in the bread's crumb hardness. This trend was observed after both 24 and 72 h of bread storage. However, storage itself had a relatively minor impact on this parameter. The crumb hardness of the bread changed from 7.77 to 21.36 N and from 8.43 to 22.54 N after one and three days of storage, respectively (Table 3). It should be noted that the addition of 6% CP caused an approximately threefold increase in this parameter compared to the control bread. These changes are attributed to a decrease in volume and an increase in crumb density due to the addition of CP, resulting in a more compact and harder crumbs. Enriching the bread had a similar effect on the chewiness of the crumbs. CP influenced the increase in these parameter, primarily due to the increase in crumb hardness. On the other hand, bread storage generally did not have a significant impact on gumminess or chewiness. However, the elasticity and cohesiveness of the crumbs decreased under the influence of CP, although the differences between the means were mostly insignificant. The highest values of these parameters were obtained for the control bread. After 72 h of storage, both the control and enriched bread showed significantly lower elasticity and cohesiveness compared to the values after 24 h since baking. The replacement of wheat flour with 6% CP addition, similar to storage, did not have a significant impact on the crumb springiness. Several authors have suggested that partial substitution of wheat flour with powdered fruits and other plant additives weakens the dough structure, resulting in decreased volume due to reduced carbon dioxide retention during dough fermentation [42], which consequently negatively affects the bread texture, particularly in terms of crumb hardness [3].

Table 3. Texture parameters of control and enriched bread samples.

Sample	Time after Baking [h]	Hardness [N]	Elasticity [−]	Springiness [−]	Cohesiveness [−]	Chewiness [N]
C	24	7.77 ± 0.30^{a}	0.31 ± 0.09^{e}	0.94 ± 0.03^{a}	0.70 ± 0.11^{e}	5.08 ± 0.72^{abc}
CP1		9.33 ± 0.23^{ab}	0.30 ± 0.01^{de}	0.90 ± 0.02^{a}	0.68 ± 0.05^{de}	5.70 ± 0.34^{abcd}
CP2		11.80 ± 0.36^{c}	0.27 ± 0.06^{cde}	0.90 ± 0.02^{a}	0.62 ± 0.07^{cde}	6.65 ± 1.05^{cde}
CP3		12.59 ± 0.51^{c}	0.25 ± 0.03^{bcde}	0.89 ± 0.01^{a}	0.59 ± 0.03^{bcde}	6.60 ± 0.39^{cde}
CP4		16.78 ± 0.60^{d}	0.22 ± 0.05^{abcde}	0.85 ± 0.04^{a}	0.56 ± 0.06^{abcde}	8.03 ± 0.92^{ef}
CP5		20.51 ± 0.36^{ef}	0.24 ± 0.01^{abcde}	0.87 ± 0.03^{a}	0.60 ± 0.02^{bcde}	10.65 ± 0.78^{g}
CP6		21.36 ± 0.65^{e}	0.21 ± 0.01^{abcde}	0.81 ± 0.02^{a}	0.56 ± 0.01^{abcde}	9.56 ± 0.22^{fg}
C	72	8.43 ± 0.72^{a}	0.19 ± 0.04^{abcd}	0.86 ± 0.02^{a}	0.53 ± 0.04^{abcd}	3.84 ± 0.35^{a}
CP1		11.08 ± 0.31^{bc}	0.15 ± 0.02^{abc}	0.82 ± 0.04^{a}	0.45 ± 0.02^{ab}	4.10 ± 0.17^{ab}
CP2		15.47 ± 0.80^{d}	0.13 ± 0.02^{a}	0.87 ± 0.01^{a}	0.41 ± 0.02^{a}	5.50 ± 0.50^{abc}
CP3		20.12 ± 1.23^{e}	0.16 ± 0.05^{abc}	0.86 ± 0.02^{a}	0.45 ± 0.07^{ab}	7.80 ± 1.35^{def}
CP4		17.34 ± 0.71^{d}	0.13 ± 0.03^{a}	0.86 ± 0.03^{a}	0.42 ± 0.05^{a}	6.27 ± 0.61^{bcde}
CP5		20.71 ± 0.62^{ef}	0.15 ± 0.04^{ab}	0.87 ± 0.13^{a}	0.44 ± 0.08^{ab}	7.93 ± 1.38^{def}
CP6		22.54 ± 0.64^{f}	0.15 ± 0.02^{abc}	0.85 ± 0.11^{a}	0.48 ± 0.04^{abc}	8.99 ± 0.63^{fg}

C—control bread, CP1, CP2, CP3, CP4, CP5, CP6—bread enriched with 1, 2, 3, 4, 5, and 6% of chokeberry powder, respectively; the values designated by the different letters $^{a-g}$ are significantly different ($p < 0.05$).

3.4. Color Coordinates

The color of food products is a crucial determinant of consumer acceptability, including for bread. The lightness values (L*) were the highest for the control sample, while those for bread enriched with 5 and 6% CP were the lowest and not significantly different from each other. Specifically, the L* value ranged from 72.00 for the control sample to 37.53 for the bread enriched with 6% CP (Table 4). The color of the bread gradually became darker with increasing levels of fruit supplementation, and the brightness was influenced by the extent of enrichment. Petković et al. [43] reported a significant decrease in bread brightness only at 10% chokeberry fruit powder supplementation, while concentrations of 1, 2.5, and 5% CP did not cause a significant change, compared to the control. The discrepancies in the results could potentially be attributed to variations in amount of used additive. Specifically, in the cited study, the fruits were subjected to convection drying, while in this research, freeze-dried CP was used. Furthermore, the addition of chokeberry led to a significant increase in redness (a*), which was dependent on the concentration of the additive. The sample containing 6% CP had the highest a* value, while the control sample had the lowest, with values of 12.74 and 1.92, respectively. In general, the decline in b* values was associated with the concentration of CP. However, there were no significant differences in b* values between breads enriched with 4 and 5% additive or between those enriched with 5 and 6% additive. Therefore, changes in yellowness were more evident at lower (1–3%) enrichment levels. The control and bread enriched with 6% CP exhibited the highest and lowest b* values, respectively, with values of 16.83 and 0.97. In another study by Petković et al. [43], it was observed that the b* value of bread decreased with the addition of powdered chokeberry fruit. However, the significant reduction was only observed at higher substitution levels (10% for powder dried at 50, 60, and 70 °C, and 5% for powder dried at 70 °C), indicating that the incorporation of freeze-dried chokeberry pomace in our study had a greater effect on the change in yellowness than the addition of convection-dried chokeberry fruits. The total color difference (ΔE) ranged from 15.22 to 37.88 for the control and the bread enriched with 5% CP, respectively. This demonstrates that the incorporation of CP into bread had a strong influence on the color of enriched bread. The red pigment fraction, composed of cyanidin derivatives, is mainly responsible for these color changes [44].

Table 4. Results of colorimetric measurements of control and enriched bread samples.

Sample	L*	a*	b*	ΔE
C	72.00 ± 0.38^{e}	1.92 ± 0.19^{a}	16.83 ± 0.53^{f}	-
CP1	59.67 ± 0.86^{d}	5.39 ± 0.27^{b}	8.63 ± 0.12^{e}	15.22 ± 0.76^{a}
CP2	51.12 ± 1.11^{c}	6.84 ± 0.04^{c}	6.42 ± 0.26^{d}	23.85 ± 0.87^{b}
CP3	44.31 ± 1.13^{b}	9.59 ± 0.24^{d}	4.12 ± 0.27^{c}	31.43 ± 1.06^{c}
CP4	42.03 ± 1.63^{b}	10.56 ± 0.10^{e}	2.31 ± 0.59^{b}	34.41 ± 1.66^{d}
CP5	38.81 ± 0.75^{a}	11.77 ± 0.09^{f}	1.47 ± 0.62^{ab}	37.88 ± 0.41^{e}
CP6	37.53 ± 0.29^{a}	12.74 ± 0.22^{g}	0.97 ± 0.01^{a}	36.84 ± 0.24^{de}

C—control bread, CP1, CP2, CP3, CP4, CP5, CP6—bread enriched with 1, 2, 3, 4, 5, and 6% of chokeberry powder, respectively; the values designated by the different letters $^{a-g}$ are significantly different ($p < 0.05$).

3.5. Basic Chemical Composition of Raw Materials and Bread

CP enriched the bread in terms of mineral, TDF, and fat content. In particular, TDF content increased almost threefold, from 3.56 (C) to 9.27% (CP6), whereas fat and ash content changed from 1.81 (C) to 2.56% (CP6) and from 0.928 (C) to 0.974% (CP6), respectively (Table 5). Conversely, increasing the percentage of CP in the bread tends to decrease the protein and carbohydrate content. However, the differences in protein content between the samples were small, and the addition of up to 3% CP was not statistically influential on this component. The correlation analysis revealed significant linear relations ($p < 0.05$) between pomace content in bread and ash, TDF, fat, protein, and carbohydrate content (r = 0.998; 0.972; 0.998, −0.975 and −0.998, respectively). Bread enriched with chokeberry pomace can be a valuable source of fiber and minerals. CP is an especially rich source of such macroelements (K, Ca, Mg and Na) and microelements (B, Cr, Cu, Fe, Mn, Ni, Se, Sr and Zn) [27]. Moreover, CP is a also abundant in insoluble dietary fiber, which makes up 76% of its total fiber content. It also contains about 20% of high molecular weight soluble fraction and 4% low molecular weight soluble fiber fraction [45].

Table 5. Basic composition [% DM] of raw materials and bread samples.

Sample	Ash	Protein	Total Dietary Fiber	Fat	Carbohydrates
CP	1.62 ± 0.010 B	6.76 ± 0.01 A	57.20 ± 0.24 B	3.72 ± 0.02 B	30.7 ± 0.08 A
WF	0.73 ± 0.001 A	12.74 ± 0.02 B	2.83 ± 0.10 A	1.69 ± 0.02 A	83.01 ± 0.12 B
C	0.928 ± 0.003 a	12.98 ± 0.01 c	3.56 ± 0.16 a	1.81 ± 0.12 a	80.73 ± 0.27 a
CP1	0.935 ± 0.002 ab	12.91 ± 0.01 bc	4.59 ± 0.09 b	2.06 ± 0.05 b	79.51 ± 0.11 b
CP2	0.942 ± 0.008 bc	12.86 ± 0.02 bc	5.69 ± 0.16 c	2.22 ± 0.03 c	78.30 ± 0.16 c
CP3	0.952 ± 0.003 cd	12.82 ± 0.04 bc	6.44 ± 0.07 d	2.29 ± 0.03 cd	77.50 ± 0.08 d
CP4	0.956 ± 0.002 de	12.72 ± 0.02 ab	7.37 ± 0.12 e	2.42 ± 0.02 de	76.52 ± 0.10 e
CP5	0.964 ± 0.002 ef	12.61 ± 0.17 a	8.52 ± 0.20 f	2.48 ± 0.01 e	73.43 ± 0.13 f
CP6	0.974 ± 0.003 f	12.64 ± 0.09 a	9.27 ± 0.11 g	2.56 ± 0.02 e	72.56 ± 0.16 g

CP—chokeberry pomace, WF—wheat flour, C—control bread, CP1, CP2, CP3, CP4, CP5, CP6—bread enriched with 1, 2, 3, 4, 5, and 6% of chokeberry powder, respectively; the values designated by the different letters A,B or a–g are significantly different ($p < 0.05$).

3.6. Phenolics Content and Antioxidant Capacity

Chokeberry pomace is known to be abundant in phenolic compounds, especially anthocyanins, which have antioxidant properties and may provide various health benefits [46]. The content of TPC and AA of raw materials and enriched breads is presented in Table 6. CP contained the highest phenolics content with the strongest antioxidant activity. Most phenolic compounds are primarily found in the fruit skin. As a result, a considerable amount of these compounds are retained in the CP even after the extraction of juice from the fruit [47]. Kapci et al. [48] found about 63 mg GAE g DM^{-1} of TPC in chokeberry pomace. This effect was visible in the enrichment of bread. With the addition of 1% CP, the TPC doubled; with 6% addition, it was about 4.5 times higher compared to the control sample (Table 6). The relationship between the percentage of CP and TPC in bread samples was linear (r = 0.987, p = 0.001). Consequently, antioxidant capacity of enriched bread increased (decrease in EC_{50} index). Both ABTS and DPPH antiradical activity increased with the percentage of CP in the bread recipe. However, the relations were non-linear, and the highest decrease in EC_{50} was observed when wheat flour was replaced with 1% of CP from 552.47 to 202.33 mg DM mL^{-1} and from 194.37 to 115.89 mg DM mL^{-1}, for ABTS and DPPH, respectively. Other authors also found that CP significantly increased TPC and antioxidant capacity of shortbread cookies [49].

Table 6. Phenolics content and antioxidant capacity of raw materials, control, and enriched bread samples.

Sample	TPC [mg GAE g DM^{-1}]	EC_{50DPPH} [mg DM mL^{-1}]	EC_{50ABTS} [mg DM mL^{-1}]
CP	54.04 B	43.33 A	26.07 A
WF	0.52 A	585.60 B	190.25 B
C	0.46 ± 0.03 a	552.47 ± 24.93 d	194.37 ± 3.30 f
CP1	0.86 ± 0.05 b	202.33 ± 32.40 c	115.89 ± 5.18 e
CP2	1.03 ± 0.05 bc	187.03 ± 11.10 c	90.24 ± 4.01 d
CP3	1.25 ± 0.08 cd	140.42 ± 4.04 b	62.64 ± 1.62 c
CP4	1.34 ± 0.06 e	110.73 ± 4.36 ab	51.67 ± 1.28 b
CP5	1.76 ± 0.06 e	88.84 ± 8.08 a	38.65 ± 1.52 a
CP6	2.07 ± 0.18 f	75.31 ± 1.81 a	35.62 ± 2.01 a

CP—chokeberry pomace, WF—wheat flour, C—control bread, CP1, CP2, CP3, CP4, CP5, CP6—bread enriched with 1, 2, 3, 4, 5, and 6% of chokeberry powder, respectively; the values designated by the different letters A,B or a–f are significantly different ($p < 0.05$).

3.7. Sensory Evaluation Results

The acceptance of food products by consumers is a very important element of their presence on the market. The final acceptability of bread depends on both the type and quality of the raw materials used, as well as the applied technological process. When evaluating the overall appearance of a loaf, consumers paid attention to the degree of its rising and shape, as well as the appearance of the crust surface. The picture of the control and enriched bread is presented in Figure 2. The addition of CP caused the crust color to become darker and gradually take on a purple hue. The control sample and bread with the addition of 1 to 4% pomace were characterized by a shiny crust, while the crust of bread with 5 and 6% CP addition was matte, with visible cracks and unevenness on its surface. The highest scores for appearance acceptability were given by evaluators to the control bread, which had the most uniform shape, was the best risen, and had no visible mechanical damage. The addition of CP caused the bread loaves to have a lower degree of rising and an irregular shape. The appearance of bread with more than 3% CP was generally not accepted by consumers, and all attempts with a higher content of this addition received a rating below 5 points on a 9-point hedonic scale (Table 7). Five points on the hedonic scale (neither like or dislike) can be considered as the threshold of consumer acceptability for a product [50]. However, it is worth remembering that the threshold of acceptability may vary depending on the context, product, or target group. The final assessment of consumer satisfaction depends on various factors such as expectations, preferences, and individual experiences. The taste and aroma of the control bread were rated the highest. The introduction of CP into the bread recipe up to 2% allowed for the production of products with good acceptability according to these indicators. A higher addition of CP was poorly accepted by consumers. Panelists found that, as the pomace content in the bread recipe increased, its taste became too acidic and tart, and the smell was unique. When evaluating the bread texture, evaluators observed that the crumbs were more compact and hard above 3% CP addition. As a result, all breads above this enrichment level were already unacceptable to consumers. In summary, the overall acceptability rating showed that the addition of dried and powdered pomace to wheat bread should not be higher than 3% relative to the amount of flour used. Many authors have shown that replacement of wheat flour with different powdered and dried plant materials in wheat bread production should not exceed a few percentage points [3,51], whereas, in the case of others cereal products such pasta or cookies, this level can be higher [52,53]. The main reason for this is the unfavorable effect of the plant additives used on the volume of bread. These additives weaken the structure of the gluten network, and as a result, the bread loses its ability to retain carbon dioxide, which often leads not only to a decrease in volume but also to deformation of the shape of the loaves.

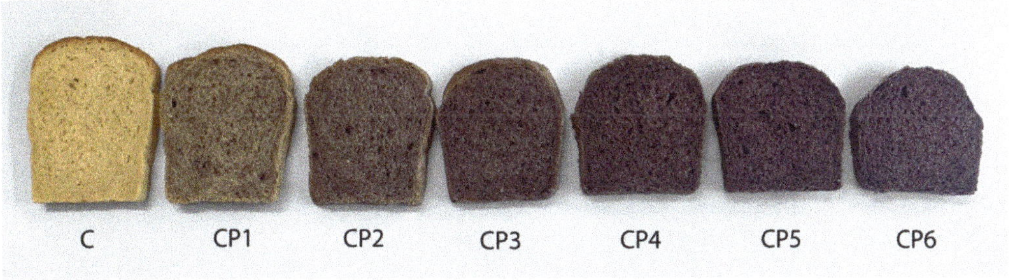

Figure 2. Picture of control and CP-enriched bread samples: C—control bread, CP1, CP2, CP3, CP4, CP5, CP6—bread enriched with 1, 2, 3, 4, 5, and 6% of chokeberry powder.

Table 7. Results of sensory evaluation of control and enriched bread samples.

Sample	Appearance	Smell	Taste	Texture	Color	OA
C	7.9 ± 0.6 [d]	8.3 ± 0.4 [e]	7.7 ± 0.5 [e]	8.3 ± 0.2 [d]	8.1 ± 0.5 [e]	7.9 ± 0.5 [g]
CP1	7.5 ± 0.6 [d]	7.5 ± 0.5 [de]	7.2 ± 0.7 [de]	7.8 ± 0.3 [d]	7.3 ± 0.6 [d]	7.6 ± 0.5 [f]
CP2	6.3 ± 0.4 [c]	6.9 ± 0.4 [d]	6.8 ± 0.4 [cd]	7.4 ± 0.3 [c]	6.5 ± 0.6 [c]	6.8 ± 0.5 [e]
CP3	5.6 ± 0.6 [c]	5.8 ± 0.4 [d]	5.4 ± 0.6 [bc]	6.3 ± 0.4 [c]	5.7 ± 0.4 [c]	5.7 ± 0.3 [d]
CP4	4.2 ± 0.7 [b]	5.3 ± 0.3 [c]	4.1 ± 0.4 [b]	6.3 ± 0.2 [b]	4.5 ± 0.3 [b]	4.3 ± 0.3 [c]
CP5	3.3 ± 0.5 [ab]	4.8 ± 0.2 [ab]	4.2 ± 0.3 [a]	5.2 ± 0.3 [a]	4.1 ± 0.3 [a]	3.8 ± 0.2 [b]
CP6	2.5 ± 0.4 [a]	4.6 ± 0.2 [a]	3.2 ± 0.2 [a]	4.8 ± 0.3 [a]	3.6 ± 0.2 [a]	3.4 ± 0.2 [a]

C—control bread, CP1, CP2, CP3, CP4, CP5, CP6—bread enriched with 1, 2, 3, 4, 5, and 6% of chokeberry powder, respectively; OA—overall acceptability; the values designated by the different letters [a–g] are significantly different ($p < 0.05$).

4. Conclusions

The addition of powdered chokeberry pomace to wheat flour resulted in a deterioration in the bread's quality. There was a decrease in volume and an increase in crumb hardness. Additionally, a decrease in consumer acceptance of the enriched bread was observed, especially when the wheat flour was replaced with pomace in a proportion higher than 3%. On the other hand, the chokeberry pomace enriched the bread with dietary fiber, minerals, and phenolic compounds, and increased its antioxidant activity. Notably, the total phenolic content in the CP3 bread increased by about threefold compared to the control bread. Future studies should focus on the possibility of improving the sensory characteristics of bread enriched with chokeberry pomace by testing different types of wheat flour, dough fermentation methods, and raw material selection in order to enhance its consumer acceptability at higher levels of pomace inclusion in the bread recipe.

Author Contributions: Conceptualization: G.C.-P. and D.D.; methodology: G.C.-P., D.D., U.G.-D., N.P-N. and S.K.; software: U.G.-D. and S.K.; validation: G.C.-P. and D.D.; formal analysis: G.C.-P., N.P.-N. and A.K.; investigation: G.C.-P., D.D., U.G.-D., N.P.-N. and S.S.; resources: G.C.-P. and D.D., data curation: G.C.-P. and D.D.; writing—original draft preparation: G.C.-P., D.D. and A.K.; writing—review and editing: G.C.-P., D.D. and A.K.; visualization: G.C.-P. and S.S.; supervision: D.D.; project administration: G.C.-P.; funding acquisition: G.C.-P., D.D. and S.K. All authors have read and agreed to the published version of the manuscript.

Funding: This research received no external funding.

Institutional Review Board Statement: Not applicable.

Informed Consent Statement: Not applicable.

Data Availability Statement: The data presented in this study are available on request from the corresponding author.

Acknowledgments: Research for this publication was conducted using research equipment purchased as part of the "Food and Nutrition Centre—modernisation of the WULS campus to create a Food and Nutrition Research and Development Centre (CZi Z)" co-financed by the European Union from the European Regional Development Fund under the Regional Operational Programme of the Mazowieckie Voivodeship for 2014–2020 (Project No. RPMA.01.01.00-14-8276/17).

Conflicts of Interest: The authors declare no conflict of interest.

References

1. Conte, P.; Fadda, C.; Piga, A.; Collar, C. Techno-functional and nutritional performance of commercial breads available in Europe. *Food Sci. Technol. Int.* **2016**, *22*, 621–633. [CrossRef] [PubMed]
2. Capurso, A.; Capurso, C. The mediterranean way: Why elderly people should eat whole wheat sourdough bread—A little known component of the mediterranean diet and healthy food for elderly adults. *Aging Clin. Exp. Res.* **2020**, *32*, 1–5. [CrossRef] [PubMed]
3. Cacak-Pietrzak, G.; Dziki, D.; Gawlik-Dziki, U.; Sułek, A.; Wójcik, M.; Krajewska, A. Dandelion flowers as an additive to wheat bread: Physical properties of dough and bread quality. *Appl. Sci.* **2023**, *13*, 477. [CrossRef]

4. Pilipović, K.; Jurišić Grubešić, R.; Dolenec, P.; Kučić, N.; Juretić, L.; Mršić-Pelčić, J. Plant-Based Antioxidants for Prevention and Treatment of Neurodegenerative Diseases: Phytotherapeutic Potential of *Laurus nobilis*, *Aronia melanocarpa*, and *Celastrol*. *Antioxidants* **2023**, *12*, 746. [CrossRef] [PubMed]
5. Romankiewicz, D.; Hassoon, W.H.; Cacak-Pietrzak, G.; Sobczyk, M.; Wirkowska-Wojdyła, M.; Ceglińska, A.; Dziki, D. The effect of chia seeds (*Salvia hispanica* L.) addition on quality and nutritional value of wheat bread. *J. Food Qual.* **2017**, *2017*, 7352631. [CrossRef]
6. Sluková, M.; Levková, J.; Michalcová, A.; Horáčková, Š.; Skřivan, P. Effect of the dough mixing process on the quality of wheat and buckwheat proteins. *Czech J. Food Sci.* **2017**, *35*, 522–531. [CrossRef]
7. Cao, H.; Gao, F.; Shen, H.; Su, Q.; Guan, X.; Sun, Z.; Yu, Z. Influence of partial substitution of wheat flour with sprouted oat flours on physicochemical and textural characteristics of wheat bread. *J. Cereal Sci.* **2023**, *110*, 103649. [CrossRef]
8. Cotovanu, I.; Ungureanuluga, M.; Mironeasa, S. Investigation of quinoa seeds fractions and their application in wheat bread production. *Plants* **2021**, *10*, 2150. [CrossRef] [PubMed]
9. Yamsaengsung, R.; Schoenlechner, R.; Berghofer, E. The effects of chickpea on the functional properties of white and whole wheat bread. *Int. J. Food Sci. Technol.* **2010**, *45*, 610–620. [CrossRef]
10. Bojňanská, T.; Musilová, J.; Vollmannová, A. Effects of Adding Legume Flours on the rheological and breadmaking properties of dough. *Foods* **2021**, *10*, 1087. [CrossRef] [PubMed]
11. Dziki, D.; Cacak-Pietrzak, G.; Hassonn, W.H.; Gawlik-Dziki, U.; Sułek, A.; Różyło, R.; Suger, D. The fruit of sumac (*Rhus coriaria* L.) as a functional additive and salt replacement to wheat bread. *LWT Food Sci. Technol.* **2021**, *136*, 110346. [CrossRef]
12. Dziki, D.; Cacak-Pietrzak, G.; Gawlik-Dziki, U.; Sułek, A.; Kocira, S.; Biernacka, B. Effect of Moldavian dragonhead (*Dracocephalum moldavica* L.) leaves on the baking properties of wheat flour and quality of bread. *CyTA—J. Food* **2019**, *17*, 536–543. [CrossRef]
13. Wójcik, M.; Różyło, R.; Łysiak, G.; Kulig, R.; Cacak-Pietrzak, G. Textural and sensory properties of wheat bread fortified with nettle (*Urtica dioica* L.) produced by scalded flour method. *J. Food Process. Preserv.* **2021**, *45*, e15851. [CrossRef]
14. Cacak-Pietrzak, G.; Dziki, D.; Gawlik-Dziki, U.; Sułek, A.; Kalisz, S.; Sujka, K. Effect of the addition of dried dandelion roots (*Taraxacum officinale* F.H. Wigg.) on wheat dough and bread properties. *Molecules* **2021**, *26*, 7564. [CrossRef] [PubMed]
15. Korus, A.; Witczak, M.; Korus, J.; Juszczak, L. Dough rheological properties and characteristics of wheat bread with the addition of lyophilized kale (*Brassica oleracea* L. var. *sabellica*) powder. *Appl. Sci.* **2023**, *13*, 29. [CrossRef]
16. Czubaszek, A.; Czaja, A.; Sokół-Letowska, A.; Kolniak-Ostek, J.; Kucharska, A. Changes in antioxidant properties and amounts of bioactive compounds during simulated in vitro digestion of wheat bread enriched with plant extracts. *Molecules* **2021**, *26*, 6292. [CrossRef]
17. Baiano, A.; la Gatta, B.; Rutigliano, M.; Fiore, A. Functional bread produced in a circular economy perspective: The use of brewers' spent grain. *Foods* **2023**, *12*, 834. [CrossRef]
18. Valková, V.; Ďúranová, H.; Havrlentová, M.; Ivanišová, E.; Mezey, J.; Tóthová, Z.; Gabríny, L.; Kačániová, M. Selected physico-chemical, nutritional, antioxidant and sensory properties of wheat bread supplemented with apple pomace powder as a by-product from juice production. *Plants* **2022**, *11*, 1256. [CrossRef]
19. Czubaszek, A.; Wojciechowicz-Budzisz, A.; Spychaj, R.; Kawa-Rygielska, J. Effect of added brewer's spent grain on the baking value of flour and the quality of wheat bread. *Molecules* **2022**, *27*, 1624. [CrossRef]
20. Granucci, N.; Harris, P.J.; Villas-Boas, S.G. Chemical Compositions of Fruit and Vegetable Pomaces from the Beverage Industries. *Waste Biomass Valor.* **2023**, *early access*. [CrossRef]
21. European Commision. *A Sustainable Bioeconomy for Europe: Strengthening the Connection between Economy, Society and the Environment: Updated Bioeconomy Strategy*; Publications Office of the European Union: Brussels, Belgium, 2018.
22. Ochmian, I.; Grajkowski, J.; Smolik, M. Comparsion of some morphological features, quality and chemical content of four cultivars of chokeberry fruits (*Aronia melanocarpa*). *Not. Bot. Horti. Agrobot.* **2012**, *40*, 253–260. [CrossRef]
23. Mayer-Miebach, E.; Adamiuk, M.; Behsnilian, D. Stability of chokeberry bioactive polyphenols during juice processing and stabilization of a polyphenol-rich material from the by-product. *Agriculture* **2012**, *2*, 244–258. [CrossRef]
24. Tanaka, T.; Tanaka, A. Chemical components and characteristics of black chokeberry. *J. Jpn. Soc. Food Sci. Technol.* **2001**, *8*, 606–610. [CrossRef]
25. Samoticha, A.J.; Wojdyło, A.; Lech, K. The influence of different the drying methods on chemical composition and antioxidant activity in chokeberries. *LWT—Food Sci. Technol.* **2016**, *66*, 484–489. [CrossRef]
26. Pavlović, A.N.; Brcanović, J.M.; Veljković, J.N.; Mitić, S.S.; Tošić, S.B.; Kaličanin, B.M.; Kostić, D.A.; Dordević, M.S.; Velimirović, D.S. Characterization of commercially available products of aronia according to their metal content. *Fruits* **2015**, *70*, 385–393. [CrossRef]
27. Kaličanin, B.; Velimirović, D.; Nešić, I. A study on mineral composition of *Aronia melanocarpa* as a functional food with potential therapeutic properties. *J. Food Nutr. Res.* **2022**, *61*, 53–60.
28. Kitryte, V.; Kraujaliene, V.; Sulniute, V.; Pukalskas, A.; Rimantas Venskutonis, P. Chokeberry pomace valorization into foodingredients by enzyme-assisted extraction: Process optimization and product characterization. *Food Bioprod. Process.* **2017**, *105*, 36–50. [CrossRef]
29. AACC. *American Association of Cereal Chemistry Approved Methods*, 11th ed.; AACC: Greensboro, NC, USA, 2009.
30. Belyaev, A.G.; Kovaleva, A.E.; Pyanikova, E.A. The influence of fireweed powder on the quality of wheat bread. *Proc. Vor. State Univ. Eng. Technol.* **2019**, *80*, 254–258. [CrossRef]

31. Almero, E.; Collar, C. Texture properties of formulated wheat doughs: Relationship with dough and bread technological quality. *Z. Lebensm. Forsch. A* **1997**, *204*, 136–145. [CrossRef]
32. Różyło, R.; Wójcik, M.; Dziki, D.; Biernacka, B.; Cacak-Pietrzak, G.; Gawłowski, S.; Zdybel, A. Freeze-dried elderberry and chokeberry as natural colorants for gluten-free wafer sheets. *Int. Agrophys.* **2018**, *33*, 217–225. [CrossRef]
33. Gawlik-Dziki, U.; Kaszuba, K.; Piwowarczyk, K.; Świeca, M.; Dziki, D.; Czyż, J. Onion skin—Raw material for the production of suplement that enhances the health beneficial properties of wheat bread. *Food Res. Int.* **2015**, *73*, 97–106. [CrossRef]
34. Singleton, V.L.; Rossi, J. Colorimetry of total phenolics with phosphomolybdic. *Am. J. Enol. Vitic.* **1965**, *6*, 144–158. [CrossRef]
35. Brand-Williams, W.; Cuvelier, E.; Berset, C. Use of a free radical method to evaluate antioxidant activity. *LWT—Food Sci. Technol.* **1995**, *28*, 25–30. [CrossRef]
36. Re, R.; Pellegrini, N.; Proteggente, A.; Pannala, A.; Yang, M.; Rice-Evans, C. Antioxidant activity applying an improved ABTS radical cation decolorization assay. *Free Radic. Biol. Med.* **1999**, *26*, 1231–1237. [CrossRef] [PubMed]
37. García-Gómez, B.; Fernández-Canto, N.; Vázquez-Odériz, M.L.; Quiroga-García, M.; Muñoz-Ferreiro, N.; Romero-Rodríguez, M.Á. Sensory descriptive analysis and hedonic consumer test for Galician type breads. *Food Control* **2022**, *134*, 108765. [CrossRef]
38. Biel, W.; Jaroszewska, A.; Stankowski, S.; Sobolewska, M.; Kępińska-Pacelik, J. Comparison of yield, chemical composition and farinograph properties of common and ancient wheat grains. *Eur. Food Res. Technol.* **2021**, *247*, 1525–1538. [CrossRef]
39. Sudha, M.L.; Vetrimani, R.; Leelavathi, K. Influence of fibre from different cereals on the rheological characteristics of wheat flour dough and on biscuit quality. *Food Chem.* **2007**, *100*, 1365–1370. [CrossRef]
40. Jaskulska, I.; Jaskulski, D.; Gałęzewski, L.; Knapowski, T.; Kozera, W.; Wacławowicz, R. Mineral composition and baking value of the winter wheat grain under varied environmental and agronomic conditions. *J. Chem.* **2018**, *2018*, 5013825. [CrossRef]
41. Kotoki, D.; Deka, S.C. Baking loss of bread with special emphasis on increasing water holding capacity. *J. Food Sci. Technol.* **2010**, *47*, 128–131. [CrossRef]
42. Seguchi, B.M.; Tabara, A.; Fukawa, I.; Ono, H.; Kumashiro, C.; Yoshino, Y.; Kusunose, C.; Yamane, C. Effects of size of cellulose granules on dough rheology, microscopy, and breadmaking properties. *J. Food Sci.* **2007**, *72*, E79–E84. [CrossRef]
43. Petković, M.; Filipović, V.; Filipović, J.; Đurović, I.; Miletić, N.; Radovanović, J. Chemical, antioxidative, and sensory characteristics of wheat bread partially substituted with black chokeberry (*Aronia melanocarpa* L.) powder. *J. Food Process. Preserv.* **2021**, *45*, e15027. [CrossRef]
44. Matsumoto, M.; Hara, H.; Chiji, H.; Kasai, T. Gastroprotective effect of red pigments in black chokeberry fruit (*Aronia melanocarpa Elliot*) on acute gastric hemorrhagic lesions in rats. *J. Agric. Food Chem.* **2004**, *52*, 2226–2229. [CrossRef]
45. Schmid, V.; Steck, J.; Mayer-Miebach, E.; Behsnilian, D.; Bunzel, M.; Karbstein, H.P.; Emin, M.A. Extrusion processing of pure chokeberry (*Aronia melanocarpa*) pomace: Impact on dietary fiber profile and bioactive compounds. *Foods* **2021**, *10*, 518. [CrossRef] [PubMed]
46. Woźniak, L.; Marszalek, K.; Skapska, S.; Jedrzejczak, R. The application of supercritical carbon dioxide and ethanol for the extraction of phenolic compounds from chokeberry pomace. *Appl. Sci.* **2017**, *7*, 322. [CrossRef]
47. Sójka, M.; Kołodziejczyk, K.; Milala, J. Polyphenolic and basic chemical composition of black chokeberry industrial by-products. *Ind. Crops Prod.* **2013**, *51*, 77–86. [CrossRef]
48. Kapci, B.; Neradová, E.; Čížková, H.; Voldřich, M.; Rajchl, A.; Capanoglu, E. Investigating the antioxidant potential of chokeberry (*Aronia melanocarpa*) products. *J. Food Nutr. Res.* **2013**, *52*, 219–229.
49. Raczkowska, E.; Nowicka, P.; Wojdyło, A.; Styczyńska, M.; Lazar, Z. chokeberry pomace as a component shaping the content of bioactive compounds and nutritional, health-promoting (anti-diabetic and antioxidant) and sensory properties of shortcrust pastries sweetened with sucrose and erythritol. *Antioxidants* **2022**, *11*, 190. [CrossRef]
50. Wichchukit, S.; O'Mahony, M. The 9-point hedonic scale and hedonic ranking in food science: Some reappraisals and alternatives. *J. Sci. Food Agric.* **2015**, *95*, 2167–2178. [CrossRef]
51. Betoret, E.; Rosell, C.M. Enrichment of bread with fruits and vegetables: Trends and strategies to increase functionality. *Cereal Chem.* **2020**, *97*, 9–19. [CrossRef]
52. Dziki, D. Current trends in enrichment of wheat pasta: Quality, nutritional value and antioxidant properties. *Processes* **2021**, *9*, 1280. [CrossRef]
53. das Chagas, E.G.L.; Vanin, F.M.; dos Santos Garcia, V.A.; Yoshida, C.M.P.; de Carvalho, R.A. Enrichment of antioxidants compounds in cookies produced with camu-camu (*Myrciaria dubia*) coproducts powders. *LWT—Food Sci. Technol.* **2021**, *137*, 110472. [CrossRef]

applied sciences

MDPI

Review

Exploring the Advantages and Disadvantages of a Whole Foods Approach for Elevating Dietary Nitrate Intake: Have Researchers Concentrated Too Much on Beetroot Juice?

Alex Griffiths [1], Shatha Alhulaefi [2,3], Eleanor J. Hayes [2], Jamie Matu [1], Kirsten Brandt [2], Anthony Watson [2], Mario Siervo [4] and Oliver M. Shannon [2,*]

1 School of Health, Leeds Beckett University, Leeds LS1 3HE, UK; a.griffiths@leedsbeckett.ac.uk (A.G.); j.matu@leedsbeckett.ac.uk (J.M.)
2 Human Nutrition & Exercise Research Centre, Centre for Healthier Lives, Population Health Sciences Institute, Newcastle University, Newcastle upon Tyne NE2 4HH, UK; s.alhulaefi2@newcastle.ac.uk (S.A.); e.j.hayes2@newcastle.ac.uk (E.J.H.); kirsten.brandt@newcastle.ac.uk (K.B.); anthony.watson@newcastle.ac.uk (A.W.)
3 Department of Nutrition, Taif University, Taif 21944, Saudi Arabia
4 School of Population Health, Curtin University, Perth, WA 6845, Australia; mario.siervo@curtin.edu.au
* Correspondence: oliver.shannon@newcastle.ac.uk; Tel.: +44-1912081140

Abstract: In recent years, a number of studies have explored the potential salutary effects of dietary nitrate, with promising findings emerging. Indeed, numerous investigations have now demonstrated that increasing intake of dietary nitrate can reduce blood pressure, improve endothelial function, decrease platelet aggregation, increase cognitive function and brain perfusion, and enhance exercise performance. Most researchers have explored the health and/or performance effects of dietary nitrate by providing participants with concentrated beetroot juice, which is rich in this compound. Another strategy for increasing/optimising dietary nitrate intake, which could be embraced alongside or instead of nitrate-rich supplements in research and non-research settings, is the consumption of whole nitrate-rich vegetables. In this review, we explore the potential advantages and disadvantages of increasing consumption of various whole nitrate-rich vegetables to augment dietary nitrate intake. We compare the cost, convenience, availability, feasibility/acceptability, and efficacy of consumption of nitrate via whole nitrate-rich vegetables against concentrated beetroot juice 'shots' as defined supplements. We also discuss possible strategies that could be used to help individuals maximise their intake of nitrate via whole vegetables, and outline potential avenues for future research.

Keywords: nitrate; beetroot juice; health; exercise performance; supplementation; whole foods

1. Introduction

Dietary nitrate is an inorganic anion which, in the human body, serves as a substrate for the multifunctional signalling molecule nitric oxide (NO) [1]. Dietary nitrate is available from a variety of food sources, but most nitrate in the diet comes from vegetables, with smaller contributions coming from certain herbs, fruits, processed meat, and drinking water [2–4]. Although dietary nitrate intake can vary considerably between individuals, the average intake tends to be just over 100 mg/d in both healthy (median intake: 108 mg/d) and patient (median intake: 110 mg/d) populations [2]. Early data from observational and animal studies suggested that a high intake of dietary nitrate could be detrimental to health, increasing the risk of methaemoglobinaemia and certain types of cancer [5–7]. Consequently, an acceptable daily intake (ADI) of nitrate was established at 3.7 mg/kg/d [8]. However, other studies have questioned this view [9–11]. Indeed, recent studies have suggested that consumption of dietary nitrate alone does not cause methaemoglobinaemia [4], which, instead, is typically caused following ingestion or inhalation of an oxidising agent [12]. Likewise, some [13], although not all [14,15], recent investigations have refuted the notion that nitrate intake increases cancer risk. In contrast to this early body of evidence, which tended to focus on nitrate as a potentially harmful dietary compound, research emerging in recent years suggests that increasing dietary nitrate intake (and thus enhancing NO bioavailability) can improve markers of cardiovascular and cognitive function. For example, dietary nitrate supplementation has been shown to reduce blood pressure (BP) by between ~2 and 10 mmHg [16–20], and could represent a potential adjunct to traditional antihypertensive medications. In addition, dietary nitrate supplementation has been shown to improve endothelial function [17,21], decrease platelet aggregation [17,19], and, albeit less reliably, enhance cerebral blood flow and cognitive function [22–24]. Over 100 studies have also been conducted exploring the potential benefits of dietary nitrate for sports performance, with findings from both a recent meta-analysis [25] and Delphi expert consensus [26] suggesting that nitrate can improve performance across a range of exercise modalities and in various populations. Several mechanisms have been proposed to explain these effects, including improvements in the efficiency of muscle contraction [27,28] and mitochondrial respiration [29], and modulation of tissue blood flow [30,31]. Interestingly, the health and performance benefits of nitrate are not apparent in individuals who use antibacterial mouthwash, which destroys bacteria in the mouth that play a crucial role in the processing of nitrate in the body [32,33].

Most studies investigating the health and/or exercise performance effects of dietary nitrate have administered this compound via nitrate-rich supplements, especially concentrated beetroot juice 'shots', which are highly enriched in nitrate and for which a nitrate-depleted placebo product is available [17,25,34,35]. For mechanistic research studies, the advantages to the provision of nitrate via concentrated beetroot juice are obvious, which may explain the ubiquity and dominance of this nitrate source in the literature. However, the translation of research that predominantly relies on the use of beetroot juice into the 'real world' could be limited if individuals are not willing or able to incorporate this particular product into their habitual diet. Another strategy for increasing/optimising dietary nitrate intake to influence human health and/or exercise performance, which could be used alongside or instead of nitrate-rich supplements in both research and 'real world' settings, is the consumption of whole nitrate-rich vegetables [36–38]. In this review, we explore the potential advantages and disadvantages of increasing consumption of various whole nitrate-rich vegetables to augment dietary nitrate intake, by focusing on the following questions:

1. Is it more cost-effective to consume dietary nitrate in the form of nitrate-rich vegetables or concentrated beetroot juice 'shots'?
2. Is it more convenient to consume dietary nitrate in the form of nitrate-rich vegetables or concentrated beetroot juice 'shots'?
3. Which is the more readily available source of dietary nitrate—nitrate-rich vegetables or concentrated beetroot juice 'shots'?
4. Is it more feasible/acceptable to consume dietary nitrate in the form of nitrate-rich vegetables or concentrated beetroot juice 'shots'?
5. What is the efficacy (in terms of eliciting health and/or performance effects) of dietary nitrate in the form of nitrate-rich vegetables versus concentrated beetroot juice 'shots'?

We also discuss some potential barriers to consumption of nitrate-rich vegetables, as well as opportunities for helping individuals optimise their intake of nitrate-rich vegetables, which may be of value in research and 'real world' settings.

2. Cost

Individual dietary choices are influenced by a number of factors, one of which is food pricing/cost [39]. Commercially available concentrated beetroot juice 'shots' are typically priced around GBP 2 or ~USD 2.50 each and contain a standardised nitrate dose of at least 400 mg for the most used product (Beet It Sport Shot, James White Drinks LTD, Ashbocking, UK). This may be prohibitively expensive for some individuals, if purchased/consumed daily. For example, in the United Kingdom (UK), the average household spend on food and non-alcoholic drink in the financial year ending in March 2021 was GBP 69.20 per week [40] (equivalent to ~USD 86/week). The purchase of one beetroot juice 'shot' per day would, therefore, constitute roughly 20% of the average household spend on food and non-alcoholic drinks every week, and a much higher percentage in lower-income households. Two examples of how the cost might impact the intake of nitrate-containing foods are provided below.

2.1. Example 1: Vegetables Purchased from General Retailers in the UK

In developed countries and some developing countries, most consumers obtain vegetables from supermarkets supplied from large retailers with long-distance supply chains providing year-round supply. Table 1 shows the average dose of nitrate in some commonly consumed nitrate-rich vegetables, alongside the estimated cost to achieve a nitrate dose equivalent to the beetroot juice 'shot' specified above. As can be seen, it is possible to obtain 400 mg of nitrate at a relatively low cost via different whole vegetables. For example, for whole beetroot, the cost per 400 mg nitrate, based around the average nitrate content in this vegetable reported by Blekkenhorst et al. [41], is around $\frac{1}{4}$ the cost of a beetroot juice 'shot'.

Table 1. Quantity and cost of consuming a standardised nitrate dose (400 mg, equivalent to one beetroot juice 'shot') via nitrate-containing vegetables widely available in the UK.

Food	Average Nitrate Content per 100 g of Each Vegetable [a]	Quantity Required to Consume 400 mg Nitrate [b]	Cost per 100 g of Each Vegetable [c]	Cost per 400 mg Nitrate [c]
Spinach	193 mg	208 g	GBP 0.43 (USD 0.53)	GBP 0.89 (USD 1.11)
Beetroot	158 mg	253 g	GBP 0.20 (USD 0.25)	GBP 0.51 (USD 0.63)
Iceberg lettuce	94 mg	424 g	GBP 0.22 (USD 0.27)	GBP 0.93 (USD 1.16)
Rhubarb	110 mg	364 g	GBP 0.75 (USD 0.93)	GBP 2.73 (USD 3.39)
Watercress	232 mg	172 g	GBP 1.45 (USD 1.80)	GBP 2.49 (USD 3.09)
Radish	166 mg	240 g	GBP 0.23 (USD 0.29)	GBP 0.55 (USD 0.68)
Arugula (Rocket)	348 mg	115 g	GBP 1.64 (USD 2.04)	GBP 1.89 (USD 2.35)
Celery	187 mg	214 g	GBP 0.32 (USD 0.40)	GBP 0.69 (USD 0.86)
Pak choi	249 mg	161 g	GBP 0.56 (USD 0.70)	GBP 0.90 (USD 1.12)
Kale	175 mg	229 g	GBP 0.42 (USD 0.52)	GBP 0.96 (USD 1.19)
Romaine Lettuce	129 mg	310 g	GBP 0.38 (USD 0.47)	GBP 1.18 (USD 1.46)
Swiss chard	208 mg	192 g	GBP 0.84 (USD 1.04)	GBP 1.61 (USD 2.00)
Turnip	171 mg	234 g	GBP 0.21 (USD 0.26)	GBP 0.49 (USD 0.61)
Fennel	131 mg	305 g	GBP 0.50 (USD 0.62)	GBP 1.52 (USD 1.89)

[a] Nitrate content per 100 g of each vegetable adapted from [41], [b] Aggregate cost represents cost in 3 UK supermarkets (Sainsburys, Tesco, and Asda) per 100 g as of May 2023. Data rounded to three significant figures. [c] US dollar equivalent presented in parentheses and based on the available exchange rate in May 2023.

One major challenge is that the nitrate content of vegetables can be highly variable, depending upon factors such as plant genetics [42] and the way in which they were grown (e.g., soil conditions, sunlight, fertilisation) and stored (e.g., storage duration and temperature) [4,43]. Indeed, in the case of spinach—a commonly consumed nitrate-rich vegetable—Hord et al. [4] report a measured nitrate content in raw leaves varying between 23.9 and 387.2 mg/100 g. Thus, whilst the estimated cost of consuming 400 mg nitrate via spinach based around the average nitrate dose of 193 mg/100 g might be GBP 0.89 (~USD 1.11), the actual cost for this nitrate dose could be as little as GBP 0.44 (~USD 0.55) and as much as GBP 7.20 (~USD 8.94). After purchase, it also matters how the vegetables are prepared (e.g., washed, peeled, pickled) and cooked (e.g., boiled, steamed, fried).

2.2. Example 2: Vegetables Grown in Own Garden or Purchased from Local Markets

In developing countries, particularly in rural areas, the supply of perishable foods, such as leafy vegetables, tends to be more local; vegetables are grown in the consumer's family garden or peri-urban vegetable farms, and harvested on the day they will be used, or dried in the sun for use in soups and stews [44]. While drying drastically reduces vitamin C content, it probably does not affect the nitrate content (per leaf). The cost is difficult to quantify, but is likely to be much lower than purchasing an imported product, such as beetroot juice.

In Table 2, we provide an outline of the nitrate content of some common vegetables grown or collected and available in markets in tropical and subtropical regions. It should be noted that many other vegetables sources of nitrate are likely to exist in these settings. However, for some of the most widely consumed leafy vegetables, such as sweet potato leaves, cassava leaves, Indian pennywort, and baobab leaves, few, if any, values on nitrate content have been published. Therefore, as discussed later, more research is needed to understand key vegetable sources of nitrate in different settings.

Table 2. Examples of nitrate-containing vegetables grown or collected and available in markets in tropical and subtropical regions, alongside the quantity required to consume a standardised nitrate dose (400 mg, equivalent to one beetroot juice 'shot').

Food	Average Nitrate Content per 100 g	Quantity Required to Consume 400 mg Nitrate
Mustard greens [a]	260 mg	153 g
Swiss chard [a]	208 mg	192 g
Water spinach [a]	145 mg	276 g
Fenugreek leaves [a]	137 mg	292 g
Amaranth [a]	112 mg	357 g
Malabar spinach [a]	102 mg	392 g
Jute mallow [b]	311 mg	128 g
Fluted pumpkin [b]	280 mg	143 g
Bitter leaf [b]	135 mg	296 g
Roselle [b]	128 mg	313 g

[a] From [41], and [b] from [45].

3. Convenience

A key strength of the commercially available beetroot 'shots' is their convenience. They are compact and easy to transport, which may be beneficial for individuals looking to 'top up' their nitrate intake away from home (e.g., when travelling) [46]. They also contain a specified nitrate dose, which reduces the burden on individuals. In contrast, the variable nitrate content of vegetables could make achieving a specific dose challenging (and, potentially, costly) on a case-by-case/day-by-day basis [4], as individuals will not know exactly how much nitrate is being consumed for each food/meal they eat. However, this may be less of an issue when looking at average nitrate intake over a period of weeks or months, if the nitrate content of vegetables is normally distributed, such that the variability may 'average out' over time.

From a research perspective, the standardised nature of the beetroot juice 'shots' allows researchers to ensure all participants receive the same nitrate dose [47]. This may be challenging with a whole-vegetable approach, given the variability in vegetable nitrate content could result in slightly different doses administered to each participant [4]. Similarly, the availability of a nitrate-depleted beetroot juice, which tastes, looks, and smells the same as the nitrate-rich equivalent, but is devoid of nitrate—something achieved by passing the juice through an ion-exchange resin that selectively extracts the nitrate—is particularly valuable as a placebo to researchers [48]. Not only does this allow researchers to isolate the impact of the nitrate on the research outcome of interest, but it also allows proper blinding of participants who may be able to discern their experimental condition with less refined placebos, such as fruit squash drinks.

Improved knowledge of the dietary sources of inorganic nitrate could make it easier for individuals to optimise their intake of dietary nitrate. For example, awareness of the types of vegetable that are typically high in nitrate may allow individuals looking to boost their nitrate intake to prioritise the consumptions of such foods by swapping lower-nitrate vegetables with ones they know to be rich in this compound. The effectiveness of this strategy may be augmented by a recent development in the field by Qadir and colleagues [49], who demonstrated that by tightly controlling and manipulating the growing conditions of different vegetables, it is possible to produce particularly high- or low-nitrate vegetables. Specifically, differential fertilisation allowed the researchers to grow lettuce with higher (1060 mg nitrate/100 g) and lower (6 mg nitrate/100 g) nitrate concentrations. The growth of high-nitrate vegetables with a standardised minimum nitrate dose (if approved for commercial sale, e.g., as a high-nitrate salad bag) could be useful for individuals looking to boost their habitual nitrate intake. Similarly, this could be valuable for researchers as an alternative to the beetroot juice (active and placebo) 'shots'.

One final consideration from a convenience standpoint relates to the consumption of dietary nitrate prior to or during exercise to elicit ergogenic effects. In the days leading up to a competitive sporting event, research indicates that it is possible to elevate habitual nitrate intake and achieve ergogenic effects using both concentrated beetroot juice and whole vegetables [50–53]. Nevertheless, consumption of a nitrate-rich salad in the hours prior to exercise (e.g., to provide a final 'top up' dose of nitrate pre-competition), and during exercise, may be logistically difficult, due to the challenges of transporting enough of the vegetables to, or around, a sporting event. Similarly, the amount that must be consumed may be undesirably high for some individuals due to the perceived risk of stomach upset. On the other hand, beetroot juice 'shots', alongside other nitrate-containing supplements, such as nitrate-rich gels [54], may be particularly useful for consumption immediately pre- and/or during exercise, where they can be easily transported and consumed in relatively small amounts to achieve a potentially beneficial effect [46]. For example, a recent study from Tan and colleagues [46] showed that consumption of concentrated beetroot juice both before and during exercise helped to preserve elevated plasma nitrite concentrations, spare muscle glycogen, and limit the gradual rise in oxygen uptake during two hours of cycling compared with pre-exercise beetroot juice ingestion alone or placebo ingestion.

4. Availability

A range of nitrate-rich vegetables, such as those outlined in Table 1, are widely available in supermarkets and grocery stores in high-income countries, such as the UK and United States (US). Meanwhile, beetroot juice 'shots' can be purchased in many health food shops, in some supermarkets, and via the internet in both the UK and US. For individuals in high-income countries actively trying to increase their nitrate intake, availability is unlikely to be a major barrier to consumption for either nitrate vehicle. However, this may not be the case in other settings, such as low- and middle-income countries, where the intake of vegetables is often much less than 100 g per day [44].

High rates of hypertension are observed in Africa, where almost 50% of adults aged 25 years and above have elevated blood pressure [55]. Interestingly, Siervo et al. [56,57] demonstrated considerable potential for dietary nitrate to help combat hypertension in Tanzania, a country in Sub-Saharan Africa experiencing a hypertension epidemic. Specifically, 60 days of dietary nitrate supplementation reduced 24 h ambulatory systolic BP by around 10 mmHg, suggesting considerable potential for dietary nitrate as an anti-hypertension strategy in this setting. Although this study administered nitrate in the form of beetroot juice 'shots', which were provided to participants via the research team, this strategy may not be viable for many individuals in Tanzania or other lower- or middle-income countries outside of a research setting, given limited availability (alongside cost restrictions). On the other hand, the growth and purchase of native vegetables rich in nitrate (Table 2) could represent a vital tool to help combat the high rates of hypertension in such settings.

5. Feasibility and Acceptability

In contrast to the wealth of research exploring the efficacy of dietary nitrate (discussed further below), only a handful of studies have reported data on the feasibility and/or acceptability of different strategies for augmenting dietary nitrate intake [34]. In one study by Ormesher et al. [58], who provided pregnant women with a single 70 mL concentrated beetroot juice 'shot' every day for 8 days, almost all of the participants (97%) indicated that they would consume beetroot juice 'shots' again, if they were experiencing benefits. However, only 62% reported finding it easy to consume this supplement. In addition, only 54% reported the 'shots' to be palatable. In another study, Babateen et al. [59] examined the feasibility of consuming low (1 beetroot juice 'shot' every second day), medium (1 beetroot juice 'shot' every day), and high (2 beetroot juice 'shots' every day) doses of concentrated beetroot juice in overweight/obese older adults in a 13-week intervention trial. Self-reported compliance was >90%. However, around 20% of the participants in that study withdrew prior to the end of the trial, with most of the dropouts coming from the moderate- or high-dose groups. Although some of the participants reported finding the taste of the beetroot juice 'pleasant', this was not a universal view, with others noting that the taste was 'bad' and that they had to 'hold [their] nose to gulp it down'. It is noteworthy that three of the participants who withdrew cited GI discomfort as their main reason for dropping out of the study, whilst three additional participants reported withdrawing because they disliked the smell/taste of beetroot juice.

In another study conducted in Tanzania, participants who consumed beetroot juice 'shots' for 60 days reported the taste to be 'good' or 'very good', and compliance with the intervention was >90% [60]. Nevertheless, the findings from both Ormesher et al. [58] and Babateen et al. [59] suggest that, at least for some individuals, more palatable alternatives to beetroot juice may be required to help some individuals elevate their nitrate intake. Other nitrate-containing supplements, such as beetroot-containing pills/capsules or nitrate-rich gels, could represent a more acceptable strategy to increase nitrate intake in individuals who do not like the taste of beetroot juice. In addition, the variety of options available with nitrate-rich vegetables (see Table 1 for some examples) could help reduce the risk of taste fatigue/monotony and allow individuals to self-select vegetables that fit better with their own food/taste preferences. This could, theoretically, help improve longer-term compliance.

Very high compliance (97–98%) was reported in 2 studies lasting 4–5 weeks, which used nitrate-rich vegetables as a nitrate source [61,62]. Meanwhile, in a study by Jonvik et al. [63], in which participants consumed four different nitrate-containing beverages—a sodium nitrate solution, concentrated beetroot juice, a rocket salad beverage, and a spinach beverage—ingestion of all four beverages was generally well tolerated. However, GI complaints were reported by two participants following consumption of the rocket salad beverage and by one participant following consumption of the spinach beverage, which was believed to be related to the high volume of these supplements provided (which was greater than for the beetroot juice and sodium nitrate conditions). In another study from the same group, participants increased their intake of dietary nitrate through either concentrated beetroot juice or whole vegetables over a 1-week period [64]. Both interventions were well tolerated, with no adverse events reported. Participants reported 100% compliance with the beetroot juice intervention, and consumed an average of 241 g/d nitrate-rich vegetables, which was close to their daily target of 250 g/d, suggesting that it is feasible to increase nitrate intake through both vegetables and nitrate-rich beetroot juice [64]. However, as noted by the authors, it is important to acknowledge that food and supplements were provided to participants in this study, such that different results may occur in the 'real world', where participants must purchase their own products [64]. It should also be noted that there may be some selection bias in the participants involved in these studies (i.e., individuals who like beetroot juice or vegetables may be more inclined to take part), such that these nitrate sources may be appraised less favourably in other cohorts.

One final point to note is that those choosing to use beetroot juice as a nitrate source should be aware of the potential for a slight discolouration of the urine, termed 'beeturia'. Although harmless, beeturia could cause shock/surprise (and, potentially, unnecessary visits to a clinician) in some individuals if they are not forewarned about this side effect. Regardless, overall, it is clear that, at least in a research setting, it is feasible to elevate habitual nitrate intake using both whole vegetables and beetroot juice 'shots' over at least several weeks, and individuals looking to boost their nitrate intake could adopt either strategy, based around their own personal preference.

6. Efficacy

Dietary nitrate in the form of unconcentrated and concentrated beetroot juice [16,18,28,53,65,66], alongside whole vegetables—including, but not restricted to, whole beetroot [67], spinach [68], lettuce [69], mixed green leafy vegetables [70], and traditional Japanese vegetables, such as ta cai, chin gin cai, and garland chrysanthemum [71] (Table 3)—has been demonstrated to improve health parameters (especially blood pressure) and/or enhance exercise performance. This suggests that both juiced and whole-vegetable sources of nitrate can be effective at eliciting potentially beneficial effects, if a sufficient nitrate dose is administered [26].

Interestingly, several studies have compared the health and/or exercise performance effects of nitrate-rich beetroot juice against nitrate salts, and have demonstrated superior BP-lowering effects [63], greater reductions in oxygen consumption during steady-state exercise [72], and more pronounced attenuation of muscle pain after damaging exercise [73,74] with beetroot juice. Here, it is argued that the presence of compounds such as polyphenols and antioxidants in the beetroot juice may elicit effects synergistically to the nitrate (e.g., by potentiating nitrite reduction into NO and/or reducing NO degradation by mitigating superoxide scavenging) [75,76]. Similarly, McDonagh et al. [77] demonstrated greater BP-lowering effects of concentrated beetroot juice versus unconcentrated beetroot juice, a beetroot flapjack, and a drink containing nitrate crystals, providing further evidence that the form in which nitrate is consumed could impact its effects in the body. In contrast, there are few head-to-head comparisons between beetroot juice 'shots' and other vegetable sources of nitrate, which make it difficult to conclusively state whether one is superior (in terms of efficacy) to the other.

Some relevant evidence comes from a study by Jonvik and colleagues [63], who contrasted the effects of acute ingestion of concentrated beetroot juice against a rocket salad beverage, spinach beverage, and sodium nitrate solution (all containing 800 mg nitrate). In this study, all of the vegetable sources of nitrate were more effective at reducing BP than the sodium nitrate supplement, with no appreciable difference between the BP-lowering effects of concentrated beetroot juice and the spinach or rocket beverages. Interestingly, there were some differences in the plasma nitrate and nitrite response between supplements, which might suggest small differences in the optimal timing for ingesting each of the different nitrate sources, and requires further investigation. In another study by the same group, van der Avoort et al. [64] explored the BP-lowering effects of concentrated beetroot juice versus a variety of whole nitrate-rich vegetables (e.g., spinach, lettuce, arugula, bok choy) over a 1-week period. Although the concentrated beetroot juice resulted in greater increases in plasma nitrate and nitrite concentrations, both interventions were similarly effective at reducing systolic and diastolic BP (both by ~5 mmHg) [64]. Although additional head-to-head comparisons are required, from the limited evidence available, it is clear that both concentrated beetroot juice and vegetables can be effective at increasing nitrate intake and, consequently, impacting health parameters. Assuming similar doses are administered, it appears likely that the effects of dietary nitrate via concentrated beetroot juice and other vegetable sources are broadly comparable [26].

Table 3. Key intervention studies exploring the impact of nitrate-rich vegetables (excepting beetroot juice) on health and/or exercise performance parameters.

Participants	Experimental Conditions	Duration	Key Findings	Reference
19 healthy women aged 20 ± 2 years	Intervention: Diet rich in nitrate-containing vegetables, including lettuce, rocket, celery, leeks, fennel, and mixed salad leaves (nitrate dose: ~339 mg/d) Control: Diet low in nitrate-containing vegetables (nitrate dose: ~8 mg/d)	1 week	Diet rich in nitrate-containing vegetables increased plasma nitrate and nitrite concentrations and decreased systolic BP by ~4 mmHg.	Ashworth et al. [78]
30 (20 M, 10 F) participants with prehypertension or untreated grade 1 hypertension aged 63 (range: 56–71) years	Intervention 1: Increased intake of high-nitrate vegetables (~200 g/d, nitrate dose: ~150 mg) Intervention 2: Increased intake of low-nitrate vegetables (~200 g/d, nitrate dose: ~22 mg) Control: No increase in vegetables	4 weeks	Intake of high-nitrate vegetables increased plasma nitrate and nitrite concentrations, but did not significantly impact BP or arterial stiffness.	Blekkenhorst et al. [62]
30 (6 M, 24 F) healthy participants aged 47 ± 14 years	Condition 1 (control): Apple flesh (low in flavonoids, low in nitrate) Condition 2: Apple flesh and skin (high in flavonoids, low in nitrate) Condition 3: Apple flesh plus spinach (low in flavonoids, high in nitrate) Condition 4: Apple flesh and skin plus spinach (high in flavonoids, high in nitrate)	Acute intervention	Compared to the low-flavonoid, low-nitrate condition (condition 1), all treatments elevated plasma nitrite concentration, increased flow-mediated dilation, and reduced pulse pressure. Systolic BP was reduced by ~3 mmHg only in the high-flavonoid, low-nitrate (condition 2) and low-flavonoid, high-nitrate (condition 3) conditions.	Bondonno et al. [68]
38 (12 M, 26 F) participants with normal to high BP aged 61 ± 7 years	Intervention: Diet rich in nitrate-containing vegetables, including frozen spinach and fresh lettuce, spinach, rocket, and other leafy green vegetables (nitrate dose: ~400 mg/d) Control: Diet low in nitrate-containing vegetables (nitrate dose: <100 mg/d)	1 week	The high-nitrate diet increased plasma and salivary nitrate and nitrite. However, ambulatory, home, and clinic BP alongside arterial stiffness were unchanged with both the high- and low-nitrate diets.	Bondonno et al. [70]
36 (14 M, 22 F) healthy participants aged 67 ± 6 years,	Intervention: 150 g whole cooked beetroot (nitrate dose: ~145 mg/d), plus 1 medium-sized banana every second day Control: 1 medium-sized banana every second day	8 weeks	Beetroot consumption reduced systolic BP by ~8 mmHg and increased diversity (Shannon index) of the gut microbiota.	Capper et al. [67]

Table 3. *Cont.*

Participants	Experimental Conditions	Duration	Key Findings	Reference
12 (6 M, 6 F) younger adults aged 27 ± 4 years, and 12 (3 M, 9 F) older adults aged 64 ± 5 years	Condition 1: 100 g cooked beetroot (nitrate dose: ~272 mg) Condition 2: 200 g cooked beetroot (nitrate dose: ~544 mg) Condition 3: 300 g cooked beetroot (nitrate dose: 816 mg) Condition 4: Potassium nitrate solution (nitrate dose: 1000 mg)	Acute	Plasma nitrate and nitrite concentrations increased in a dose-dependent manner, with more pronounced responses in younger vs. older adults. All interventions reduced systolic and diastolic BP (both by ~2–5 mmHg) compared to baseline in younger adults, whereas only the high-dose potassium nitrate intervention (condition 4) reduced systolic and diastolic BP (both by ~5 mmHg) in the older adults.	Capper et al. [79]
18 (11 M, 7 F) healthy participants aged 18 ± 1 year	Condition 1: Sodium nitrate solution (nitrate dose: 800 mg) Condition 2: Concentrated beetroot juice (nitrate dose: 800 mg) Condition 3: Spinach beverage (nitrate dose: 800 mg) Condition 4: Rocket salad beverage (nitrate dose: 800 mg)	Acute	All conditions increased plasma nitrate and nitrite concentrations. The vegetable nitrate sources (conditions 2–4) reduced systolic (~5–7 mmHg) and diastolic (~4–8 mmHg) BP by a similar amount. The sodium nitrate did not impact systolic BP and led to smaller reductions in diastolic BP (~2–4 mmHg).	Jonvik et al. [63]
27 healthy participants aged 24.5 ± 11 years	Condition 1: Spinach soup (845 mg/d nitrate) Condition 2: Asparagus soup (0.6 mg/d nitrate)	1 week	The high-nitrate spinach soup reduced the augmentation index, lowered central systolic (−3.39 ± 5.6 mmHg) and diastolic BP (−2.60 ± 5.8 mmHg), and also reduced brachial systolic BP (−3.48 ± 7.4 mmHg) compared with the low-nitrate asparagus soup.	Jovanovski et al. [80]
26 (6 M, 20 F) participants aged 59 ± 8 years	Intervention: High-nitrate meal (nitrate dose: 220 mg from spinach) Control: Low-nitrate meal	Acute	Consumption of the high-nitrate meal elevated salivary nitrate and nitrite, increased the large artery elasticity index, reduced pulse pressure, and decreased systolic BP (up to ~7.5 mmHg).	Liu et al. [81]

Table 3. *Cont.*

Participants	Experimental Conditions	Duration	Key Findings	Reference
8 (5 M, 3 F) normotensive adults aged 73 ± 5 years	Condition 1 (control): 3 days low-nitrate diet (nitrate dose: 43.4 mg/d) Condition 2: 3 days low-nitrate diet + acute ingestion of nitrate-rich beetroot juice supplement (nitrate dose: 527 mg) Condition 3: 3 days high-nitrate diet (nitrate dose: 155 mg/d) Condition 4: 3 days high-nitrate diet + acute ingestion of nitrate-rich beetroot juice supplement	3 days	Plasma nitrate and nitrite were significantly increased with the low-nitrate diet + supplement (condition 2) and high-nitrate diet + supplement (condition 4), but were unchanged in the other conditions. Systolic and diastolic BP were lower, compared to pre-breakfast levels, in all conditions. However, there was no difference in response between groups.	Miller et al. [82]
7 healthy men aged 25 ± 2 years	Intervention: High-nitrate diet (nitrate dose: ~512 mg/d) including spinach and collard greens Control: Low-nitrate diet (nitrate dose: ~181 mg/d)	6 days	The high-nitrate diet significantly increased plasma nitrate and nitrite concentrations, reduced the oxygen cost of moderate intensity cycling, increased total muscle work during intermittent leg-extension exercise, and improved repeated sprint ability test performance.	Porcelli et al. [51]
25 (10 M, 15 F) healthy volunteers aged 36 ± 10 years	Intervention: Diet rich in traditional Japanese vegetables, such as ta cai, chin gin cai, and garland chrysanthemum (nitrate dose: 18.8 mg/kg/d) Control: Diet low in traditional Japanese vegetables	10 days	The high-nitrate diet increased plasma and salivary nitrate and nitrite concentrations and reduced diastolic BP (~4.5 mmHg). Participants reported finding it easier to consume the diet rich in Japanese vegetables compared with the control diet.	Sobko et al. [71]
231 (109 M, 122 F) adults with elevated BP aged 63 ± 6 years	Intervention 1: Low-nitrate vegetables + potassium nitrate pills (nitrate dose: 300 mg/d) Intervention 2: High-nitrate vegetables (nitrate dose: ~300 mg/d) + placebo pills Control: Low-nitrate vegetables + placebo pills	5 weeks	Plasma nitrate concentrations increased with intake of potassium nitrate pills (intervention 1) and high-nitrate vegetables (intervention 2), whereas plasma nitrite only increased with the potassium nitrate pills. Ambulatory BP was no different between conditions, whereas with the potassium nitrate pills, there was a significant decrease in clinic diastolic BP post intervention.	Sundqvist et al. [61]

Table 3. *Cont.*

Participants	Experimental Conditions	Duration	Key Findings	Reference
30 (15 M, 15 F) recreationally active participants aged 24 ± 6 years	Condition 1: High intake of nitrate-rich vegetables (nitrate dose: ~400 mg), including salad, lettuce, arugula, and beetroot Condition 2: Concentrated beetroot juice (nitrate dose: 400 mg)	1 week	Plasma nitrate and nitrite concentrations were elevated above baseline in both conditions, with greater increases in the concentrated beetroot juice condition. However, similar changes in systolic and diastolic BP (~5 mmHg reductions in both) were observed with vegetables and concentrated beetroot juice.	van der Avoort et al. [64]

7. Other Benefits of Consuming Whole Nitrate-Rich Vegetables

For individuals who may be considering increasing their intake of dietary nitrate, doing so via consumption of a variety of whole nitrate-rich vegetables could have health benefits in addition to those associated with the nitrate, such as via the provision of various vitamins, minerals, and phytochemicals alongside dietary fibre [75,83] (an in-depth outline of the chemical/nutritional properties of beetroot juice can be found in [84,85]). Indeed, in a recent study by Capper et al. [67], ingestion of whole beetroot for 8 weeks reduced systolic BP by ~8 mmHg, and also increased the diversity of the gut microbiota and stool short-chain fatty acid concentrations. The increased (by ~7 g/d) intake of fibre with whole beetroot is likely to have contributed towards these effects [67,86,87]. Similarly, a recent animal model investigation suggested that intake of spinach, which is rich in nitrate, fibre, and a range of phenolics, can mitigate the impact of high-fat overfeeding on the gut microbiome and improve circulating glucose and lipid profiles [88].

8. Potential Barriers to Consuming Whole Nitrate-Rich Vegetables

To date, no studies have specifically investigated the barriers to dietary nitrate intake in the general public. However, it seems likely that barriers to general consumption of vegetables, such as convenience, availability, cost, knowledge, and time [89], may also be evident for consumption of nitrate-rich vegetables. Although concentrated beetroot juice overcomes some of these barriers (e.g., convenience and time), consumption of nitrate-rich vegetables may still represent a viable, cheaper, more palatable, and, therefore, potentially effective public health strategy for increasing dietary nitrate intake for a range of individuals.

Lack of education and knowledge about vegetable sources of dietary nitrate could be one potential barrier limiting intake of nitrate-rich vegetables. Increasing awareness of which vegetables are rich in nitrate may facilitate elevated intake in a range of individuals [90]. For example, it has previously been shown that better nutritional knowledge leads to better diet quality in both athletes [91] and the general population [92]. However, education alone is likely not sufficient to induce long-term lifestyle behaviour change [93,94], as dietary and other lifestyle changes can be incredibly difficult to make and maintain, and this must be considered at an individual, environmental, social, and policy level [95]. Whilst consideration for these broad strategies for behaviour change in this context are beyond the scope of this review, we consider herein some specific strategies that may be useful in optimising the intake of nitrate-rich vegetables.

9. Strategies to Help Optimise Intake of Nitrate-Rich Vegetables

For individuals considering increasing their intake of nitrate via whole vegetables, below we discuss some potential strategies that may be helpful and warrant further exploration in research and practical settings. Note, however, that the present paper exclusively addresses the scientific aspects of what is feasible and may be useful from a technical perspective and on a global scale. Many countries and regions have detailed rules and regulations on food health claims and marketing of foods with functional constituents, such as nitrate, and all countries tend to use the WHO guidance defining the ADI of nitrate at 3.7 mg/kg bodyweight [4], which (depending upon body weight) is close to the 400 mg nitrate per day used as a reference value throughout the present paper. The authors are aware that this affects, and in most cases limits, whether and how food producers and retailers can and should inform their customers about the nitrate contents of products, not to mention actively advertise the potentially associated benefits briefly reviewed in the Introduction. Due to the substantial differences in the relevant legislation across the world, any changes or harmonisations will require a combination of global and local initiatives in collaboration between scientific and legislative communities, which are outside the scope of the present paper. Therefore, these aspects are not specified or discussed at all.

9.1. Self-Monitoring Tools

Self-monitoring tools could be a useful strategy to allow individuals to monitor and, if necessary, adjust their intake of nitrate via vegetables and other whole-food sources. This could take the form of an app or other computerised tool, which could be used by individuals to estimate their intake of nitrate—something which many individuals do already for other nutrients/energy intake [96]. Comprehensive databases have now been developed on food sources of nitrate, which are beginning to be incorporated into commercially available dietary analysis software [97], which make this a realistic possibility in the near future. The integration of these tools with supermarket websites or online recipe books could be helpful for those who are not currently achieving their target nitrate intakes. Another self-monitoring strategy could include measuring biomarkers, which provide a more objective measure of nitrate intake and subsequent 'activation' into nitrite in the body. For example, relatively inexpensive test strips are now available, which allow individuals to monitor their salivary nitrite, and could be used to help 'fine tune' dietary nitrate intake [32].

Around 65–70% of ingested nitrate is excreted in the urine within 24 h of nitrate intake [98]. As such, urinary nitrate can also be used as an objective marker of habitual nitrate intake [3,99]. Although commercially available test strips are available, which can be used to measure urinary nitrate and the nitrite concentration in a free-living setting, individuals should be aware that measuring the nitrate concentration of spot urine samples may be less informative than collecting 24 h measurements (something that is likely to be burdensome and unrealistic for many individuals outside of a research setting). Moreover, urinary nitrate/nitrite concentrations can be altered as a consequence of different health conditions, such as urinary tract infections and kidney injury [100], which could confound interpretation of these values.

9.2. Growth of High-Nitrate Vegetables

As noted previously, a recent study by Qadir et al. [49] has demonstrated the potential to manipulate growing conditions to produce lettuce with a controlled high- and low-nitrate content. These products have potential for use as an alternative nitrate source and placebo (to nitrate-rich and -depleted beetroot juice) in a research setting [69]. In addition, if the high-nitrate vegetables (50 g of which would provide >500 mg nitrate) are mass produced at a relatively low cost, and approved for commercial sale, then they could represent one useful strategy to help individuals elevate their nitrate intake in a range of settings. Similarly, if the same growing methods could be applied to other vegetables, then this would open the possibility to provide a range of different high-nitrate vegetable sources that individuals could consume based on their preferred taste profile.

Urban agriculture (i.e., the growth of foods, such as fruits and vegetables, in urban areas) has attracted recent attention as a potential strategy to increase fruit and vegetable intake, reduce transportation costs, reduce stress, and increase the connectivity between people and the food system [44,101]. In densely populated cities, vertical farming units are one tool being employed in urban agriculture initiatives that allow highly standardised growing conditions [102], and could also be used to help ensure regular access to high-nitrate vegetables in specific community settings (e.g., work places, schools, retirement homes, prisons, and other locations where the provision of fresh vegetables poses logistical or economic challenges and intake is typically low).

9.3. Incorporation of Nitrate-Rich Vegetables into Existing Meals

Some individuals may find it difficult or objectionable to make major changes to the types of meals they consume daily (for example, swapping a favourite burger, pasta dish, or casserole for a nitrate-rich salad). In such cases, the incorporation of nitrate-rich vegetables into an individual's favourite foods or recipes (so called dietary 'hacks') could be a useful strategy to help 'boost' their habitual nitrate intake without requiring substantial modifications to the types of food they are eating. Covert incorporation of vegetables into meals has been shown to be an effective strategy to enhance vegetable intake and can also have the parallel benefit of lowering the energy density of the meal [103]. This could be as simple as adding nitrate-rich vegetables, such as spinach, to a favourite soup or casserole. For example, one study demonstrated beneficial haemodynamic effects of a soup enriched with nitrate via spinach versus a lower-nitrate equivalent [80]. Another approach, which could be commercially exploited, is the development of baked products, in which nitrate-rich vegetables have been incorporated. The proof of concept for this idea was demonstrated in a series of studies by Hobbs et al. [104,105], who showed that beetroot-enriched bread was effective at elevating nitrate intake and reducing BP, particularly in T carriers of the Glu298Asp polymorphism in the eNOS gene. Importantly, the incorporation of beetroot into the bread did not appear to negatively impact the acceptability or palatability of the product [104]. A final dietary 'hack' that could help to increase nitrate intake could be the selection of cooking methods that minimise nitrate degradation/loss from foods, such as frying instead of boiling, as was employed in the DePEC trial in Malaysia [106].

9.4. Integration into Dietary Guidelines

Awareness of nitrate-rich vegetables could be increased via the incorporation of specific recommendations into relevant public health policies. For example, the UK-based Eatwell Guide recommends consumption of five portions of fruit and vegetables per day [107]; however, no consideration for the specific type of vegetables is provided. Specifying the number of portions that should come from nitrate-rich vegetables could be a useful strategy to increase public awareness and, therefore, intake of dietary nitrate. Similar concepts are currently adopted within the Eatwell Guide around the intake of fish, for which individuals are recommended to consume two portions per week, with one of these portions coming from an oily variety. Such recommendations could be implemented into other national policy recommendations, such as the US, in which specific recommendations for 'nitrate rich vegetables' could be added as a subgroup under the 'vegetables' food group [108]. It is important to note that it is still likely too early to make such recommendations, given the lack of long-term safety data regarding dietary nitrate intake, but this may be a feasible option in the future.

An alternative strategy to increasing consumption of nitrate-rich vegetables may be via a 'vegetables on prescription' approach. With this strategy, healthcare providers identify individuals or groups of individuals with low intake of fruit and vegetables (e.g., males, young people, low-socioeconomic-status groups, etc.), and prescribe fruit and vegetable intake through discounts on fruit and vegetable purchases [109]. This prescription is often delivered alongside messaging regarding the importance of fruit and vegetable intake. These interventions have had some success previously, improving knowledge related to the importance of fruit and vegetable intake [110]. This concept could be adopted for nitrate-rich vegetables, but the feasibility and efficacy of such interventions warrants investigation before implementation.

10. Conclusions

In this review, we have discussed the potential to use whole nitrate-rich vegetables as a strategy to increase dietary nitrate intake. We suggest that increasing consumption of whole nitrate-rich vegetables could be a useful and effective strategy for individuals seeking to elevate their habitual nitrate intake, either alone or alongside the beetroot juice 'shots' most commonly investigated in research. Whole vegetables are more cost effective than beetroot juice 'shots' in most contexts, and the potential to choose from a variety of vegetables to elevate nitrate intake could help avoid taste fatigue. In addition, numerous studies have shown that, providing a sufficient dose of nitrate is administered, whole vegetables can be an effective strategy to augment circulating NO metabolites and concomitantly improve health/exercise performance parameters. Nevertheless, it is our view that whole-vegetable sources of nitrate may not be a perfect replacement for beetroot juice 'shots' or equivalent concentrated and standardised products in all conditions, such as pre- or during exercise, where the size (both for portability and in terms of volume required) and assured provision of a set nitrate dose may be advantageous. In addition, although other vegetable-based alternatives to the high- and low-nitrate beetroot juice 'shots' are beginning to emerge, these products are not yet widely available, such that the beetroot juice 'shots' are likely to remain the preferred nitrate donor in a research setting for some time yet.

Future research needs: In developed countries, different strategies should be explored, including self-monitoring tools, which can be used to help individuals optimise their nitrate intake via whole vegetables. Meanwhile, in developing countries, while the use of locally produced vegetables is more advantageous than imported beetroot concentrate, the most urgent need is to obtain more and better data on nitrate contents in the available vegetables. Together with the results from recent research on the benefits of dietary nitrate, such data can help to support producers and consumers to understand that high-nitrate content in vegetables may be an innocuous and even potentially beneficial trait, particularly when the overall vegetable intake is much lower than recommended.

Author Contributions: Conceptualisation, O.M.S.; methodology, A.G., S.A., E.J.H., J.M., K.B., A.W., M.S. and O.M.S.; software, not applicable; validation, not applicable; formal analysis, not applicable; investigation, A.G., S.A., E.J.H., J.M., K.B., A.W., M.S. and O.M.S.; resources, A.G., S.A., E.J.H., J.M., K.B., A.W., M.S. and O.M.S.; data curation, A.G. and O.M.S.; writing—original draft preparation, A.G., J.M., K.B., A.W. and O.M.S.; writing—review and editing, A.G., S.A., E.J.H., J.M., K.B., A.W., M.S. and O.M.S.; visualisation, E.J.H. and O.M.S.; supervision, O.M.S., K.B. and A.W.; project administration, O.M.S.; funding acquisition, not applicable. All authors have read and agreed to the published version of the manuscript.

Funding: This research received no external funding.

Institutional Review Board Statement: Not applicable.

Informed Consent Statement: Not applicable.

Data Availability Statement: Not applicable.

Conflicts of Interest: The authors declare no conflict of interest.

References

1. Lundberg, J.O.; Govoni, M. Inorganic Nitrate Is a Possible Source for Systemic Generation of Nitric Oxide. *Free Radic. Biol. Med.* **2004**, *37*, 395–400. [CrossRef]
2. Babateen, A.M.; Fornelli, G.; Donini, L.M.; Mathers, J.C.; Siervo, M. Assessment of Dietary Nitrate Intake in Humans: A Systematic Review. *Am. J. Clin. Nutr.* **2018**, *108*, 878–888. [CrossRef]
3. McMahon, N.F.; Brooker, P.G.; Kadach, S.; Pavey, T.G.; Leveritt, M.D. Estimating Nitrate Intake in the Australian Diet: Design and Validation of a Food Frequency Questionnaire. *J. Hum. Nutr. Diet.* **2023**, *36*, 169–180. [CrossRef]
4. Hord, N.G.; Tang, Y.; Bryan, N.S. Food Sources of Nitrates and Nitrites: The Physiologic Context for Potential Health Benefits. *Am. J. Clin. Nutr.* **2009**, *90*, 1–10. [CrossRef]
5. Winter, J.W.; Paterson, S.; Scobie, G.; Wirz, A.; Preston, T.; McColl, K.E.L. N-Nitrosamine Generation from Ingested Nitrate via Nitric Oxide in Subjects with and without Gastroesophageal Reflux. *Gastroenterology* **2007**, *133*, 164–174. [CrossRef]
6. Addiscott, T.M.; Benjamin, N. Nitrate and Human Health. *Soil Use Manag.* **2004**, *20*, 98–104. [CrossRef]
7. Powlson, D.S.; Addiscott, T.M.; Benjamin, N.; Cassman, K.G.; de Kok, T.M.; van Grinsven, H.; L'hirondel, J.-L.; Avery, A.A.; Van Kessel, C. When Does Nitrate Become a Risk for Humans? *J. Environ. Qual.* **2008**, *37*, 291–295. [CrossRef]
8. Scientific Committee for Food Nitrates and Nitrites. *Reports of the Scientific Committee for Food*; European Commission: Brussels, Belgium, 1997.
9. McKnight, G.M.; Duncan, C.W.; Leifert, C.; Golden, M.H. Dietary Nitrate in Man: Friend or Foe? *Br. J. Nutr.* **1999**, *81*, 349–358. [CrossRef]
10. Milkowski, A.; Garg, H.K.; Coughlin, J.R.; Bryan, N.S. Nutritional Epidemiology in the Context of Nitric Oxide Biology: A Risk-Benefit Evaluation for Dietary Nitrite and Nitrate. *Nitric Oxide Biol. Chem.* **2010**, *22*, 110–119. [CrossRef]
11. Khatri, J.; Mills, C.E.; Maskell, P.; Odongerel, C.; Webb, A.J. It Is Rocket Science—Why Dietary Nitrate Is Hard to 'Beet'! Part I: Twists and Turns in the Realization of the Nitrate-Nitrite-NO Pathway. *Br. J. Clin. Pharmacol.* **2017**, *83*, 129–139. [CrossRef]
12. Methemoglobinemia Following Unintentional Ingestion of Sodium Nitrite—New York. 2002. Available online: https://www.cdc.gov/mmwr/preview/mmwrhtml/mm5129a2.htm (accessed on 8 March 2022).

13. Xie, L.; Mo, M.; Jia, H.-X.; Liang, F.; Yuan, J.; Zhu, J. Association between Dietary Nitrate and Nitrite Intake and Sitespecific Cancer Risk: Evidence from Observational Studies. *Oncotarget* **2016**, *7*, 56915–56932. [CrossRef] [PubMed]
14. Schullehner, J.; Hansen, B.; Thygesen, M.; Pedersen, C.B.; Sigsgaard, T. Nitrate in Drinking Water and Colorectal Cancer Risk: A Nationwide Population-Based Cohort Study. *Int. J. Cancer* **2018**, *143*, 73–79. [CrossRef]
15. Espejo-Herrera, N.; Gràcia-Lavedan, E.; Boldo, E.; Aragonés, N.; Pérez-Gómez, B.; Pollán, M.; Molina, A.J.; Fernández, T.; Martín, V.; La Vecchia, C.; et al. Colorectal Cancer Risk and Nitrate Exposure through Drinking Water and Diet. *Int. J. Cancer* **2016**, *139*, 334–346. [CrossRef]
16. Larsen, F.J.; Ekblom, B.; Sahlin, K.; Lundberg, J.O.; Weitzberg, E. Effects of Dietary Nitrate on Blood Pressure in Healthy Volunteers. *N. Engl. J. Med.* **2006**, *355*, 2792–2793. [CrossRef]
17. Jackson, J.K.; Patterson, A.J.; MacDonald-Wicks, L.K.; Oldmeadow, C.; McEvoy, M.A. The Role of Inorganic Nitrate and Nitrite in Cardiovascular Disease Risk Factors: A Systematic Review and Meta-Analysis of Human Evidence. *Nutr. Rev.* **2018**, *76*, 348–371. [CrossRef]
18. Bailey, S.J.; Winyard, P.; Vanhatalo, A.; Blackwell, J.R.; DiMenna, F.J.; Wilkerson, D.P.; Tarr, J.; Benjamin, N.; Jones, A.M. Dietary Nitrate Supplementation Reduces the O2 Cost of Low-Intensity Exercise and Enhances Tolerance to High-Intensity Exercise in Humans. *J. Appl. Physiol.* **2009**, *107*, 1144–1155. [CrossRef]
19. Siervo, M.; Scialò, F.; Shannon, O.M.; Stephan, B.C.M.; Ashor, A.W. Does Dietary Nitrate Say NO to Cardiovascular Ageing? Current Evidence and Implications for Research. *Proc. Nutr. Soc.* **2018**, *77*, 112–123. [CrossRef]
20. Babateen, A.M.; Shannon, O.M.; O'Brien, G.M.; Olgacer, D.; Koehl, C.; Fostier, W.; Mathers, J.C.; Siervo, M. Moderate Doses of Dietary Nitrate Elicit Greater Effects on Blood Pressure and Endothelial Function than a High Dose: A 13-Week Pilot Study. *Nutr. Metab. Cardiovasc. Dis. NMCD* **2023**, *33*, 1263–1267. [CrossRef]
21. Hobbs, D.A.; George, T.W.; Lovegrove, J.A. The Effects of Dietary Nitrate on Blood Pressure and Endothelial Function: A Review of Human Intervention Studies. *Nutr. Res. Rev.* **2013**, *26*, 210–222. [CrossRef]
22. Shannon, O.M.; Gregory, S.; Siervo, M. Dietary Nitrate, Aging and Brain Health: The Latest Evidence. *Curr. Opin. Clin. Nutr. Metab. Care* **2022**, *25*, 393–400. [CrossRef]
23. Siervo, M.; Babateen, A.; Alharbi, M.; Stephan, B.; Shannon, O. Dietary Nitrate and Brain Health. Too Much Ado about Nothing or a Solution for Dementia Prevention? *Br. J. Nutr.* **2022**, *128*, 1130–1136. [CrossRef]
24. Clifford, T.; Babateen, A.; Shannon, O.M.; Capper, T.; Ashor, A.; Stephan, B.; Robinson, L.; O'Hara, J.P.; Mathers, J.C.; Stevenson, E.; et al. Effects Of Inorganic Nitrate And Nitrite Consumption On Cognitive Function And Cerebral Blood Flow: A Systematic Review And Meta-Analysis Of Randomised Clinical Trials. *Crit. Rev. Food Sci. Nutr.* **2019**, *59*, 2400–2410. [CrossRef]
25. Senefeld, J.W.; Wiggins, C.C.; Regimbal, R.J.; Dominelli, P.B.; Baker, S.E.; Joyner, M.J. Ergogenic Effect of Nitrate Supplementation: A Systematic Review and Meta-Analysis. *Med. Sci. Sports Exerc.* **2020**. *ahead of print*. [CrossRef]
26. Shannon, O.M.; Allen, J.D.; Bescos, R.; Burke, L.; Clifford, T.; Easton, C.; Gonzalez, J.T.; Jones, A.M.; Jonvik, K.L.; Larsen, F.J.; et al. Dietary Inorganic Nitrate as an Ergogenic Aid: An Expert Consensus Derived via the Modified Delphi Technique. *Sports Med.* **2022**, *52*, 2537–2558. [CrossRef] [PubMed]
27. Hernandez, A.; Schiffer, T.A.; Ivarsson, N.; Cheng, A.J.; Bruton, J.D.; Lundberg, J.O.; Weitzberg, E.; Westerblad, H. Dietary Nitrate Increases Tetanic [Ca^{2+}]i and Contractile Force in Mouse Fast-Twitch Muscle. *J. Physiol.* **2012**, *590*, 3575–3583. [CrossRef] [PubMed]
28. Bailey, S.J.; Fulford, J.; Vanhatalo, A.; Winyard, P.G.; Blackwell, J.R.; DiMenna, F.J.; Wilkerson, D.P.; Benjamin, N.; Jones, A.M. Dietary Nitrate Supplementation Enhances Muscle Contractile Efficiency during Knee-Extensor Exercise in Humans. *J. Appl. Physiol.* **2010**, *109*, 135–148. [CrossRef]
29. Larsen, F.J.; Schiffer, T.A.; Borniquel, S.; Sahlin, K.; Ekblom, B.; Lundberg, J.O.; Weitzberg, E. Dietary Inorganic Nitrate Improves Mitochondrial Efficiency in Humans. *Cell Metab.* **2011**, *13*, 149–159. [CrossRef]
30. Ferguson, S.K.; Hirai, D.M.; Copp, S.W.; Holdsworth, C.T.; Allen, J.D.; Jones, A.M.; Musch, T.I.; Poole, D.C. Impact of Dietary Nitrate Supplementation via Beetroot Juice on Exercising Muscle Vascular Control in Rats. *J. Physiol.* **2013**, *591*, 547–557. [CrossRef] [PubMed]
31. Jones, A.M.; Ferguson, S.K.; Bailey, S.J.; Vanhatalo, A.; Poole, D.C. Fiber-Type Specific Effects of Dietary Nitrate. *Exerc. Sport Sci. Rev.* **2016**, *44*, 53–60. [CrossRef]
32. Babateen, A.M.; Shannon, O.M.; Mathers, J.C.; Siervo, M. Validity and Reliability of Test Strips for the Measurement of Salivary Nitrite Concentration with and without the Use of Mouthwash in Healthy Adults. *Nitric Oxide Biol. Chem.* **2019**, *91*, 15–22. [CrossRef]
33. Woessner, M.; Smoliga, J.M.; Tarzia, B.; Stabler, T.; Van Bruggen, M.; Allen, J.D. A Stepwise Reduction in Plasma and Salivary Nitrite with Increasing Strengths of Mouthwash Following a Dietary Nitrate Load. *Nitric Oxide* **2016**, *54*, 1–7. [CrossRef] [PubMed]
34. Shannon, O.M.; Easton, C.; Shepherd, A.I.; Siervo, M.; Bailey, S.J.; Clifford, T. Dietary Nitrate and Population Health: A Narrative Review of the Translational Potential of Existing Laboratory Studies. *BMC Sport. Sci. Med. Rehabil.* **2021**, *13*, 65. [CrossRef] [PubMed]
35. Jones, A.M. Dietary Nitrate Supplementation and Exercise Performance. *Sport. Med.* **2014**, *44*, 35–45. [CrossRef] [PubMed]
36. Avoort, C.M.T.; Loon, L.J.C.; Hopman, M.T.E.; Verdijk, L.B. Increasing Vegetable Intake to Obtain the Health Promoting and Ergogenic Effects of Dietary Nitrate. *Eur. J. Clin. Nutr.* **2018**, *72*, 1485–1489. [CrossRef]

37. Griffiths, A.; Matu, J.; Whyte, E.; Akin-Nibosun, P.; Clifford, T.; Stevenson, E.; Shannon, O.M. The Mediterranean Dietary Pattern for Optimising Health and Performance in Competitive Athletes: A Narrative Review. *Br. J. Nutr.* **2022**, *128*, 1285–1298. [CrossRef]
38. Shannon, O.M.; Stephan, B.C.M.; Minihane, A.-M.; Mathers, J.C.; Siervo, M. Nitric Oxide Boosting Effects of the Mediterranean Diet: A Potential Mechanism of Action. *J. Gerontol. Ser. A* **2018**, *73*, 902–904. [CrossRef]
39. French, S.A. Pricing Effects on Food Choices. *J. Nutr.* **2003**, *133*, 841S–843S. [CrossRef]
40. Office for National Statistics Family Spending in the UK: April 2020 to March 2021. Available online: https://www.ons.gov.uk/peoplepopulationandcommunity/personalandhouseholdfinances/expenditure/bulletins/familyspendingintheuk/april2020tomarch2021#:~:text=Households%20spent%20an%20average%20of,%2C%20and%20transport%20(13%25) (accessed on 1 April 2023).
41. Blekkenhorst, L.C.; Prince, R.L.; Ward, N.C.; Croft, K.D.; Lewis, J.R.; Devine, A.; Shinde, S.; Woodman, R.J.; Hodgson, J.M.; Bondonno, C.P. Development of a Reference Database for Assessing Dietary Nitrate in Vegetables. *Mol. Nutr. Food Res.* **2017**, *61*, 1600982. [CrossRef]
42. Wang, X.; Cai, X.; Xu, C.; Wang, S.; Dai, S.; Wang, Q. Nitrate Accumulation and Expression Patterns of Genes Involved in Nitrate Transport and Assimilation in Spinach. *Molecules* **2018**, *23*, 2231. [CrossRef]
43. Shannon, O.M.; Babateen, A.; Grisotto, G.; Mathers, J.C.; Siervo, M. No Effect of 4 Wk of Nitrate-Rich Vegetable Consumption on Blood Pressure: Reflections for Future Research. *Am. J. Clin. Nutr.* **2018**, *108*, 1352–1353. [CrossRef]
44. Diouf, M.; Ba, C.O. Contribution of African leafy vegetables to food safety and income generation in Senegal. *Acta Hortic.* **2014**, *1021*, 53–58. [CrossRef]
45. Musa, A.; Ogbadoyi, E.O. Determination of Anti-Nutrients and Toxic Substances of Selected Fresh Leafy Vegetables Obtained from Minna Town, Nigeria. *Niger. J. Basic Appl. Sci.* **2014**, *22*, 79–83.
46. Tan, R.; Wylie, L.J.; Thompson, C.; Blackwell, J.R.; Bailey, S.J.; Vanhatalo, A.; Jones, A.M. Beetroot Juice Ingestion during Prolonged Moderate-Intensity Exercise Attenuates Progressive Rise in O2 Uptake. *J. Appl. Physiol.* **2018**, *124*, 1254–1263. [CrossRef]
47. James, P.E.; Willis, G.R.; Allen, J.D.; Winyard, P.G.; Jones, A.M. Nitrate Pharmacokinetics: Taking Note of the Difference. *Nitric Oxide* **2015**, *48*, 44–50. [CrossRef]
48. Lansley, K.E.; Winyard, P.G.; Fulford, J.; Vanhatalo, A.; Bailey, S.J.; Blackwell, J.R.; DiMenna, F.J.; Gilchrist, M.; Benjamin, N.; Jones, A.M. Dietary Nitrate Supplementation Reduces the O2 Cost of Walking and Running: A Placebo-Controlled Study. *J. Appl. Physiol.* **2011**, *110*, 591–600. [CrossRef] [PubMed]
49. Qadir, O.; Siervo, M.; Seal, C.J.; Brandt, K. Manipulation of Contents of Nitrate, Phenolic Acids, Chlorophylls, and Carotenoids in Lettuce (*Lactuca sativa* L.) via Contrasting Responses to Nitrogen Fertilizer When Grown in a Controlled Environment. *J. Agric. Food Chem.* **2017**, *65*, 10003–10010. [CrossRef] [PubMed]
50. Shannon, O.M.; Duckworth, L.; Barlow, M.J.; Deighton, K.; Matu, J.; Williams, E.L.; Woods, D.; Xie, L.; Stephan, B.C.M.; Siervo, M.; et al. Effects of Dietary Nitrate Supplementation on Physiological Responses, Cognitive Function, and Exercise Performance at Moderate and Very-High Simulated Altitude. *Front. Physiol.* **2017**, *8*, 401. [CrossRef]
51. Porcelli, S.; Pugliese, L.; Rejc, E.; Pavei, G.; Bonato, M.; Montorsi, M.; La Torre, A.; Rasica, L.; Marzorati, M. Effects of a Short-Term High-Nitrate Diet on Exercise Performance. *Nutrients* **2016**, *8*, 534. [CrossRef]
52. Lansley, K.E.; Winyard, P.G.; Bailey, S.J.; Vanhatalo, A.; Wilkerson, D.P.; Blackwell, J.R.; Gilchrist, M.; Benjamin, N.; Jones, A.M. Acute Dietary Nitrate Supplementation Improves Cycling Time Trial Performance. *Med. Sci. Sport. Exerc.* **2011**, *43*, 1125–1131. [CrossRef]
53. Cermak, N.M.; Gibala, M.J.; van Loon, L.J. Nitrate Supplementation's Improvement of 10-Km Time-Trial Performance in Trained Cyclists. *Int. J. Sport Nutr. Exerc. Metab.* **2012**, *22*, 64–71. [CrossRef]
54. Muggeridge, D.J.; Sculthorpe, N.; James, P.E.; Easton, C. The Effects of Dietary Nitrate Supplementation on the Adaptations to Sprint Interval Training in Previously Untrained Males. *J. Sci. Med. Sport* **2017**, *20*, 92–97. [CrossRef] [PubMed]
55. WHO | A Global Brief on Hypertension. Available online: https://www.who.int/cardiovascular_diseases/publications/global_brief_hypertension/en/ (accessed on 14 May 2019).
56. Siervo, M.; Shannon, O.; Kandhari, N.; Prabhakar, M.; Fostier, W.; Köchl, C.; Rogathi, J.; Temu, G.; Stephan, B.C.M.; Gray, W.K.; et al. Nitrate-Rich Beetroot Juice Reduces Blood Pressure in Tanzanian Adults with Elevated Blood Pressure: A Double-Blind Randomized Controlled Feasibility Trial. *J. Nutr.* **2020**, *150*, 2460–2468. [CrossRef] [PubMed]
57. Kandhari, N.; Prabhakar, M.; Mmbaga, B.T.; Rogathi, J.; Temu, G.; Gray, W.K.; Paddick, S.-M.; Walker, R.; Siervo, M. Effects of Dietary Nitrate and Folate Supplementation on Blood Pressure in Hypertensive Tanzanians: Design and Baseline Characteristics of a Feasibility Trial. *Contemp. Clin. Trials Commun.* **2019**, *16*, 100472. [CrossRef] [PubMed]
58. Ormesher, L.; Myers, J.E.; Chmiel, C.; Wareing, M.; Greenwood, S.L.; Tropea, T.; Lundberg, J.O.; Weitzberg, E.; Nihlen, C.; Sibley, C.P.; et al. Effects of Dietary Nitrate Supplementation, from Beetroot Juice, on Blood Pressure in Hypertensive Pregnant Women: A Randomised, Double-Blind, Placebo-Controlled Feasibility Trial. *Nitric Oxide Biol. Chem.* **2018**, *80*, 37–44. [CrossRef]
59. Babateen, A.M.; Shannon, O.M.; O'Brien, G.M.; Okello, E.; Khan, A.A.; Rubele, S.; Wightman, E.; Smith, E.; McMahon, N.; Olgacer, D.; et al. Acceptability and Feasibility of a 13-Week Pilot Randomised Controlled Trial Testing the Effects of Incremental Doses of Beetroot Juice in Overweight and Obese Older Adults. *Nutrients* **2021**, *13*, 769. [CrossRef]
60. Kandhari, N.; Prabhakar, M.; Shannon, O.; Fostier, W.; Koehl, C.; Rogathi, J.; Temu, G.; Stephan, B.C.M.; Gray, W.K.; Haule, I.; et al. Feasibility and Acceptability of a Nutritional Intervention Testing the Effects of Nitrate-Rich Beetroot Juice and Folic Acid on Blood Pressure in Tanzanian Adults with Elevated Blood Pressure. *Int. J. Food Sci. Nutr.* **2021**, *72*, 195–207. [CrossRef]

61. Sundqvist, M.L.; Larsen, F.J.; Carlström, M.; Bottai, M.; Pernow, J.; Hellénius, M.-L.; Weitzberg, E.; Lundberg, J.O. A Randomized Clinical Trial of the Effects of Leafy Green Vegetables and Inorganic Nitrate on Blood Pressure. *Am. J. Clin. Nutr.* **2020**, *111*, 749–756. [CrossRef]
62. Blekkenhorst, L.C.; Lewis, J.R.; Prince, R.L.; Devine, A.; Bondonno, N.P.; Bondonno, C.P.; Wood, L.G.; Puddey, I.B.; Ward, N.C.; Croft, K.D.; et al. Nitrate-Rich Vegetables Do Not Lower Blood Pressure in Individuals with Mildly Elevated Blood Pressure: A 4-Wk Randomized Controlled Crossover Trial. *Am. J. Clin. Nutr.* **2018**, *107*, 894–908. [CrossRef]
63. Jonvik, K.L.; Nyakayiru, J.; Pinckaers, P.J.; Senden, J.M.; van Loon, L.J.; Verdijk, L.B. Nitrate-Rich Vegetables Increase Plasma Nitrate and Nitrite Concentrations and Lower Blood Pressure in Healthy Adults. *J. Nutr.* **2016**, *146*, 986–993. [CrossRef]
64. van der Avoort, C.M.T.; Jonvik, K.L.; Nyakayiru, J.; van Loon, L.J.C.; Hopman, M.T.E.; Verdijk, L.B. A Nitrate-Rich Vegetable Intervention Elevates Plasma Nitrate and Nitrite Concentrations and Reduces Blood Pressure in Healthy Young Adults. *J. Acad. Nutr. Diet.* **2020**, *120*, 1305–1317. [CrossRef]
65. Vanhatalo, A.; Bailey, S.J.; Blackwell, J.R.; DiMenna, F.J.; Pavey, T.G.; Wilkerson, D.P.; Benjamin, N.; Winyard, P.G.; Jones, A.M. Acute and Chronic Effects of Dietary Nitrate Supplementation on Blood Pressure and the Physiological Responses to Moderate-Intensity and Incremental Exercise. *AJP Regul. Integr. Comp. Physiol.* **2010**, *299*, R1121–R1131. [CrossRef] [PubMed]
66. Shannon, O.M.; Barlow, M.J.; Duckworth, L.; Williams, E.; Wort, G.; Woods, D.; Siervo, M.; O'Hara, J.P. Dietary Nitrate Supplementation Enhances Short but Not Longer Duration Running Time-Trial Performance. *Eur. J. Appl. Physiol.* **2017**, *117*, 775–785. [CrossRef]
67. Capper, T.E.; Houghton, D.; Stewart, C.J.; Blain, A.P.; McMahon, N.; Siervo, M.; West, D.J.; Stevenson, E.J. Whole Beetroot Consumption Reduces Systolic Blood Pressure and Modulates Diversity and Composition of the Gut Microbiota in Older Participants. *NFS J.* **2020**, *21*, 28–37. [CrossRef]
68. Bondonno, C.P.; Downey, L.A.; Croft, K.D.; Scholey, A.; Stough, C.; Yang, X.; Considine, M.J.; Ward, N.C.; Puddey, I.B.; Swinny, E.; et al. The Acute Effect of Flavonoid-Rich Apples and Nitrate-Rich Spinach on Cognitive Performance and Mood in Healthy Men and Women. *Food Funct.* **2014**, *5*, 849. [CrossRef] [PubMed]
69. Qadir, O.K.; Seal, C.J.; Ashor, A.W.; Tassotti, M.; Mena, P.; Del Rio, D.; Siervo, M.; Brandt, K. Double-Blind Controlled Dietary Cross-over Intervention with Differentially Fertilised Intact Lettuce Leaves Shows Acute Reduction in Blood Pressure in Young Adults, Associated with Faster Uptake of Nitrate than of Phenolics. *Eur. J. Nutr.* **2022**, *61*, 4191–4203. [CrossRef]
70. Bondonno, C.P.; Liu, A.H.; Croft, K.D.; Ward, N.C.; Yang, X.; Considine, M.J.; Puddey, I.B.; Woodman, R.J.; Hodgson, J.M. Short-Term Effects of Nitrate-Rich Green Leafy Vegetables on Blood Pressure and Arterial Stiffness in Individuals with High-Normal Blood Pressure. *Free Radic. Biol. Med.* **2014**, *77*, 353–362. [CrossRef]
71. Sobko, T.; Marcus, C.; Govoni, M.; Kamiya, S. Dietary Nitrate in Japanese Traditional Foods Lowers Diastolic Blood Pressure in Healthy Volunteers. *Nitric Oxide* **2010**, *22*, 136–140. [CrossRef]
72. Flueck, J.L.; Bogdanova, A.; Mettler, S.; Perret, C. Is Beetroot Juice More Effective than Sodium Nitrate?-The Effects of Equimolar Nitrate Dosages of Nitrate-Rich Beetroot Juice and Sodium Nitrate on Oxygen Consumption during Exercise. *Appl. Physiol. Nutr. Metab.* **2016**, *41*, 421–429. [CrossRef]
73. Clifford, T.; Howatson, G.; West, D.J.; Stevenson, E.J. Beetroot Juice Is More Beneficial than Sodium Nitrate for Attenuating Muscle Pain after Strenuous Eccentric-Bias Exercise. *Appl. Physiol. Nutr. Metab.* **2017**, *42*, 1185–1191. [CrossRef]
74. Jones, L.; Bailey, S.J.; Rowland, S.N.; Alsharif, N.; Shannon, O.M.; Clifford, T. The Effect of Nitrate-Rich Beetroot Juice on Markers of Exercise-Induced Muscle Damage: A Systematic Review and Meta-Analysis of Human Intervention Trials. *J. Diet. Suppl.* **2022**, *19*, 749–771. [CrossRef]
75. Clifford, T.; Howatson, G.; West, D.J.; Stevenson, E.J. The Potential Benefits of Red Beetroot Supplementation in Health and Disease. *Nutrients* **2015**, *7*, 2801–2822. [CrossRef] [PubMed]
76. Shannon, O.; McGawley, K.; Nybäck, L.; Duckworth, L.; Barlow, M.J.; Woods, D.; Siervo, M.; O'Hara, J.P. 'Beet-Ing' the Mountain: A Review of the Physiological and Performance Effects of Dietary Nitrate Supplementation at Simulated and Terrestrial Altitude. *Sports Med.* **2017**, *47*, 2155–2169. [CrossRef] [PubMed]
77. McDonagh, S.T.J.; Wylie, L.J.; Webster, J.M.A.; Vanhatalo, A.; Jones, A.M. Influence of Dietary Nitrate Food Forms on Nitrate Metabolism and Blood Pressure in Healthy Normotensive Adults. *Nitric Oxide* **2018**, *72*, 66–74. [CrossRef] [PubMed]
78. Ashworth, A.; Mitchell, K.; Blackwell, J.R.; Vanhatalo, A.; Jones, A.M. High-Nitrate Vegetable Diet Increases Plasma Nitrate and Nitrite Concentrations and Reduces Blood Pressure in Healthy Women. *Public Health Nutr.* **2015**, *18*, 2669–2678. [CrossRef] [PubMed]
79. Capper, T.; Clifford, T.; Taylor, G.; Iqbal, W.; West, D.; Stevenson, E.; Siervo, M. Ageing Modifies Acute Resting Blood Pressure Responses to Incremental Consumption of Dietary Nitrate: A Randomised, Cross-over Clinical Trial. *Br. J. Nutr.* **2023**, *129*, 442–453. [CrossRef]
80. Jovanovski, E.; Bosco, L.; Khan, K.; Au-Yeung, F.; Ho, H.; Zurbau, A.; Jenkins, A.L.; Vuksan, V. Effect of Spinach, a High Dietary Nitrate Source, on Arterial Stiffness and Related Hemodynamic Measures: A Randomized, Controlled Trial in Healthy Adults. *Clin. Nutr. Res.* **2015**, *4*, 160–167. [CrossRef]
81. Liu, A.H.; Bondonno, C.P.; Croft, K.D.; Puddey, I.B.; Woodman, R.J.; Rich, L.; Ward, N.C.; Vita, J.A.; Hodgson, J.M. Effects of a Nitrate-Rich Meal on Arterial Stiffness and Blood Pressure in Healthy Volunteers. *Nitric Oxide.* **2013**, *35*, 123–130. [CrossRef]

82. Miller, G.D.; Marsh, A.P.; Dove, R.W.; Beavers, D.; Presley, T.; Helms, C.; Bechtold, E.; King, S.B.; Kim-Shapiro, D. Plasma Nitrate and Nitrite Are Increased by a High Nitrate Supplement, but Not by High Nitrate Foods in Older Adults. *Nutr. Res.* **2012**, *32*, 160–168. [CrossRef]
83. Brandt, K. Chapter 50—Vegetables and Fruit in the Prevention of Chronic Age-Related Diseases. In *Molecular Basis of Nutrition and Aging*; Malavolta, M., Mocchegiani, E., Eds.; Academic Press: San Diego, CA, USA, 2016; pp. 707–722. ISBN 978-0-12-801816-3.
84. Gallardo, E.J.; Coggan, A.R. What Is in Your Beet Juice? Nitrate and Nitrite Content of Beet Juice Products Marketed to Athletes. *Int. J. Sport Nutr. Exerc. Metab.* **2019**, *29*, 345–349. [CrossRef]
85. Wootton-Beard, P.C.; Ryan, L. A Beetroot Juice Shot Is a Significant and Convenient Source of Bioaccessible Antioxidants. *J. Funct. Foods* **2011**, *3*, 329–334. [CrossRef]
86. Cronin, P.; Joyce, S.A.; O'Toole, P.W.; O'Connor, E.M. Dietary Fibre Modulates the Gut Microbiota. *Nutrients* **2021**, *13*, 1655. [CrossRef] [PubMed]
87. Kimble, R.; Gouinguenet, P.; Ashor, A.; Stewart, C.; Deighton, K.; Matu, J.; Griffiths, A.; Malcomson, F.C.; Joel, A.; Houghton, D.; et al. Effects of a Mediterranean Diet on the Gut Microbiota and Microbial Metabolites: A Systematic Review of Randomized Controlled Trials and Observational Studies. *Crit. Rev. Food Sci. Nutr.* **2022**, 1–22. [CrossRef]
88. Elvira-Torales, L.I.; Periago, M.J.; González-Barrio, R.; Hidalgo, N.; Navarro-González, I.; Gómez-Gallego, C.; Masuero, D.; Soini, E.; Vrhovsek, U.; García-Alonso, F.J. Spinach Consumption Ameliorates the Gut Microbiota and Dislipaemia in Rats with Diet-Induced Non-Alcoholic Fatty Liver Disease (NAFLD). *Food Funct.* **2019**, *10*, 2148–2160. [CrossRef] [PubMed]
89. De Leon, A.; Jahns, L.; Casperson, S.L. Barriers and Facilitators to Following the Dietary Guidelines for Vegetable Intake: Follow-up of an Intervention to Increase Vegetable Intake. *Food Qual. Prefer.* **2020**, *83*, 103903. [CrossRef]
90. Shannon, O.M.; Grisotto, G.; Babateen, A.; McGrattan, A.; Brandt, K.; Mathers, J.C.; Siervo, M. Knowledge and Beliefs about Dietary Inorganic Nitrate among UK-Based Nutrition Professionals: Development and Application of the KINDS Online Questionnaire. *BMJ Open* **2019**, *9*, e030719. [CrossRef]
91. Heaney, S.; O'Connor, H.; Michael, S.; Gifford, J.; Naughton, G. Nutrition Knowledge in Athletes: A Systematic Review. *Int. J. Sport Nutr. Exerc. Metab.* **2011**, *21*, 248–261. [CrossRef]
92. Spronk, I.; Kullen, C.; Burdon, C.; O'Connor, H. Relationship between Nutrition Knowledge and Dietary Intake. *Br. J. Nutr.* **2014**, *111*, 1713–1726. [CrossRef]
93. Kelly, M.P.; Barker, M. Why Is Changing Health-Related Behaviour so Difficult? *Public Health* **2016**, *136*, 109–116. [CrossRef]
94. Michie, S.; van Stralen, M.M.; West, R. The Behaviour Change Wheel: A New Method for Characterising and Designing Behaviour Change Interventions. *Implement. Sci.* **2011**, *6*, 42. [CrossRef]
95. Hu, F.B. Globalization of Diabetes: The Role of Diet, Lifestyle, and Genes. *Diabetes Care* **2011**, *34*, 1249–1257. [CrossRef]
96. Levinson, C.A.; Fewell, L.; Brosof, L.C. My Fitness Pal Calorie Tracker Usage in the Eating Disorders. *Eat. Behav.* **2017**, *27*, 14–16. [CrossRef] [PubMed]
97. McMahon, N.F.; Brooker, P.G.; Pavey, T.G.; Leveritt, M.D. Nitrate, Nitrite and Nitrosamines in the Global Food Supply. *Crit. Rev. Food Sci. Nutr.* **2022**, 1–22. [CrossRef] [PubMed]
98. Bartholomew, B.; Hill, M.J. The Pharmacology of Dietary Nitrate and the Origin of Urinary Nitrate. *Food Chem. Toxicol.* **1984**, *22*, 789–795. [CrossRef] [PubMed]
99. Wu, Z.; Tian, T.; Ma, W.; Gao, W.; Song, N. Higher Urinary Nitrate Was Associated with Lower Prevalence of Congestive Heart Failure: Results from NHANES. *BMC Cardiovasc. Disord.* **2020**, *20*, 498. [CrossRef] [PubMed]
100. Mian, A.I.; Du, Y.; Garg, H.K.; Caviness, A.C.; Goldstein, S.L.; Bryan, N.S. Urinary Nitrate Might Be an Early Biomarker for Pediatric Acute Kidney Injury in the Emergency Department. *Pediatr. Res.* **2011**, *70*, 203–207. [CrossRef]
101. Audate, P.P.; Fernandez, M.A.; Cloutier, G.; Lebel, A. Scoping Review of the Impacts of Urban Agriculture on the Determinants of Health. *BMC Public Health* **2019**, *19*, 672. [CrossRef]
102. Al-Kodmany, K. The Vertical Farm: A Review of Developments and Implications for the Vertical City. *Buildings* **2018**, *8*, 24. [CrossRef]
103. Blatt, A.D.; Roe, L.S.; Rolls, B.J. Hidden Vegetables: An Effective Strategy to Reduce Energy Intake and Increase Vegetable Intake in Adults. *Am. J. Clin. Nutr.* **2011**, *93*, 756–763. [CrossRef]
104. Hobbs, D.A.; Kaffa, N.; George, T.W.; Methven, L.; Lovegrove, J.A. Blood Pressure-Lowering Effects of Beetroot Juice and Novel Beetroot-Enriched Bread Products in Normotensive Male Subjects. *Br. J. Nutr.* **2012**, *108*, 2066–2074. [CrossRef]
105. Hobbs, D.A.; George, T.W.; Lovegrove, J.A. Differential Effect of Beetroot Bread on Postprandial DBP According to Glu298Asp Polymorphism in the ENOS Gene: A Pilot Study. *J. Hum. Hypertens.* **2014**, *28*, 726–730. [CrossRef]
106. Lee, S.S.; McGrattan, A.; Soh, Y.C.; Alawad, M.; Su, T.T.; Palanisamy, U.D.; Hussin, A.M.; Kassim, Z.B.; Mohd Ghazali, A.N.B.; Christa Maree Stephan, B.; et al. Feasibility and Acceptability of a Dietary Intervention to Reduce Salt Intake and Increase High-Nitrate Vegetable Consumption in Malaysian Middle-Aged and Older Adults with Elevated Blood Pressure: Findings from the DePEC-Nutrition Trial. *Nutrients* **2022**, *14*, 430. [CrossRef] [PubMed]
107. Public Health England. *The Eatwell Guide: How Does It Differ to the Eatwell Plate and Why?* Public Health England: London, UK, 2016.
108. US Department of Agriculture. *Dietary Guidelines for Americans, 2020–2025*; US Department of Agriculture: Washington, DC, USA, 2020.

109. Kearney, M.; Bradbury, C.; Ellahi, B.; Hodgson, M.; Thurston, M. Mainstreaming Prevention: Prescribing Fruit and Vegetables as a Brief Intervention in Primary Care. *Public Health* **2005**, *119*, 981–986. [CrossRef] [PubMed]
110. Buyuktuncer, Z.; Kearney, M.; Ryan, C.L.; Thurston, M.; Ellahi, B. Fruit and Vegetables on Prescription: A Brief Intervention in Primary Care. *J. Hum. Nutr. Diet.* **2014**, *27*, 186–193. [CrossRef] [PubMed]

applied sciences

MDPI

Review

Old-Fashioned, but Still a Superfood—Red Beets as a Rich Source of Bioactive Compounds

Aleksandra Sentkowska [1,*] and Krystyna Pyrzyńska [2]

[1] Heavy Ion Laboratory, University of Warsaw, Pasteur 5A, 02-093 Warsaw, Poland
[2] Faculty of Chemistry, University of Warsaw, Pasteur 1, 02-093 Warsaw, Poland; kryspyrz@chem.uw.edu.pl
* Correspondence: sentkowska@slcj.uw.edu.pl

Abstract: Beetroot (*Beta vulgaris* L.) is a vegetable that is consumed worldwide in the form of juices, soups, or salads. It is also known for its high content of biologically active substances such as betalains, polyphenolic compounds, vitamins, carotenoids, and other nutrients including, sodium, potassium, and magnesium. The distribution of these compounds in the plant is diverse, some occur in greater amounts in the leaves (e.g., vitamin A, B6) and others are in the tubers (e.g., folate, lycopene). The concentration of bioactive compounds in beetroot also depends on its variety and growing conditions. Recent studies have reported on the beneficial effect of beetroot juice and beetroot products on the body's efficiency during prolonged physical exercise. The purpose of this review is to discuss the content of biologically active compounds in beetroot and the impact of beetroot product consumption on the human body, based on the latest literature.

Keywords: red beets; bioactive compounds; betalains; phenolic compounds; antioxidants

1. Introduction

Beta vulgaris (beet) is an edible plant in the subfamily *Betoideae* of the family *Amaranthaceae*. It has several cultivar groups such as sweet beet, garden beet, chard, and mangoldwurzel. Red beetroots (*Beta vulgaris*) have been cultivated for many hundreds of years in all temperate climates. Surprisingly, the green parts of the beetroot were consumed initially in the form of chard [1]. The dish was so popular that the ancient Greeks and Romans developed a method of growing beetroot also during the hot summer months. It should be emphasized that this variety had only thin and fibrous roots, which were occasionally used in medicine. Consumption of underground beet parts was first reported in Germany and Italy in 1542 [2]. However, that beet variety was significantly different from the one we know today and rather closely resembled a parsnip. The beetroot variety that is still cultivated today appeared in the late 16th century, and is believed to have originated from a prehistoric North African root vegetable. Currently, most beetroots are consumed as a vegetable and also as a juice. Beetroots can also be processed into a powder for use as a food colorant E162 and additive to cosmetics [3–6].

Although beets have been known for centuries, understanding of their value is still limited. New reports about bioactive compounds in beets suggest that they fit the definition of a functional food, as they offer health benefits that extend beyond their nutritional value [7–9]. In addition to proteins, carbohydrates, fat, amino acids, fatty acids, phytosterols, minerals, and fibres, beetroot contains also vitamins, nitrate, polyphenolic compounds, and betalains, a class of red and yellow natural pigments found in plants. Betalains are considered to be strong antioxidant agents as they can decrease oxidative stress by efficiently removing the reactive oxygen species [9–11]. Other health benefits of betalains include antimicrobial, antiviral, and anti-inflammatory activities [12–14]. Several studies showed the tumour-chemopreventive effects of beetroot extracts in laboratory animals [15–17]. This is probably a synergistic effect of both betalains and other compounds present in beetroot, e.g., polyphenolic compounds. The scheme of the main bioactive compounds present in beetroot is depicted in Figure 1.

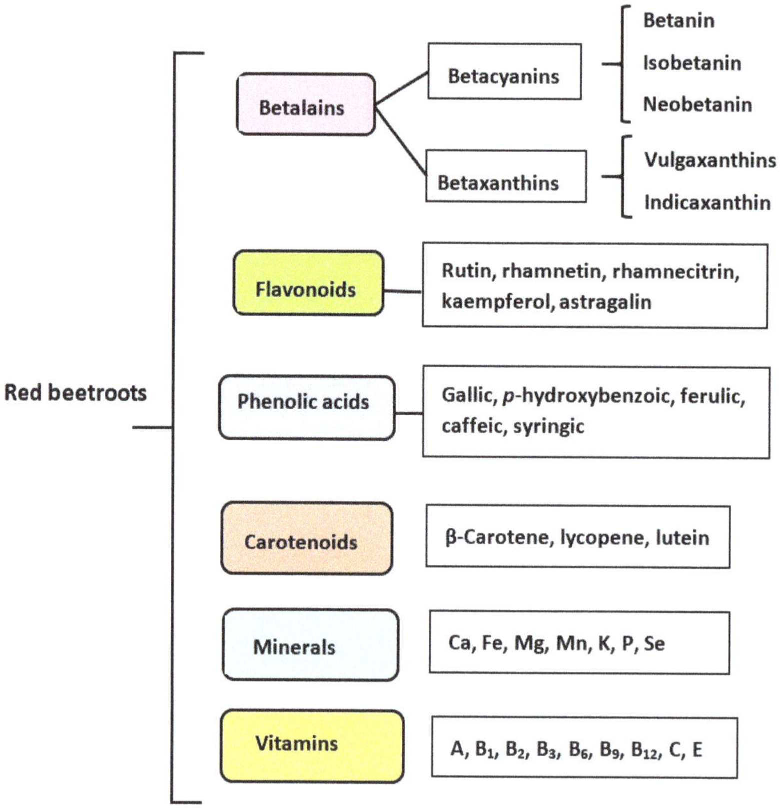

Figure 1. Main bioactive compounds present in red beetroots.

This review aims to collect the latest literature on the content of biologically active compounds in beets and beet products. It also provides an overview of the positive health effect associated with beet consumption.

2. Betalains

Betalains are water-soluble secondary metabolites that occur in many plants of the order *Caryphyllales*. They have a higher solubility in water compared to anthocyanins, and are stable in the pH range of 3–7, thus, can be used in low-acid and neutral foods [18,19]. Betalamic acid is a component of all betalains and its substituent type defines the class of betalain [20]. Betalamic acid can spontaneously condense with cyclo-DOPA (3,4-dihydroxy-phenylalanine, a precursor of betalain pigment in plants) yielding the violet betacyanins, or with amines or amino acids to produce yellow betaxanthins [21]. Their general chemical structures are presented in Figure 2 [18]. Red beets are the main edible source of these compounds. According to Delgado-Vargas et al., betacyanins content in beetroot accounts for 75–95% of the total pigments, the remaining 5–25% corresponds to betaxanthins [22]. The violet betacyanin is the most studied of all the betalain groups. Beetroot contains also its epimer, isobetanin, as well as other pigments like betaxanthins vulgaxanthin I and miraxanthin V [23]. Four betacyanins (betanin, prebetanin, isobetanin, neobetanin) and three betaxanthins (vulgaxanthin I, vulgaxanthin II, indicaxanthin) were identified in the water fractions of crude extracts of beetroot peel [24]. In beetroot juice samples, the presence of red pigments was reported in ranges of 797.5–421.7 mg·L^{-1} for betanin and 321.7–423.1 mg·L^{-1} for vulgaxanthin [25]. Concentrations of both dyes in purchased organic beetroot juice were higher than those in freshly squeezed juice [26]. However, the characteristic color quickly lost its intensity after storage at 20 °C, and after some time it was possible to notice the presence of a flocculent precipitate. Degradation products of betanin are isobetanin, decarboxylated betanin, betalamic acid, and cyclodopa glucoside [27]. Kazimierczak et al. determined higher concentrations of betalains in pure beetroot juices in comparison to those combined with other components (e.g., apple, lemon) [28].

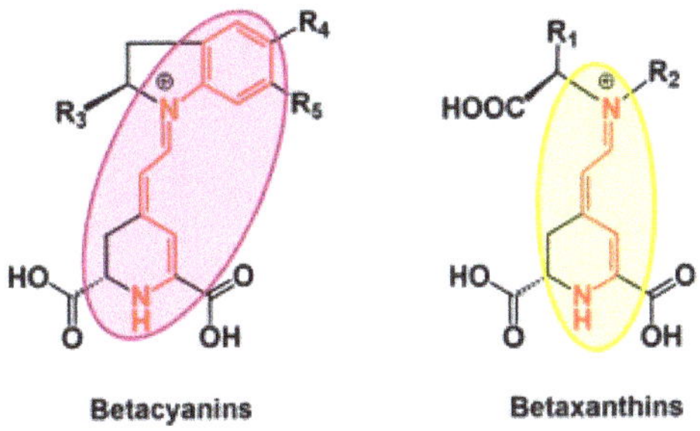

Figure 2. General chemical structures for betacyanins and betaxanthins. Reprinted with permission from [18].

Betalains can exert their action by different mechanisms, such as scavenging free radicals, regulating their content produced during biochemical reactions, and protecting cells from oxidative stress and damage. In food science, the term "antioxidant" is commonly used to describe compounds that prevent oxidative reactions, such as lipid peroxidation, maintaining the freshness of food products. In health science, this term includes free-radical scavenging of reactive oxygen species, reducing agents, and chelators of metal ions; thus, they have roles in preventing different chronic diseases [29]. However, many antioxidants at high concentrations could act as prooxidants and induce toxicity. Several studies confirmed the high radical-scavenging activity of betalains [30–32]. While betanin exhibits a high radical-scavenging activity at pH > 4, the highest activity of betanidin was reported at a pH range of 2–4 [30]. In betacyanins, acylation raises the antioxidant potential, whereas glycosylation decreases their activity. The highest potential was reported for betanin, followed by neobetanin, vulgaxanthin I, and indicaxanthin [32].

The losses of red pigments in the betalain solution (heated at 90 °C for 30 min in a water bath without access to light) increased, along with rising pH levels, due to their degradation [33]. The opposite correlation was observed for yellow pigments; the most pronounced increase in their levels was observed at a pH of 6.5–7.0. However, during heating, the antioxidant capacity of the solution was subject to only minor changes in the pH range of 4–9.

Many studies demonstrated the health-promoting effects of betalains. Most of these reports focus on their antioxidant, anti-inflammatory, and anti-carcinogenic properties, as was mentioned earlier [9,10,12–14]. The first studies investigating the chemopreventive potential of beetroot extract were carried out in mice in 1996 [34]. This study proved that oral administration of beetroot extract contributed to the reduction of the number of skin papillomas. Zhang et al. also proved the chemopreventive effect of betanin in mice [35]. In their study, two carcinogenic substances, vinyl carbamate and benzopyrene, were used to induce lung tumours, and then an oral treatment with an aqueous solution of betanin was administered. As a result, a reduction of cancer cells was observed, explained by the increased expression of caspase-3, a proapoptotic enzyme belonging to the cysteine-protease group, which increases cell apoptosis levels. The anticancer properties of betalains in mice with xenografts from human lung tumour cells were also proved (A549) [36].

According to Martinez et al., a more complex mechanism of betalain action is probably responsible for their antitumor effects [18]. The mechanism of action of *Beta vulgaris* betanin on tumour cells is present in Figure 3. First, the antioxidant properties of betanin increase the activity of antioxidant enzymes that neutralize ROS produced by immune cells. Then, the red pigments, due to their anti-inflammatory capacity, are able to reduce the production of pro-inflammatory cytokines that increase the migration of leukocytes to the tumour area [37]. At the final stage, betanin causes the formation of blood vessels (angiogenesis) around the tumour area, while increasing levels of apoptosis.

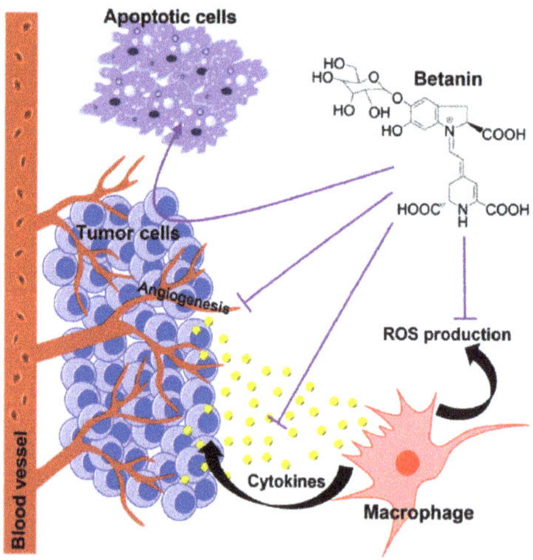

Figure 3. The mechanism of action of *Beta vulgaris* betanin on tumor cells. Reprinted with permission from [18].

The anti-inflammatory properties of betalains are obvious from the use of *B. vulgaris* betanin-rich pigments used as additives in the food industry. Several studies have attempted to explain the mechanism behind the mechanism of action of betalains [13,38,39]. The mechanism that could be responsible for observed anti-inflammatory effects may be the demonstrated in vitro inhibition that betalains exert on the enzymes lipoxygenase and cyclooxygenase, which are involved in the formation of the inflammation-mediating compounds leukotrienes and prostaglandins. The cardioprotective effect of betalains was also proved. Consumption of beetroot chips by rats resulted in a decrease in cholesterol and triglyceride levels in their blood [40]. Novel studies have evaluated the protective effect of betalain against heart failure [41], and the potential impacts of betalain supplementation on some atherosclerotic risk factors in coronary artery disease patients [42]. The effects of betalains on hyperglycemia in diabetic rats have also been reported [43,44].

Even though the results of animal studies clearly show the health-promoting effect of betalains, only a few similar studies on humans have been described. These examined the impact of betalain intake on the body's efficiency and cardiovascular system [42] and the impact of beet juice supplementation in reducing muscle damage following eccentric exercises [45]. The intake of red beetroot improves glucose metabolism and can help in the treatment of people with type 2 diabetes [46].

3. Flavonoids and Polyphenolic Acids

The high antioxidant capacity of beets is due not only to the high content of betalains, but also other phenolic compounds—flavonoids and polyphenolic acids. Significant amounts of catechins and polyphenolic acid (ferulic, protocatechuic, vanillic, p-coumaric, p-HBA, syringic, and caffeic) have been detected in beets [47–50]. However, the composition of plant extract depends on many factors, including the original plant (variety, maturation state), conditions of its cultivation (climate, harvesting), as well as its storage condition [51].

Kavalcowa et al. showed how much the total content of polyphenolic compounds depends on the beet variety and the place of its cultivation [50]. The highest content of total polyphenols was found in the variety of Renova from Sliač, in Slovakia (1280.56 $\mu g \cdot g^{-1}$), followed by red beetroot from Zohor. In addition, the variety of Zohor, obtained from Sihelné, had the lowest concentration of polyphenols (820 $\mu g \cdot g^{-1}$). In the case of the variety Monorubrawe, the highest value of total polyphenols was determined in samples from the area of Sihelné (1201.6 $\mu g \cdot g^{-1}$), followed by red beetroot from Sliač (1023.21 $\mu g \cdot g^{-1}$) and Zohor (988.66 $\mu g \cdot g^{-1}$). The variety-specific differences between the highest and lowest polyphenol contents were 460.56 $\mu g \cdot g^{-1}$ in the Renova variety and 212.94 $\mu g \cdot g^{-1}$ in the Monorubrawe variety. The authors concluded that the observed differences may be due to the high humus (4.0%) and potassium (520 $\mu g \cdot g^{-1}$) content in the soil. Studies conducted by Kujala et al. clearly showed that the total phenolics content decreases in order: peel (50%), crown (37%), and flesh (13%) [52]. The comparison of the phenolic composition in betalain extracts from intact *B. vulgaris* cv. *Detroit Dark Red* plants and hairy root cultures was described by Georgiev et al. [11]. The authors proved that the concentrations of selected polyphenolic compounds were even more than 100 times higher in extracts from hairy root cultures than in extracts obtained from intact plants. Moreover, rutin was present only in hairy root cultures, while chlorogenic acid was found only in intact plants. While quercetin was not reported in this study, Pratimasari and Puspitasari proved its presence in the purified extract of beetroot leaf [53]. In the study performed by Płatosz et al., quercetin was detected in fresh red beet, fermented red beet, and commercial fermented beet juice, in concentrations of 0.023 $\mu g \cdot g^{-1}$, 0.002 $\mu g \cdot g^{-1}$, and 0.009 $\mu g \cdot mL^{-1}$, respectively [47]. The results obtained by Georgiev indicated that the extract of the hairy roots had substantially higher concentrations of the identified phenolic compounds than did the intact plants, as noted above [11]. Moreover, there were differences in the phenolic profiles of the studied extracts. Chlorogenic acid was present only in the extract from intact plants, while rutin was found only in the extract from hairy root cultures. As the authors emphasize, it is a valuable discovery as rutin interacts synergistically with many other biologically active compounds present in the plant material enhancing its antioxidant properties [54]. The fermentation process caused an increase in the content of phenolic acids and a reduction in the content of their conjugated forms, while the same process caused a decrease in the content of free flavonoids [47]. In the case of the studies described by Płatosz, significant differences in the concentrations of polyphenolic acids were observed between the extracts from fresh and fermented beetroot [47]. Such large differences were not observed in the case of flavonoids, their concentrations in most cases were rather similar.

Attention to possible interactions between phenolic compounds in beetroot juice has been also drawn [25]. The content of the main active compounds in beetroot juice was determined by HPLC analysis before and after the addition of these compounds to the study juice, and the antioxidant activities of the obtained solutions were examined. Each of the studied additives had a great impact on the scavenging of the DPPH radicals in the juice. The addition of caffeic acid, rhamnetin, and vitamin B5 resulted in a decrease in this ability, which was shown in Figure 4. Antioxidant interactions can be synergistic, additive, or antagonistic, depending on the specific substances and their concentration ratios. In the case of examined additives, only betanin caused an increase in the ability to neutralize the DPPH radical, in comparison to beetroot juice without any additives. There is a synergistic effect here, which is particularly visible in the initial stage of the kinetic curve. In contrast, an antagonistic effect is observed for the remaining additives. The kind of interaction depends on the specific antioxidants interacting in the system and the condition behind the evaluation, as noted above, however, the knowledge about the interaction between polyphenolic compounds and vitamins as well as non-antioxidant compounds is still poor. A common theme in the scientific literature is that interactions between antioxidant molecules do occur, but a mechanism that allows a prediction of synergistic and antagonistic interactions is not clear [55].

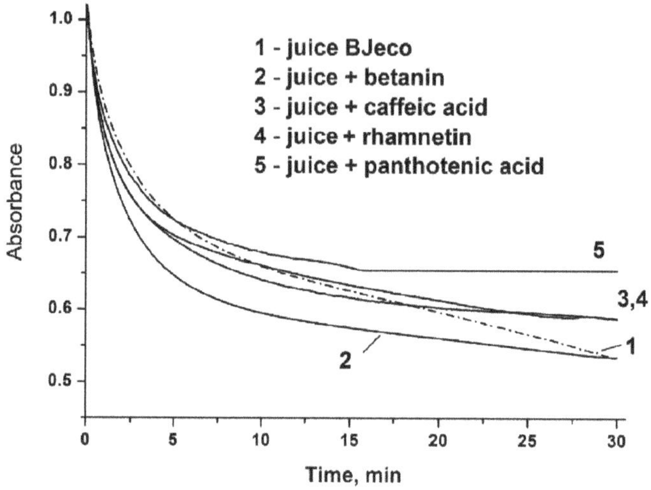

Figure 4. The kinetic curves for DPPH radical scavenging by beetroot juice from ecological cultivar (BJeco) with different additives (concentration of each 10 mg·L^{-1}). The measurements were carried out at λ = 539 nm. Reprinted with permission from [25].

Vulic et al. studied the presence of bioactive compounds in industrial beetroot pomace extract obtained from different cultivars grown in Serbia [56]. The concentrations of total phenolics (45.68 mg gallic acid equivalents per gram), flavonoids (25.89 mg rutin equivalents per gram), and betalains (betanin, 4.09 mg·g^{-1}; vulgaxanthin I, 7.32 mg·g^{-1}) were determined.

Some studies reported that beets from organic cultivation are characterized by a higher content of bioactive compounds [57–59], while others reported the opposite relationship [60,61]. According to Szopińska and Gawęda, these inconsistent results may be due to the differences in the form and amount of fertilizer used, weather conditions, soil type, cultivation, and genetic factors [62]. It is generally accepted that plants from organic farming contain more polyphenolic compounds than those from conventional cultivation. This, in turn, translates into the antioxidant properties of such foods, and their usefulness in preventing oxidative stress by neutralizing free radicals. Different reaction mechanisms between the polyphenolic compounds and radicals have been proposed. They can be divided into two groups: H-atom abstraction and radical adduct formation. Moreover, H-abstraction processes can be divided into four subgroups of mechanisms: hydrogen atom transfer (HAT), proton coupled electron transfer (PCET), sequential proton loss electron transfer (SPLET), and single electron transfer followed by proton transfer (SET-PT). It should be mentioned that all of these mechanisms have the same result [63]. Tosovic et al. mentioned that the mechanisms of action of many important food and beverage ingredients, such as chlorogenic acid, are not fully clarified [64]. Based on the electron-spin resonance spectroscopy [ESR] and density-functional theory [DFT], authors have concluded that in acidic and neutral media, chlorogenic acid rather takes the HAT or RAF pathway. On the other hand, Dimić et al. proved that SPLET is the dominant mechanism for the radical activity towards DPPH radicals in polar solvents [63]. Based on our experience in working with beetroot juice, we know that its pH, depending on the method of preparation and the variety of beetroot from which it was obtained, is in the range of 4–5. This may suggest that, in the case of beetroot juice, the mechanism of free radical neutralization is based on HAT or RAF mechanisms, as was proposed by Tosovic et al. [64].

4. Carotenoids

Carotenoids are another group of bioactive compounds present in beetroots; however, they are present rather in small amounts [65]. The content of these compounds strictly depends on the part of the plant that is being tested; the highest content of carotenoids is found in beetroots peel, followed by pulp, leaves, and stalks [8]. This is because these compounds are accumulated in the green parts of the plants, namely in chloroplasts, as a mixture of α- and β-carotene, β-cryptoxanthin, lutein, zeaxantin, violaxanthin, and neoxanthin [66]. The main carotenoids in red beets are β-carotene and lutein, which pose strong anticancer properties [16]. Red beets are not a very good source of red carotenoid lycopene in comparison to tomatoes, watermelon, or papaya [8]. The lycopene content was 30.0 ± 0.3 μg per 100 g of beet tubers and was not detected in the leaves. α-Carotene was detected in the leaves and tubers of the beetroot, and its concentration in tubers was almost seven times higher than in leaves. However, the leaves are a good source of β-carotene (11.64 μg per 100 g) and lutein (1.503 μg together with zeaxanthin) [67]. The presence of these compounds was not confirmed in beet tubers [8].

A higher concentration of carotenoids was found in organic stalks and cooked pulp in comparison to the conventionally grown plants [68]. This is in line with reports in the literature showing the content of β-carotene is higher in organically cultivated plants in comparison to conventionally grown plants [69]. However, significant differences were only observed between these two parts. After cooking the beet peel and pulp, a decrease in the carotene content was observed; however, its content was still higher in organic beets after this process [68].

5. Minerals

Beetroot contains natural minerals that are essential for the proper functioning of the human body. Wruss et al. analyzed the mineral composition of seven different popular beetroot varieties grown in Upper Austria [49]. Obtained values were in a range similar to those reported in previous studies [70]. However, as the authors highlighted, data regarding the concentration of trace elements and minerals in beet juices are still limited. It is also crucial to note which part of the plant is tested, as the content of individual elements varies among leaves and tubers [8]. The concentration of copper ions was more than two times higher in leaves, in comparison to tubers (0.191 and 0.075 mg per 100 g of beets, respectively) [8]. A greater difference was observed for iron, for which 2.57 mg was found in leaves and 0.80 mg in tubers (values per 100 g of beets). Also, calcium, sodium, potassium, and magnesium were detected in much higher concentrations in leaves. The concentrations of zinc, phosphorus, and manganese did not significantly differ from each other.

Recent studies showed that beetroots and beet products are a great source of selenium [71,72]. Selenium is an important trace element for humans and animals, as it plays a key role in several major metabolic pathways, such as thyroid hormone metabolism and immune functions [73,74]. Its deficiency has been linked with cardiovascular and inflammatory diseases and cancer [71]. The major selenium-containing amino acids occurring naturally in dietary sources are selenomethionine (SeMet) and methyloselenocysteine (MeSeCys), which were also detected in beetroot juice [26,72]. A study involving five juices from conventionally grown beetroot and one from an ecological cultivar reported that all juices differed from each other in terms of selenium content and antioxidant activity, which is shown in Figure 5. Overall, the squeezed juice was lower in selenium compounds and polyphenols, perhaps because commercially available juice is made with a higher weight of beetroot.

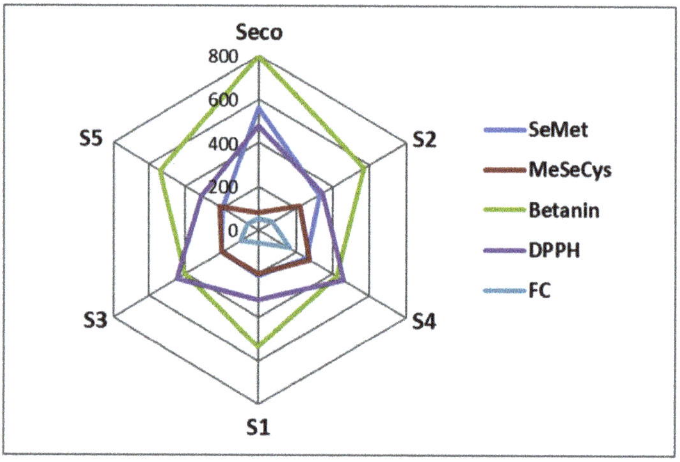

Figure 5. The comparison of selenium species and betanin contents, and antioxidant properties of beetroot juice. Reprinted with permission from [71].

An important aspect is the instability of selenium compounds, which is particularly visible in freshly squeezed juice that has not been pasteurized, especially SeMet, which was oxidized to selenomethionine oxide (SeMetO) [26]. This process resulted in a decrease in the antioxidant activity of the studied juices. The antioxidant activity of selenium species in the DPPH assay decreases in the following order: MeSeCys > SeMet ~ Se(VI) > Se(IV). The values of all species were lower than that of α-tocopherol at the same concentration [75]. From the consumer's point of view, the best way to consume beetroot juices is to drink them immediately after preparation or opening the bottle.

6. Vitamins

Beetroots and beet products are also great sources of vitamin C, vitamin A, and vitamins from the B group. The B-group vitamins are a collection of 8 water-soluble vitamins including thiamine (B1), riboflavin (B2), niacin (B3), pantothenic acid (B5), pyridoxine (B6), biotin (B7), folate (B9), and cyanocobalamin (B12). According to Odoh and Okoro 100 g of this plant contains: vitamin A (2.6 mcg), K, (3.2 mcg), C (4.36 mg), E (0.18 mg) B3 (0.03 mg), B6 (90 mg), B2 (0.034 mg), and B5 (0.151 mg) [67]. These values are in agreement with findings published by Neha et al. [76]. Similar concentrations of vitamins mentioned above were reported by the US Department of Agriculture in the most recent release of the USDA National Nutrient Database for Standard Reference, a major source of food composition data [77]. Analysis of vitamin C in different parts of the beetroot was conducted by Rosseto et al. [68]. According to this study, the highest concentration of ascorbic acid was found in the pulp of the beet and the lowest in its leaf. The concentrations of vitamin C, a well-known antioxidant, vary by plant part; in decreasing order of concentration, the parts are: pulp > skin > leaf > stalk for organic beet and pulp > stalk > skin > leaf for conventionally grown beet. This suggests that the method of cultivation influences the concentration of ascorbic acid in a specific part of the plant. Part-specific concentrations were observed in beets from organic and conventional cultivation. However, the highest level of vitamin C was detected in each part of plants fertilized with organic manure [68].

Ceclu and Nistor also showed the differences in the concentration of vitamins between beet leaves and tubers [8]. In the case of niacin, higher concentrations of vitamins A, B6, and C were detected in leaves, while folate was identified in much higher concentrations in tubers. Vitamins from the B group were also detected in beetroot juices [25]. This study also showed that organic juice is a better source of vitamins in comparison to those obtained from beets from conventional cultivation. It can be seen in the example of pyridoxal (B6), whose concentration in organic juice was about three times higher in comparison to other studied juices. Also, nicotinamide (B3) was detected in organic juice and in one juice from conventionally grown beetroots. In other juices, this vitamin was below the limit of detection (0.03 mg·L^{-1}). It is difficult to compare the obtained results with others in the literature as there is no information from what vegetables the juice was made. However, the published results focused on the analysis of vitamins in the vegetable rather than in the juice made from it. It should also be noted that possible discrepancies in the content of biologically active compounds and vitamins may result from different varieties or growing conditions.

7. Other Compounds

Nitrate (NO_3^-) is one of the most important inorganic compounds in red beets, and their ingestion provides a natural means of increasing in vivo nitric oxide (NO) [8]. This can prevent several pathologies such as hypertension and endothelial dysfunction [78,79]. For a long time, nitrates, which can be converted into nitrosamines, were thought to be harmful substances. These, in turn, lead to endocrinological diseases, defects in human fetuses or cancer [80]. Today it is known that red beetroot nitrates are one of the most important nutrients. In general, nitrates provide ergogenic and cardio-protective properties. Nitrates present in beetroots are converted into nitrite and nitric oxide, which are responsible for lowering blood pressure and vasoprotection [81]. Jonvik et al. reported that beetroot juice lowers blood pressure to a greater extent than sodium nitrate [82]. Many studies have focused on the impact of beetroot juice consumption on the body's efficiency, mainly during intense physical exercise.

Volino-Souza et al. performed a randomized, clinical, crossover, double-blind study which has shown that the consumption of 140 mL of juice from red beets improved macrovascular endothelial function [83]. However, no oxygen saturation parameters in muscle tissue of pregnant women were observed. Other studies showed that one week of supplementation with beetroot juice improved submaximal endurance and reduced blood pressure in older patients with heart failure and preserved ejection fraction [84]. Subsequent works emphasize the beneficial effect of beetroot supplementation on the body's efficiency during physical exertion, as well as its impact on the speed of regeneration [85,86].

The content of nitrates in beetroot depends on its variety. Wruss et al. analyzed nitrates content in seven beetroot varieties grown in Upper Austria [49]. The highest content was detected in the Mona Lisa variety (4626 $\pm$ 568 mg·L^{-1}), while the lowest was found in Robuschka (564 $\pm$ 129 mg·L^{-1}). The mean value across all analyzed cultivars was 1970 mg·L^{-1}, but the established standard deviation was 1395 mg·L^{-1}, which demonstrates the variability of nitrate content among beet varieties. Gallardo and Coggan studied the nitrate and nitrite content of beet juice products marketed to athletes [87]. The NO_3^- concentration of powders was significantly higher than that of concentrates (174 $\pm$ 63 µmol·g^{-1} and 70 $\pm$ 39 µmol·mL^{-1}, respectively). A lower concentration of nitrates was found in mixed drinks (13 $\pm$ 5 µmol·mL^{-1}) and bulk juices (18 $\pm$ 11 µmol·mL^{-1}). Authors highlighted that regardless of the type of product, there was considerable variability in NO_3^- concentration/content between products and, often, between samples of the same product [86]. The evaluation of nitrate and nitrite contents of beetroot from different regions of Brazil and the USA showed the highest nitrate (31.2 $\pm$ 0.010 mmol·L^{-1}) and nitrite (0.45 $\pm$ 0.005 mmol·L^{-1}) contents in US beets when compared to beetroots from Brazil [88]. In Brazil, Rio de Janeiro was the region that showed the highest nitrate content (17.1 $\pm$ 0.020 mmol·L^{-1}), while Rio Grande do Norte presented the highest nitrite content (0.13 $\pm$ 0.010 mmol·L^{-1}).

Red beetroots have more sugar than many other vegetables. According to Wruss et al., its average total content was 77.5 $\pm$ 10.2 g·L $^{-1}$ (7.8%) [49]. In the roots, sucrose was the most commonly identified sugar (94.8%), followed by glucose (3.3%) and fructose (1.9%). However, glucose was the major sugar in the shoots [89]. Red beetroot can be considered as a part of a healthy diet for diabetic patients. It was found that raw, red beetroot consumption by diabetes mellitus type 2 patients for 8 weeks, has beneficial impacts on cognitive function, glucose metabolism, and other metabolic markers [46]. The list of biologically active compounds in beets and beet products is presented in Table 1.

Unfortunately, in contrast to its health benefits, beetroots contain also oxalic acid; high levels of which lead to the production of kidney stones and their related negative health outcomes [90]. Moreover, oxalate, by chelating metal ions (such as magnesium and calcium), may inhibit their absorption. Thus, the consumption of red beetroots in large amounts may lead to negative health effects.

Table 1. The content of bioactive compounds in red beets and beet juices.

Compound	Concentration *	Sample	References
Betalains			
Betanin	128.7 ± 22.0	beet	[8]
	797 − 421.7	beet juices	[25]
	797 ± 24.0	Organic beet juice	[71]
	406 ± 17.0	Conventional beet juice	[71]
	705 ± 156	Beet extract	[49]
Vulgahantin I	321 − 432.1	Beet juices	[25]
	424 ± 16.0	Organic beet juice	[71]
	311 ± 13.0	Conventional beet juice	[71]
	397 ± 100	Beet extract	[49]
Flavonoids			
Myricetin	0.27 ± 0.091	Organic beets	[28]
	0.30 ± 0.109	Conventional beets	[28]
Luteolin	0.14 ± 0.004	Organic beets	[28]
	0.13 ± 0.003	Conventional beets	[28]
Quercetin	0.13 ± 0.017	Organic beets	[28]
	0.010 ± 0.009	Conventional beets	[28]
	0.0023	Fresh red beets	[47]
	0.009	Commercial juice	[47]
Epicatechin	3.20	Intact beet	[11]
	2.1 ± 0.100	Commercial juice	[25]
	0.253	Fresh beet	[47]
	0.202	Fermented beet	[47]
	0.034	Commercial juice	[47]
Catechin	0.715 ± 0.018	Commercial juice	[25]
	6.73 ± 0.031	Organic juice	[25]
Polyphenolic acids			
Gallic acid	36.40 ± 23.77	Organic beet	[28]
	65.93 ± 45.38	Conventional beet	[28]
	0.147± 0.008	Commercial juice	[25]
	1.24 ± 0.054	Organic juice	[25]
Chlorogenic acid	1.70 ± 0.55	Beet juice	[91]
	4.67 ± 3.67	Organic beet	[28]
	2.29 ± 2.09	Conventional beet	[28]
	1.80	Intact beet	[11]
Caffeic acid	2.22 ± 0.75	Beet juice	[91]
	2.40 ± 0.050	Commercial beet juice	[25]
	0.900 ± 0.008	Organic beet juice	[25]
	0.74 ± 0.40	Organic beet	[28]
	0.77 ± 0.28	Conventional beet	[28]
	3.70	Intact plant	[11]

Table 1. *Cont.*

Compound	Concentration *	Sample	References
Ferulic acid	0.120 ± 0.005	Commercial beet juice	[25]
	1.81 ± 0.062	Organic beet juice	[25]
	0.54 ± 0.37	Organic beet	[28]
	1.71 ± 0.76	Conventional beet	[28]
pHBA	1.2	Intact plant	[11]
	4.03 ± 0.053	Commercial beet juice	[25]
	6.83 ± 0.095	Organic beet juice	[25]
p-coumarc acid	5.27 ± 0.98	Beet juice	[91]
Sinapic acid	1.99 ± 0.80	Beet juice	[91]
Vitamin C	4.55 ± 2.16	Organic beet	[28]
	5.08 ± 2.10	Conventional beet	[28]
	4.36	Red beet	[67]
	7.20	Red beet	[8]
	4.90	Red beet	[76]
Vitamins B			
Riboflavin (B2)	0.034	Red beet	[67]
	0.040	Red beet	[76]
Nicotinamide (B3)	2.85 ± 0.064	Commercial beet juice	[25]
	2.43 ± 0.040	Organic beet juice	[25]
	0.334	Red beet	[76]
	0.030	Red beet	[67]
Pantothenic acid (B5)	2.49 ± 0.041	Commercial beet juice	[25]
	1.070 ± 0.047	Organic beet juice	[25]
	0.151	Red beets	[67]
Pirydoxal (B6)	1.420 ± 0.025	Commercial beet juice	[25]
	1.67 ± 0.038	Organic beet juice	[25]
	90	Red beets	[67]
	0.067	Red beet	[76]
Folate (B9)	0.109	Red beet	[8]
Selenium compounds			
Selenomethionine	0.56 ± 0.020	Organic beet juice	[71]
	0.20 ± 0.01	Conventional beet juice	[71]
Methylselenocysteine	0.08 ± 0.03	Organic beet juice	[71]
	0.20 ± 0.01	Conventional beet juice	[71]
Selenocysteine	0.27 ± 0.02	Conventional beet juice	[71]
Minerals			
Iron	2.57	Beetroot leaves	[8]
	0.80	Tubers	[8]
Copper	0.191	Leaves	[8]
	0.075	Tuber	[8]
Zinc	0.365 ± 0.015	Tubers	[8]
	0.38	Leaves	[8]
Magnesium	23.0	Tubers	[8]
	70	Leaves	[8]

Table 1. *Cont.*

Compound	Concentration *	Sample	References
Carotenoids			
A-carotene	22.0 ± 2.0	Tubers	[8]
	3.50 ± 0.5	Leaves	[8]
B-carotene	0.012	Leaves	[67]
	0.001	Leaves	[67]
Lycopene	0.030	Tubers	[8]
Lutein + zeaxanthin	0.001	Leaves	[8]

* Concentration in beets are expressed as mg per 100 g of raw material and mg·L^{-1} of juice.

8. Conclusions

Red beetroot (*Beta vulgaris*) is known and cultivated in many regions of the world. It is consumed as a component of salads, juices, and dietary supplements or as a natural food colorant. Many studies have been published on beetroots, however data on the bioactive compounds of fresh beets and beet products are still limited. The research described in this paper shows that this is a complicated issue as, depending on the variety and cultivation conditions, the content of individual compounds in beetroot can be significantly different. However, the research results show that red beetroot is a valuable source of several bioactive compounds such as betalains, flavonoids, phenolic acids, carotenoids, vitamins, and minerals, which are necessary for the proper functioning of the human body. It was proved that beetroots have a beneficial effect in the treatment of such serious diseases as diabetes, cardiovascular diseases, and cancer. All parts of the beet, such as the tuber, leaves, or seeds, have these properties. In summary, beetroot can be called a superfood and could be used as a potential material to develop innovative foods.

Author Contributions: Conceptualization, A.S. and K.P.; resources, A.S. and K.P.; writing—original draft preparation, A.S. and K.P.; writing—review and editing, A.S. and K.P.; supervision, K.P.; project administration, A.S. All authors have read and agreed to the published version of the manuscript.

Funding: This research received no external funding.

Institutional Review Board Statement: Not applicable.

Informed Consent Statement: Not applicable.

Data Availability Statement: Not applicable.

Conflicts of Interest: The authors declare no conflict of interest.

References

1. Montanari, M. *A Cultural History of Food in the Medieval Age*; Berg Publishing: London, UK, 2012.
2. Rupp, R. *How Carrots Won the Trojan War: Curious (But True) Stories of Common Vegetables*; Storey Pub.: North Adams, MA, USA, 2011.
3. Aykın-Dinçer, E.; Güngör, K.K.; Çağlar, E.; Erbaş, M. The use of beetroot extract and extract powder in sausages as natural food colorant. *Int. J. Food Eng.* **2020**, *17*, 75–82. [CrossRef]
4. Domínguez, R.; Munekata, P.E.S.; Pateiro, M.; Maggiolino, A.; Bohrer, B.; Lorenzo, J.M. Red Beetroot. A Potential Source of Natural Additives for the Meat Industry. *Appl. Sci.* **2020**, *10*, 8340.
5. Costa, A.P.D.; Hermes, V.S.; Rios, A.O.; Flôres, S.H. Minimally processed beetroot waste as an alternative source to obtain functional ingredients. *J. Food Sci. Technol.* **2017**, *54*, 2050–2058. [CrossRef] [PubMed]
6. European Food Safety Authority Panel on Food Additives and Nutrient Sources added to Food. Scientific Opinion on the re-evaluation of beetroot red (E 162) as a food additive. *EFSA J.* **2015**, *13*, 4318.
7. Chen, L.; Zhu, Y.; Hu, Z.; Wu, S.; Jin, C. Beetroot as a functional food with huge health benefits: Antioxidant, antitumor, physical function, and chronic metabolomics activity. *Food Sci. Nutr.* **2021**, *9*, 6406–6420. [CrossRef] [PubMed]
8. Ceclu, L.; Oana-Viorela, N. Red Beetroot: Composition and Health Effects—A Review. *J. Nutri. Med. Diet Care* **2020**, *6*, 043.
9. Fu, Y.; Shi, J.; Xie, S.Y.; Zhang, T.Y.; Soladoye, O.P.; Aluko, R.E. Red Beetroot Betalains: Perspectives on Extraction, Processing, and Potential Health Benefits. *J. Agric. Food Chem.* **2020**, *68*, 11595–11611. [CrossRef]

10. Teixeira da Silva, D.V.; dos Santos Baião, D.; de Oliveira Silva, F.; Genilton Alves, G.; Perrone, D.; Mere Del Aguila, E.; Flosi Paschoalin, V.M. Betanin, a Natural Food Additive: Stability, Bioavailability, Antioxidant and Preservative Ability Assessments. *Molecules* **2019**, *24*, 458. [CrossRef]
11. Georgiev, V.G.; Weber, J.; Kneschke, E.M.; Denev, P.N.; Bley, T.; Pavlov, A.I. Antioxidant activity and phenolic content of betalain extracts from intact plants and hairy root cultures of the red beetroot *Beta vulgaris* cv. Detroit dark red. *Plant Foods Hum. Nutr.* **2010**, *65*, 105–111. [CrossRef]
12. Madadi, E.; Mazloum-Ravasan, S.; Yu, J.S.; Ha, J.W.; Hamed Hamishehkar, H.; Kim, K.H. Therapeutic Application of Betalains: A Review. *Plants* **2020**, *9*, 1219. [CrossRef]
13. Moreno-Ley, C.M.; Osorio-Revilla, G.; Hernández-Martínez, D.M.; Ramos-Monroy, O.A.; Gallardo-Velázquez, T. Anti-inflammatory activity of betalains: A comprehensive review. *Hum. Nutr. Metab.* **2021**, *25*, 200126. [CrossRef]
14. Wijesinghe, V.N.; Choo, W.S. Antimicrobial Betalains. *J. Appl. Microbiol.* **2022**, *133*, 3347–3367. [CrossRef] [PubMed]
15. Tan, M.L.; Hamid, S.B.S. Beetroot as a Potential Functional Food for Cancer Chemoprevention, a Narrative Review. *J. Cancer Prev.* **2021**, *26*, 1–17. [CrossRef]
16. Lechner, J.F.; Stoner, G.D. Red Beetroot and Betalains as Cancer Chemopreventative Agents. *Molecules.* **2019**, *24*, 1602. [CrossRef] [PubMed]
17. Kapadia, G.J.; Rao, G.S.; Cheppail, R.; Akira, I.; Nobutaka, S.; Harukuni, T. Synergistic cytotoxicity of red beetroot (*Beta vulgaris* L.) extract with doxorubicin in human pancreatic, breast and prostate cancer cell lines. *J. Complement. Integr. Med.* **2013**, *10*, 113–122. [CrossRef]
18. Martinez-Rodrigez, P.; Guerrero-Rubio, M.A.; Henarejos-Escudero, P.; Garcia-Carmona, F.; Gandia-Herrero, F. Health-promoting potential of betalains in vivo and their relevance as functional ingredients: A review. *Trends Food Sci. Nutr.* **2022**, *122*, 66–82. [CrossRef]
19. Polturak, G.; Aharoni, A. "La Vie en Rose": Biosynthesis, Sources, and Applications of Betalain Pigments. *Mol. Plant* **2018**, *11*, 7–22. [CrossRef]
20. Sadowska-Bartosz, I.; Bartosz, G. Biological properties and applications of betalains. *Molecules* **2021**, *26*, 2520. [CrossRef]
21. Khan, M.I.; Giridhar, P. Plant betalains: Chemistry and biochemistry. *Phytochemistry* **2015**, *117*, 267–295. [CrossRef]
22. Delgado-Vargas, F.; Jimenez, A.R.; Paredes-Lopez, O.; Francis, F.J. Natural pigments: Carotenoids, anthocyanins, and betalains—Characteristics, biosynthesis, processing, and stability. *Crit. Rev. Food Sci. Nutr.* **2000**, *40*, 173–289. [CrossRef]
23. Sawicki, T.; Wiczkowski, W. The effects of boiling and fermentation on betalain profiles and antioxidant capacities of red beetroot products. *Food Chem.* **2018**, *259*, 292–303. [CrossRef] [PubMed]
24. Kujala, T.; Loponen, J.; Pihlaja, K. Betalains and phenolic acids in red beetroot (*Beta vulgaris*) peel extracts: Extraction and characterization. *Z. Für Nat. C* **2001**, *56*, 343–348. [CrossRef] [PubMed]
25. Sentkowska, A.; Pyrzynska, K. Zwitterionic hydrophilic interaction liquid chromatography coupled to mass spectrometry for analysis of beetroot juice and antioxidant interactions between its bioactive compounds. *LWT-Food Sci. Technol.* **2018**, *93*, 641–648. [CrossRef]
26. Sentkowska, A.; Pyrzynska, K. The influence of thiol addition on selenium stability and antioxidant activity of beetroot juice. *Appl. Sci.* **2022**, *12*, 12634. [CrossRef]
27. Herbach, K.M.; Stintzing, F.C.; Carle, R. Betalain Stability and Degradation—Structural and Chromatic Aspects. *J. Food Sci.* **2006**, *71*, R41–R50. [CrossRef]
28. Kazimierczak, R.; Siłakiewicz, A.; Hallmann, E.; Średnicka-Tober, D.; Rembiałkowska, E. Chemical composition of selected beetroot juices in relation to beetroot production system and processing technology. *Not. Bot. Horti Agrobot.* **2016**, *44*, 491–498. [CrossRef]
29. Yang, C.S.; Ho, C.T.; Zhang, J.; Wan, X.; Zhang, K.; Lim, J. Antioxidants: Different Meanings in Food Science and Health Science. *J. Agric. Food Chem.* **2018**, *66*, 3063–3068. [CrossRef]
30. Slimen, I.B.; Najar, T.I.; Abderrabba, M. Chemical and antioxidant properties of betalains. *J. Agric. Food Chem.* **2017**, *65*, 675–689. [CrossRef]
31. Taira, J.; Tsuchida, E.; Katoh, M.C.; Uehara, M.; Ogi, T. Antioxidant capacity of betacyanins as radical scavengers for peroxyl radical and nitric oxide. *Food Chem.* **2015**, *166*, 531–536. [CrossRef]
32. Fernando, G.S.N.; Sergeeva, N.N.; Frutos, M.J.; Marshall, L.J.; Boesch, C. Novel approach for purification of major betalains using flash chromatography and comparison of radical scavenging and antioxidant activities. *Food Chem.* **2022**, *385*, 132632. [CrossRef]
33. Mikołajczyk-Bator, K.; Czapski, J. Effect of pH Changes on Antioxidant Capacity and the Content of Betalain Pigments During the Heating of a Solution of Red Beet Betalains. *Pol. J. Food Nutr. Sci.* **2017**, *67*, 123–128. [CrossRef]
34. Kapadia, G.J.; Tokuda, H.; Konoshima, T.; Nishino, H. Chemoprevention of lung and skin cancer by *Beta vulgaris* (beet) root extract. *Cancer Lett.* **1996**, *100*, 211–214. [CrossRef]
35. Zhang, Q.; Pan, J.; Wang, Y.; Lubet, R.; You, M. Beetroot red (betanin) inhibits vinyl carbamate- and benzo(a)pyrene-induced lung tumorigenesis through apoptosis. *Mol. Carcinog.* **2013**, *52*, 686–691. [CrossRef] [PubMed]
36. Yin, Z.; Yang, Y.; Guo, T.; Veeraraghavan, V.P.; Wang, X. Potential chemotherapeutic effect of betalain against human non-small cell lung cancer through PI3K/Akt/mTOR signaling pathway. *Environ. Toxic* **2021**, *36*, 1011–1020. [CrossRef]

37. Martinez, R.M.; Longhi-Balbinot, D.T.; Zarpelon, A.C.; Staurengo-Ferrari, L.; Baracat, M.M.; Georgetti, S.R.; Sassonia, R.C.; Verri, W.A., Jr.; Casagrande, R. Anti-inflammatory activity of betalain-rich dye of *Beta vulgaris*: Effect on edema, leukocyte recruitment, superoxide anion and cytokine production. *Arch. Pharm. Res.* **2015**, *38*, 494–504. [CrossRef] [PubMed]
38. Macias-Ceja, D.C.; Cosín-Roger, J.; Ortiz-Masia, D.; Salvador, P.; Hernandez, C.; Esplugues, J.V.; Calatayud, S.; Barrachina, M.D. Stimulation of autophagy prevents intestinal mucosal inflammation and ameliorates murine colitis. *Br. J. Pharma.* **2017**, *174*, 2501–2511. [CrossRef]
39. Dai, R.; Wang, Y.; Wang, N. Betalain alleviates airway inflammation in an ovalbumin-induced-asthma mouse model via the TGF-β1/Smad signaling pathway. *J. Environ. Pathol. Toxicol. Oncol.* **2021**, *40*, 11–21. [CrossRef]
40. Wróblewska, M.; Juskiewicz, J.; Wiczkowski, W. Physiological properties of beetroot crisps applied in standard and dyslipidaemic diets of rats. *Lip. Health Dis.* **2011**, *10*, 178. [CrossRef]
41. Gao, Y.; Liang, X.; Tian, Z.; Ma, Y.; Sun, C. Betalain exerts cardioprotective and anti-inflammatory effects against the experimental model of heart failure. *Hum. Exp. Toxic.* **2021**, *40*, S16–S28. [CrossRef]
42. Rahimi, P.; Mesbah-Namin, S.A.; Ostadrahimi, A.; Abedimanesh, S.; Separham, M.A. Effects of betalains on atherogenic risk factors in patients with atherosclerotic cardiovascular disease. *Food Func.* **2019**, *10*, 8286–8297. [CrossRef]
43. Han, J.; Tan, C.; Wang, Y.; Yang, S.; Tan, D. Betanin reduces the accumulation and cross-links of collagen in high-fructose-fed rat heart through inhibiting non-enzymatic glycation. *Chem.-Biol. Inter.* **2015**, *227*, 37–44. [CrossRef] [PubMed]
44. Indumathi, D.; Sujithra, K.; Srinivasan, S.; Vinothkumar, V. Betanin exhibits significant potential as an antihyperglycemic and attenuating the glycoprotein components in streptozotocin–nicotinamide-induced experimental rats. *Toxic. Mech. Met.* **2018**, *28*, 547–554. [CrossRef] [PubMed]
45. Bell, T.; West, O.; Howatson, D.J.; Stevenson, E.J. The effects of beetroot juice supplementation on indices of muscle damage following eccentric exercise. *Euro. J. Appl. Phys.* **2016**, *116*, 353–362.
46. Aliahmadi, M.; Amiri, F.; Bahrami, L.S.; Hosseini, A.F.; Abiri, B.; Vafa, M. Effects of raw red beetroot consumption on metabolic markers and cognitive function in type 2 diabetes patients. *J. Diabetes Metab. Disord.* **2021**, *20*, 673–682. [CrossRef] [PubMed]
47. Płatosz, N.; Sawicki, T.; Wiczkowski, W. Profile of phenolic acids and flavonoids of red beet and its fermentation products. Does long-term consumption of fermented beetroot juice affect phenolics profile in human blood plasma and urine? *Pol. J. Food Nutr. Sci.* **2020**, *70*, 55–65. [CrossRef]
48. Guiné, R.P.F.; Gonçalves, F.; Lerat, C.; Idrissi, T.E.; Rodrigi, E.; Correia, P.M.R.; Gonçalves, J.C. Extraction of phenolic compounds with antioxidant activity from beetroot (*Beta vulgaris* L.). *Curr. Nutr. Food Sci.* **2018**, *14*, 350–357. [CrossRef]
49. Wruss, J.; Waldenberger, G.; Huemer, S.; Uygun, P.; Peter Lanzerstorfer, P.; Müller, U.; Höglinger, O.; Weghuber, J. Compositional characteristics of commercial beetroot products and beetroot juice prepared from seven beetroot varieties grown in Upper Austria. *J. Food Comp. Anal.* **2015**, *42*, 46–55. [CrossRef]
50. Kavalcová, P.; Bystrická, J.; Tomáš, J.; Kovarovič, J.; Lenková, M. The content of total polyphenols and antioxidant activity in red beetroot. *Potravinárstvo Sci. J. Food Ind.* **2015**, *9*, 77–83. [CrossRef]
51. Vulic, J.; Canadanovic-Brunet, J.; Cetkovic, G.; Tumbas, V.; Djilas, S.; Cetojevic-Simin, D.; Canadanovic, V. Antioxidant and cell growth activities of beet root pomace extracts. *J. Funct. Foods* **2012**, *4*, 670–678. [CrossRef]
52. Kujala, T.S.; Loponen, J.M.; Kika, K.D.; Pihlaja, K. Phenolics and betacyanins in red beetroot (*Beta vulgaris*) root: Distribution and effect of cold storage on the content of total phenolics and three individual compounds. *J. Agric. Food Chem.* **2000**, *48*, 5338–5342. [CrossRef]
53. Pratimasari, D.; Puspitasari, D. Identification of flavonoid compounds from purified extract of beetroot leaf (*Beta vulgaris* L.). *Indones. J. Glob. Health Res.* **2022**, *2*, 811–816.
54. Boyle, S.P.; Dobson, V.L.; Duthie, S.J.; Hinselwood, D.C.; Kyle, J.A.M.; Collins, A.R. Bioavailability and efficiency of rutin as an antioxidant: A human supplementation study. *Eur. J. Clin. Nutr.* **2000**, *54*, 774–782. [CrossRef] [PubMed]
55. Sentkowska, A.; Pyrzyńska, K. Simultaneous determination of vitamin B6 and catechins in dietary supplements by ZIC-HILIC chromatography and their antioxidant interactions. *Europ. Food Res. Technol.* **2020**, *246*, 1609–1615. [CrossRef]
56. Vulic, J.J.; Cebovic, T.N.; Canadanovic, V.M.; Cetkovic, G.S.; Djilas, S.M.; Canadanovic-Brunet, J.M.; Velicanski, A.S.; Cvetkovic, D.D.; Tumbas, V. Antiradical, antimicrobial and cytotoxic activities of commercial beetroot pomace. *Food Funct.* **2013**, *4*, 713–721. [CrossRef]
57. Kazimierczak, R.; Hallmann, E.; Treščinska, V.; Rembiałkowska, E. Estimation of the nutritive value of two red beet (*Beta vulgaris*) varieties from organic and conventional cultivation. *J. Res. Appl. Agric. Eng.* **2011**, *56*, 206–210.
58. Jabłońska-Ceglarek, R.; Rosa, R. Influence of green manures on the quantity and quality of the yield of red beet. *Hortorum Cultus* **2003**, *2*, 21–30.
59. Kazimierczak, R.; Hallmann, E.; Lipowski, J.; Drela, N.; Kowalik, A.; Püssa, T.; Matt, D.; Luik, A.; Gozdowski, D.; Rembiałkowska, E. Beetroot (*Beta vulgaris* L.) and naturally fermented beetroot juices from organic and conventional production: Metabolomics, antioxidant levels and anticancer activity. *J. Sci. Food Agric.* **2014**, *94*, 2618–2629. [CrossRef]
60. Rembiałkowska, E. Health and sensory quality of potatoes and selected vegetables from organic farms. *Acta Hortic.* **2003**, *604*, 473–477.
61. Sikora, M.; Klonowska, K.; Hallmann, E.; Rembiałkowska, E. Nutritive quality of red beet roots from organic and conventional production. In *The Impact of Organic Production Methods on the Vegetable Product Quality*; Rembiałkowska, E., Ed.; SGGW: Warsaw, Poland, 2010; pp. 209–220.

62. Szopińska, A.A.; Gawęda, M. Comparison of yield and quality of red beet roots cultivated using conventional, integrated and organic method. *J. Hortic. Res.* **2013**, *21*, 107–114. [CrossRef]
63. Dimić, D.; Milenković, D.; Marković Dimitrić, J.; Marković, Z. Antiradical activity of catecholamines and metabolites of dopamine:theoretical and experimental study. *Phys. Chem. Chem. Phys.* **2017**, *19*, 12970. [CrossRef]
64. Tosovic, J.; Marković, S.; Dimitrić Marković, J.M.; Mojović, M.; Milenković, D. Antioxidative mechanism in chlorogenic acid. *Food Chem.* **2017**, *237*, 390–398. [CrossRef]
65. Jeyanthi, R.; Sharmila, S.; Das, M.P.; Seshiah, C. Extraction and purification of carotenoids from vegetables. *J. Chem. Pharma. Res.* **2014**, *6*, 594–598.
66. Paliwal, H.; Goyal, S.; Singla, S.; Daksh, S. Pigments from natural sources: An overview. *Inter. J. Res. Pharma. Pharm. Sci.* **2016**, *3*, 1–12.
67. Odoh, U.E.; Okoro, E.C. Quantitative Phytochemical, Proximate/Nutritive Composition Analysis of *Beta vulgaris* Linnaeus (Chenopodiaceae). *Int. J. Curr. Res.* **2013**, *5*, 3723–3728.
68. Rossetto, M.R.; Vianello, F.; da Rocha, S.A.; Lima, G.P.P. Antioxidant substances and pesticide in parts of beet organic and conventional manure. *Afr. J. Plant Sci.* **2009**, *3*, 245–253.
69. Ismail, A.; Fun, C.S. Determination of vitamin C, β-carotene and riboflavin contents in five green vegetables organically and conventionally grown. *Malays. J. Nutr.* **2003**, *9*, 31–39. [PubMed]
70. Petek, M.; Custic, M.H.; Toth, N.; Slunjski, S.; Coga, L.; Pavlovic, I.; Karazija, T.; Lazarevic, B.; Cvetkovic, S. Nitrogen and crude proteins in beetroot (*Beta vulgaris* var. conditiva) under different fertilization treatments. *Not. Bot. Horti Agrobot.* **2012**, *40*, 215–219. [CrossRef]
71. Sentkowska, A.; Pyrzynska, K. Determination of selenium species in beetroot juices. *Heliyon* **2020**, *6*, e04194. [CrossRef] [PubMed]
72. Adadi, P.; Barakova, N.V.; Muravyov, K.Y.; Krivoshapkina, E.F. Designing selenium functional foods and beverages: A review. *Food Res. Inter.* **2019**, *120*, 708–772. [CrossRef] [PubMed]
73. Kieliszek, M. Selenium–fascinating microelement, properties and sources in food. *Molecules* **2019**, *24*, 1298. [CrossRef]
74. Rayman, M.P. Selenium intake, status, and health: A complex relationship. *Hormones* **2020**, *19*, 9–142. [CrossRef] [PubMed]
75. Sentkowska, A.; Pyrzynska, K. Investigation of antioxidant activity of selenium compounds and their mixtures with tea polyphenols. *Mol. Biol. Rep.* **2019**, *46*, 3019–3023. [CrossRef] [PubMed]
76. Neha, P.; Jain, S.K.; Jain, H.K.; Mittal, H.K.; doh, U.E.; Okoro, E.C. Chemical and functional properties of Beetroot (*Beta vulgaris* L.) for product development: A review. *Int. J. Chem. Stud.* **2018**, *6*, 3190–3194.
77. Haytowitz, D.B.; Ahuja Jaspreet, K.C.; Wu, X.; Somanchi, M.; Nickle, M.; Quyen, A.N.; Roseland, J.M.; Williams, J.R.; Patterson, K.Y.; Ying, L.; et al. USDA National Nutrient Database for Standard Reference, Legacy Release. Nutrient Data Laboratory, Beltsville Human Nutrition Research Center, ARS, USDA. 2019. Available online: https://data.nal.usda.gov/dataset/usda-national-nutrient-database-standard-reference-legacy-release (accessed on 30 May 2023).
78. Clifford, T.; Howatson, G.; West, D.J.; Stevenson, E.J. The potential benefits of red beetroot supplementation in health and disease. *Nutrients* **2015**, *7*, 2801–2822. [CrossRef]
79. dos Santos Baiao, D.; d'El-Rei, J.; Alves, G.; Neves, M.F.; Perrone, D.; Mere Del Aguila, E.; Flosi Paschoalin, V.M. Chronic effects of nitrate supplementation with a newly designed beetroot formulation on biochemical and hemodynamic parameters of individuals presenting risk factors for cardiovascular diseases: A pilot study. *J. Funct. Foods* **2019**, *58*, 85–94. [CrossRef]
80. Babarykin, D.; Smirnova, G.; Pundinsh, I.; Vasiljeva, S.; Krumina, G.; Agejchenko, V. Red beet (*Beta vulgaris*) impact on human health. *J. Biosci. Med.* **2019**, *7*, 61–79.
81. Webb, A.J.; Patel, N.; Loukogeorgakis, S.; Okorie, M.; Aboud, Z.; Misra, S.; Rashid, R.; Miall, P.; Deanfield, J.; Benjamin, N.; et al. Acute blood pressure lowering, vasoprotective, and antiplatelet properties of dietary nitrate via bioconversion to nitrite. *Hypertension* **2008**, *51*, 784–790. [CrossRef]
82. Jonvik, K.L.; Nyakayiru, J.; Pinckaers, P.J.M.; Senden, J.; van Loon, L.; Verdijk, L.B. Nitrate-rich vegetables increase plasma nitrate and nitrite concentrations and lower blood pressure in healthy adults. *J. Nutr.* **2016**, *146*, 986–993. [CrossRef]
83. Volino-Souza, M.; De Oliveira, G.; Alvares, T. A single dose of beetroot juice improves endothelial function but not tissue oxygenation in pregnant women: A randomised clinical trial. *Br. J. Nutr.* **2018**, *120*, 1006–1013. [CrossRef]
84. Eggebeen, J.; Kim-Shapiro, D.B.; Haykowsky, M.; Morgan, T.M.; Basu, S.; Brukaker, P.; Kitzman, D.W. One week of daily dosing with beetroot juice improves submaximal endurance and blood pressure in older patients with heart failure and preserved ejection fraction. *Heart Fail.* **2016**, *4*, 428–437. [CrossRef]
85. Woessner, M.N.; Van Bruggen, M.D.; Pieper, C.F.; Sloane, R.; Kraus, W.E.; Gow, A.J.; Allen, J.D. Beet the best? Dietary inorganic nitrate to augment exercise training in lower extremity peripheral artery disease with intermittent claudication. *Circ. Res.* **2018**, *123*, 654–659. [CrossRef] [PubMed]
86. Notay, K.; Incognito, A.V.; Millar, P.J. Acute beetroot juice supplementation on sympathetic nerve activity: A randomized, double-blind, placebo-controlled proof-of-concept study. *Am. J. Phys. Heart Circul. Physiol.* **2017**, *313*, H59–H65. [CrossRef] [PubMed]
87. Gallardo, E.J.; Coggan, A.R. What is in your beet juice? Nitrate and nitrite content of beet juice products marketed to athletes. *Int. J. Sport Nutr. Exerc. Metab.* **2019**, *29*, 345–349. [CrossRef]

88. dos Santos Baião, D.; Conte-Junior, C.A.; Paschoalin, V.M.F.; Alvares, T.S. Quantitative and comparative contents of nitrate and nitrite in *Beta vulgaris* L. by reversed-phase high-performance liquid chromatography-fluorescence. *Food Anal. Meth.* **2016**, *9*, 1002–1008. [CrossRef]
89. Oh, C.; Park, J.-E.; Son, Y.-J.; Nho, C.W.; Park, N.I.; Yoo, G. Light spectrum effects on the ions, and primary and secondary metabolites of red beets (*Beta vulgaris* L.). *Agronomy* **2022**, *12*, 1699. [CrossRef]
90. Bargagli, M.; Tio, M.C.; Waikar, S.S.; Ferraro, P.M. Dietary Oxalate Intake and Kidney Outcomes. *Nutrients* **2020**, *12*, 2673. [CrossRef]
91. Desseva, I.; Stoyanova, M.; Petkova, N.; Mihaylova, D. Red beetroot juice phytochemicals bioaccessibility: An in vitro approach. *Pol. J. Food Nutr. Sci.* **2020**, *70*, 45–53. [CrossRef]

applied sciences

MDPI

Article

Bioactive Substances and Microbiological Quality of Milk Thistle Fruits from Organic and Conventional Farming

Katarzyna Sadowska [1], Jadwiga Andrzejewska [1,*], Anna Ligocka [1], Joanna Korczyk-Szabo [2] and Miroslav Haban [2,3]

[1] Faculty of Agriculture and Biotechnology, Bydgoszcz University of Science and Technology, 7 Kaliskiego Ave., 85-796 Bydgoszcz, Poland; katarzyna.sadowska@pbs.edu.pl (K.S.); anna.ligocka@pbs.edu.pl (A.L.)
[2] Institute of Agronomic Sciences, Faculty of Agrobiology and Food Resources, Slovak University of Agriculture in Nitra, Tr. A. Hlinku 2, 94976 Nitra, Slovakia; joanna.korczyk-szabo@uniag.sk (J.K.-S.); miroslav.haban@uniag.sk (M.H.)
[3] Department of Pharmacognosy and Botany, Faculty of Pharmacy, Comenius University in Bratislava, Odbojarov 10, 83232 Bratislava, Slovakia
* Correspondence: jadwiga.andrzejewska@pbs.edu.pl

Abstract: The agricultural policy of the European Union is currently focused on increasing the area of organic farming. Medicinal plants, including milk thistle (*Silybum marianum* [L.] Gaertn.), are particularly suitable for this type of cultivation. The aim of this study was to compare milk thistle fruits from organic and conventional farming in terms of the content of silymarin and individual flavonolignans, oil content, microbiological purity, as well as antimicrobial activity of the silymarin extract, mainly in relation to microorganisms responsible for skin infections. The raw material of *Silybi mariani* fructus obtained from organic farming did not differ in terms of silymarin and oil content compared to the raw material from conventional cultivation. However, it differed in the composition of silymarin and the level of microbiological contamination. Raw material from organic farming was mostly characterized by a higher proportion of the sum of silydianin and silychristin in the silymarin complex than the sum of silybinin A and silybinin B. In the samples from conventional cultivation, only genotypes with a predominance of silybinins were present. Although the total number of microorganisms (TAMC) and yeasts and molds (TYMC) on fruit from organic farming were several times higher than on fruit from conventional farming, it was still within the standards set for food products. All raw materials were free of *Escherichia coli*, *Salmonella* spp. and *Listeria monocytogenes*. In addition, it was shown that the silymarin extract from organic farming was generally characterized by greater antimicrobial activity, especially in relation to *Staphylococcus aureus*, which is resistant and troublesome in skin infections.

Keywords: *Candida*; milk thistle; MRSA; silybinin; silydianin; *Staphylococcus aureus*

1. Introduction

Organic farming is the most pro-environmental method of agricultural production. This is a global trend that is a response to the growing awareness of consumers and the changing structure of demand on the market. Currently, the European Commission has set out an ambitious action plan for the further development of organic production by member states towards 25% of organic agricultural area by 2030 [1]. In European organic farming, the largest area of arable land is occupied by cereals [2]. Herbal (medicinal) plants, although they have a small percentage of the sown area, due to their properties and the possibility of cultivation without synthetic fertilizers and pesticides, perfectly fit into the idea of organic farming. Moreover, herb cultivation can significantly increase the economic efficiency of organic farms, especially since herbal companies are systematically expanding their offer with certified organic products.

Milk thistle (*Silybum marianum* [L.] Gaertn.) of the family Asteraceae belongs to a group of species that occupy one of the leading positions among medicinal plants, especially in Poland, Slovakia, and the Czech Republic. In Poland, the areas of milk thistle cultivation range from 1500 to 2000 ha. The herbal raw material of milk thistle are dry fruits (*Silybi mariani fructus*) which in the fruit cover accumulate unique active substances such as flavonolignans (silybinin A, silybinin B, isosilybin A, isosilybin B, silychristin, silydianin). The complex of these secondary metabolites is known as silymarin and is used in the production of medicines and dietary supplements intended mainly for people with liver ailments and diseases. However, the use of the herbal raw material of milk thistle and the use of whole plants is much wider and includes: culinary applications (cold-pressed oil rich in essential fatty acids (EFA), fruit additions to bread, edible sprouts), self-medication (consumption of defatted or non-defatted ground fruit as a source of silymarin), the use of fruits or whole plants in animal nutrition, production of cosmetics and bioenergy, soil phytoremediation [3].

A wide range of applications makes this plant very attractive for organic farms, as well as herbal companies processing such raw materials. However, the rules of organic farming differ fundamentally from the rules of conventional farming. Obtaining certification is conditioned primarily by the use of certified seeds and the non-use of synthetic mineral fertilizers and synthetic plant protection products [4]. However, the use of organic fertilizers and organic material can increase soil organic carbon and nutrients, and enhance crop yields [5]. These important elements of agricultural technology can also change the quality of herbal raw material, which is extremely important for products with medicinal or food values. The number of scientific studies on milk thistle cultivation and properties is relatively extensive, but the issues related to organic production are only fragmentary. They usually concern only selected elements of agricultural technology related to organic production, e.g., the impact of the use of organic fertilizers and some amendments on the yield and quality of the milk thistle raw material [6,7].

Scientific studies report that milk thistle fruits within the cultivars and individual plants differ in color, 1000 seed weight, germination capacity, and also in the content of fat and silymarin [8–11]. The proportion of individual flavonolignans in silymarin, similarly to the proportion of fatty acids in the oil, is genetically determined [8,11]. However, the content of silymarin, and to a lesser extent also the content of oil in milk thistle fruit is shaped by environmental and cultivation conditions. In a temperate climate, higher temperatures and a long growing season favor the accumulation of silymarin [7,12,13].

Silymarin, besides its well-known hepatoprotective properties, has also been shown to have antioxidant, antifibrotic, anti-inflammatory, choleretic, and immune-stimulating, regenerative, cytoprotective, cardioprotective, neuroprotective, anti-carcinogenic properties [14]. Recently, attention has also been paid to antimicrobial properties, especially in the context of the emergence of many antibiotic-resistant pathogens [15]. The importance of silymarin in the treatment of skin infections is also beginning to be appreciated [16–18]. The course of these infections is often long-lasting, and the use of synthetic drugs, including antibiotics, is not very effective. It appears that biologically active substances obtained from plants have as good or even better effects compared to synthetic drugs, and, in addition they do not cause microbial resistance, which often occurs during antibiotic therapy. Among the bacteria, *Staphylococcus aureus* is the most common cause of skin infections, including impetigo and cellulitis as well as ulcers and abscesses [19]. Increasingly, however, the cause of these infections is identified as a multidrug-resistant strain of MRSA (methicillin-resistant *Staphylococcus aureus*), insensitive to commonly used antibiotics [20,21]. In addition, emerging Gram-negative pathogens can cause fatal skin and soft tissue infections which is also a major threat [22]. Among the pathogenic yeasts, *Candida albicans* is most often isolated from skin lesions; however, new species are emerging among *Candida* genus that show less sensitivity and more resistance to antifungal drugs (e.g., *C. auris*, *C. parapsilosis*, *C. krusei*, *C. glabrata*, and *C. tropicalis*) [23,24]. In light of the above data, it seems reasonable to use natural antimicrobial substances, e.g., silymarin, for the production of ointments and/or creams.

The aim of this study is to compare milk thistle fruits from organic and conventional crops in terms of the content of silymarin and individual flavonolignans, oil content, microbiological purity, as well as antimicrobial activity of the silymarin extract, mainly in relation to microorganisms responsible for skin infections.

2. Materials and Method

Milk thistle fruits came from five organic and five conventional crops grown in temperate climate conditions. Three organic crops of unknown genotypes were located in Poland, the Mirel cultivar was grown in the Czech Republic, and the Silyb cultivar in Slovakia. The Silma cultivar and an unknown genotype were grown in two conventional crops in Poland, the Silma and Silyb cultivars were cultivated in Slovakia, and the Mirel cultivar in the Czech Republic (Table 1). Milk thistle was cultivated on medium-compact soils with a pH of 6.0–6.2 The average annual temperature in Poland is 9.5 °C, in Slovakia 14 °C, in the Czech Republic 13 °C, and the sum of precipitation in lowland areas in Poland is 500–600 mm, and in Slovakia and the Czech Republic it is 500–700 mm.

Table 1. Symbols, names, and places of cultivation of milk thistle.

Symbol of Genotype	Name of Genotype	Country of Cultivation	Latitude and Longitude
		organic farming	
1.OF	Unknown	Poland	50°26′ N, 16°39′ E
2.OF	Unknown	Poland	50°58′ N, 16°81′ E
3.OF	Unknown	Poland	50°58′ N, 16°39′ E
4.OF	Mirel	Czech	49°58′ N, 16°58′ E
5.OF	Silyb	Slovakia	48°19′ N, 17°43′ E
		conventional farming	
1.CF	Silma	Poland	51°17′ N, 18°48′ E
2.CF	Unknown	Poland	53°29′ N, 18°45′ E
3.CF	Silma	Slovakia	48°19′ N, 18°08′ E
4.CF	Mirel	Czech	49°58′ N, 16°58′ E
5.CF	Silyb	Czech	49°58′ N, 16°58′ E

After cleaning the fruit and drying it in a dryer at 50 °C, samples were taken in order to test the content of total silymarin and the composition of flavonolignans (100 g), fat (200 g), and microbiological purity (200 g).

2.1. *Determination of Silymarin and Oil Content*

The content of silymarin, expressed as silybinin in milk thistle fruit, was tested in accordance with the monograph of the Pharmacopea [25], *Silybi mariani fructus* (01/2014:1860) by high-performance liquid chromatography (HPLC), using a liquid chromatograph equipped with a UV detector, Agilent HP 1260 in Poznańskie Zakłady Zielarskie Herbapol S.A (Poznań Herbal Plants).

Total oil content was determined gravimetrically using cyclohexane as solvent. The analyzes were performed in the accredited Eurofins Polska laboratory.

2.2. *Dry Silymarin Extracts*

To obtain dry silymarin extracts, methanol extracts from milk thistle fruit from organic and conventional farming, with determined silymarin content, were evaporated on a Heidolph vacuum rotary evaporator. The dry extract was dissolved in 3 mL of DMSO (dimethyl sulfoxide). This procedure was carried out to demonstrate that the antimicrobial activity of silymarin is not due to the presence of methanol residues.

2.3. Microbiological Quality Assessment

Microbiological quality assessment of the raw material was carried out in the accredited GBA POLSKA laboratory and was made based on Polish or European standards:

1. Total aerobic mesophilic count (TAMC), plate method, depth inoculation [26];
2. Total yeast and mold count (TYMC), plate method, surface inoculation [27];
3. Number of ß-glucuronidase-positive *Escherichia coli*, plate method, depth inoculation; [28];
4. Number of coliforms, plate method, depth inoculation [29];
5. Detection of *Listeria monocytogenes* in 25 g, breeding method supplemented with biochemical tests [30];
6. Detection of *Salmonella* spp. in 25 g, breeding method supplemented with biochemical tests [31].

The test results were referred to the guidelines contained in the Commission Regulation (EC) [32] as amended, the Regulation of the Minister of Health [33], and to the recommendations of the World Health Organization [34].

2.4. Evaluation of the Antimicrobial Properties of Silymarin

Twenty-four common etiological agents of skin infections (three of each species) were studied: *Staphylococcus aureus*, Methicillin-Resistant *S. aureus* (MRSA), *Candida albicans*, *C. tropicalis*, *C. glabrata*, *C. krusei*, *Escherichia coli*, and *Pseudomonas aeruginosa*.

These were clinical strains obtained from the Department of Microbiology, Collegium Medicum, Nicolaus Copernicus University in Bydgoszcz. The microorganisms were grown on TSA medium for 24 h at 37 °C, then several single colonies of each strain were collected and suspensions with a density of 5×10^5 CFU/mL in BHI (Brain Heart Infusion) medium were prepared. The tests used silymarin solutions from organic (E) and conventional (K) production in DMSO (dimethylsulfoxide) at concentrations of 500 mg/mL. For tests checking their antimicrobial properties, they were diluted in BHI (at double concentration) to obtain solutions with concentrations of 250 mg/mL.

2.5. Minimum Inhibitory Concentration for Silymarin (MIC)

The minimum inhibitory concentrations of silymarin were determined by a serial broth microdilution method in BHI medium using U-bottom 96-well polystyrene microtiter plates (Profilab, 555.2.IS). A volume of 100 µL of BHI broth was added to each well of a microplate and 100 µL of the silymarin was used to conduct a two-fold serial dilution, giving concentrations of 250 to 1.8 mg/mL. Next, 10 µL of the bacterial or yeast suspension was added to the corresponding rows. Wells containing only microorganisms in BHI were used as growth control, and wells with 100 µL of DMSO in BHI with microorganisms were tested to exclude the antimicrobial effect of the solvent. As a sterility control, only growth medium without any microorganisms and silymarin was used. The cultures were incubated in a humid chamber at 37 °C for 24 h. Bacterial and fungal growth were evaluated according to turbidity. The tests were performed in triplicate. MIC was defined as the lowest concentration at which no growth was observed.

2.6. Minimum Bactericidal/Fungicidal Concentration for Silymarin (MBC/MFC)

The minimum bactericidal/fungicidal concentrations were determined based on broth dilution MIC tests by subculturing to selective agar media. Mannitol Salt Phenol Red Agar (Merck, 105404) was used to culture *S. aureus* and MRSA strains, Endo Agar (Merck, 1.04044) for *Escherichia coli*, Pseudomonas CN Selective Agar (Merck 107620) with Pseudomonas CN Selective Supplement (Merck, 1.07624) for *Pseudomonas aeruginosa*, and the Candida Chromogenic LAB-AGAR (Biomaxima, PS 665E) chromogenic medium for *Candida* spp. strains. The MBC/MFC is the lowest concentration of the agent that reduces populations by 99.9% of the tested bacteria/fungi.

2.7. Statistical Analysis

Statistical analysis was carried out for data on the content of silymarin, the percentage of individual components of silymarin (flavonolignans), and the oil content in milk thistle fruits. The obtained results were statistically processed in terms of equality of variances and normality of distributions. Differences between the means were compared using the Student's t-test at the significance level of $\alpha = 0.10$. This level was adopted due to the small sample size.

The number of microorganisms on milk thistle fruits was given as an average for organic and conventional farming.

3. Result and Discussion

The average content of silymarin in milk thistle grown in organic and conventional farming systems was very similar (1.69 and 1.82%, respectively) (Table 2). This level seems relatively low compared to the values reported in other publications [8,12]. The difference results from the method of presenting the analysis results. This study was guided by the definition of the herbal raw material of *Silybi mariani fructus* given by the Pharmacopea [25], where silymarin is expressed as silybinin, and its content should not be lower than 1.5%.

Table 2. Content of silymarin (% DM), proportion of flavonolignans (%) and oil content (% DM) in milk thistle fruits cultivated in organic and conventional farming.

Symbol of Genotype	Silymarin Expressed as Silybinin	Silychristin Silydianin	Silybinin A Silybinin B	Isosilybinin A Isosilybinin B	Oil Content
1.OF	1.88	45.9	36.3	17.8	23.10
2.OF	1.28	48.7	27.6	23.6	24.80
3.OF	1.67	53.0	24.1	22.9	23.15
4.OF	2.32	27.2	61.2	11.6	22.94
5.OF	1.29	34.6	51.8	13.6	22.60
Mean	1.69 [a]	41.9 [a]	40.2 [b]	17.9 [a]	23.32 [a]
1.CF	1.67	29.8	58.0	12.2	23.54
2.CF	1.29	30.5	58.3	11.2	24.00
3.CF	1.66	27.8	59.8	12.4	24.96
4.CF	1.88	28.3	59.1	12.6	23.25
5.CF	2.61	29.7	56.7	13.6	23.79
Mean	1.82 [a]	29.2 [b]	58.4 [a]	12.4 [b]	23.91 [a]

The means marked vertically with the same letter do not differ significantly at the level of $\alpha = 0.10$.

In silymarin from organic milk thistle fruits, a significantly higher share of silychristin and silydianin and total isosilybinins was found, and a significantly lower share of total silydianins than in silymarin from conventional milk thistle fruits. In conventionally cultivated genotypes, the dominant flavonolignans were silybinin A and B (Table 2).

In the group of genotypes cultivated in the organic system, there were two cultivars with a dominant content of silybinin (5.OF-Mirel and 6.OF-Silyb). The three remaining genotypes were characterized by a dominant content of silychristin and silydianin; they also contained more isosilybinin A and B genotypes than other genotypes. The proportion of flavonolignans in the composition of silymarin is genetically determined. Two chemotypes of *S. marinum* are distinguished in the literature—chemotype A, where the silymarin composition is dominated by silybinin, and chemotype B, where the silymarin composition is dominated by silydianin [8]. The cultivars grown in Poland, the Czech Republic, and Slovakia represent chemotype A. However, organic farms also cultivate forms representing chemotype B, including those from Ukraine and Bulgaria (personal information), which is reflected in the results of the presented study.

The cultivation system had no effect on the oil content of the milk thistle fruit (Table 2). The obtained results of the present study, in comparison with the results of other authors, also prove that the oil content in milk thistle fruit is slightly influenced by the genotype.

In temperate climates, as a rule, the oil content in milk thistle fruit is lower than in fruit obtained in warmer climates [8,11,13,35].

The therapeutic effectiveness of herbal raw materials depends on the content of bioactive ingredients. Many factors affect their concentration, including: growing conditions, processing, or storage, as well as microbiological contamination that may reduce or even inactivate their therapeutic activity [36]. Primary contamination of raw materials is caused mainly by spoiling bacteria, yeasts, and molds originating from the soil, irrigation water, air, animal waste, as well as organic fertilizers (e.g., crop residues). The latter may additionally stimulate the growth of soil microorganisms and thus increase the degree of contamination of raw materials [37]. The soil intended for plant production, especially in organic farms, is often fertilized with manure or compost, which can be a breeding ground for pathogens such as *Salmonella*, *Campylobacter*, *Listeria monocytogenes* and many others [38–40]. Secondary contamination of the raw material is associated with improper harvesting, lack of hygiene during drying and processing of harvested plants, storage, and transport, which applies to both organic and conventional farms [38,39].

Milk thistle fruits are the starting material for obtaining the extract which is the active substance in the medicinal product. In our study, organic fruit had significantly higher total microbial counts (TAMC) and yeasts and molds (TYMC) (Table 3). Bearing in mind the fulfillment of stringent standards for medicinal products, such a raw material may require additional decontamination processes before extraction [25].

Table 3. Number of microorganisms in samples of dried milk thistle fruit from organic and conventional farming.

Sample	TAMC [CFU·g^{-1}]	TYMC [CFU·g^{-1}]	Coliforms [CFU·g^{-1}]	*Escherichia coli* [CFU·g^{-1}]	*Salmonella* spp. [in 25 g]	*Listeria monocytogenes* [in 25 g]
			organic farming			
1.OF	2.5×10^6	1.0×10^4	$<1.0 \times 10^1$	$<1.0 \times 10^1$	ND	ND
2.OF	1.8×10^7	9.4×10^4	$<1.0 \times 10^1$	$<1.0 \times 10^1$	ND	ND
3.OF	9.5×10^6	6.0×10^3	$<1.0 \times 10^1$	$<1.0 \times 10^1$	ND	ND
4.OF	8.0×10^5	7.1×10^3	5.9×10^3	$<1.0 \times 10^1$	ND	ND
5.OF	5.2×10^6	8.8×10^4	$<1.0 \times 10^1$	$<1.0 \times 10^1$	ND	ND
Mean1-5.OF	72×10^5	4.1×10^4	118.8×10^1	$<1.0 \times 10^1$	ND	ND
			conventional farming			
1.CF	7.4×10^5	1.9×10^4	$<1.0 \times 10^1$	$<1.0 \times 10^1$	ND	ND
2.CF	2.4×10^6	2.0×10^4	$<1.0 \times 10^1$	$<1.0 \times 10^1$	ND	ND
3.CF	2.2×10^6	$<1.0 \times 10^1$	$<1.0 \times 10^1$	$<1.0 \times 10^1$	ND	ND
4.CF	6.4×10^5	4.1×10^2	$<1.0 \times 10^1$	$<1.0 \times 10^1$	ND	ND
5.CF	7.8×10^4	1.5×10^2	1.5×10^4	$<1.0 \times 10^1$	ND	ND
Mean1-5.CF	1.2×10^5	$7. 9 \times 10^3$	300.8×10^1	$<1.0 \times 10^1$	ND	ND

TAMC—total microbial count; TYMC—total yeasts and molds count; CFU—colony-forming unit; ND—not detected.

Whole or ground milk thistle fruits are also treated as food products with high health benefits. For food products, regulations on food safety, including microbiological quality, are applied [32–34]. This may have been due to the properly conducted process of drying the fruit right after harvesting, which inhibited the development of vegetative forms [41]. It should also be mentioned that companies or farms with organic certificates have the option of decontaminating biologically contaminated plant material, e.g., by ozonation.

Herbal plants, like agricultural and horticultural plants, can be a habitat of a large number of bacteria and fungi, including potential organisms that cause the decomposition of the raw material. The microbiological quality is evidenced by the results of the quantitative analysis of indicator microorganisms, which reflect the general condition of the raw material or the environment in which it is processed. TAMC (total aerobic mesophilic count), coliforms, and *E. coli* are commonly determined [42]. In the present study, the number of TAMCs and the number of yeasts and molds in milk thistle fruits from the organic production system were almost six times higher than in products from conventional farming (Table 3). Fungi are considered as indicators used to assess the raw material quality, in particular in the context of its spoilage and forecasting the shelf life. In scientific publications, it is often emphasized that milk thistle fruits and dietary supplements produced on their basis are inhabited by fungi, including species that accumulate mycotoxins. Mycotoxins are a threat for both production systems [14,43,44]. Excluding the use of chemicals for plant protection in organic farming may lead to increased contamination of products with fungi that produce these toxic compounds. In the present study, milk thistle fruits were not tested for the presence of mycotoxinogenic species; however, such a situation cannot be ruled out [45].

Isolation of coliforms and *E. coli*, which occur in the digestive tracts of warm-blooded animals, from food samples indicates fecal contamination and the potential presence of pathogens. In the present study, coliforms were present in only one in five samples of fruit from both organic and conventional cultivation, while the raw material from the conventional system contained 2.5 times more bacteria of this group. It is believed that organic farms using natural fertilizers are more exposed to soil and raw material contamination with fecal bacteria, while we obtained the opposite relationship. The number of *Escherichia coli* did not exceed the recommended limits in any fruit sample, and no pathogens of *Salmonella* ssp. or *Listeria monocytogenes* were found.

The assessment of the antimicrobial properties of silymarin solutions in DMSO was tested against microorganisms that are common etiological agents of skin infections—Gram-positive *Staphylococcus aureus* and MRSA, Gram-negative *Escherichia coli* and *Pseudomonas aeruginosa*, and yeasts of the genus *Candida*. The results are presented as the MIC and MBC/MFC of the compound in mg·mL^{-1} (Table 4). Silymarin is a complex of substances containing flavonolignans, mainly silybinin or silydianin. Silybinin has hepatoprotective and antimicrobial activity, especially against Gram-positive bacteria, because it inhibits the synthesis of their RNA and protein [46]. In our research, however, it was shown that against Gram-positive bacteria *S. aureus* and MRSA, a silymarin extract with a high content of silydianin and silychristin (organic raw material) was more effective than a silymarin extract with a predominance of the sum of silydianin (Tables 2 and 4). At the same time, silymarin did not inhibit the growth of Gram-negative bacteria—*E. coli* and *P. aeruginosa*, which is consistent with the reports of other researchers [46,47]. The microorganism most sensitive to its action was the methicillin-resistant strain of *S. aureus* (MRSA), for which the MIC/MBC was determined at the level of 15.6/31.25 mg·mL^{-1} for the product from the conventional system and 7.8/15.6 mg·mL^{-1} for the organic system. The antibacterial activity of silymarin against MRSA was also tested by Faezizadeh et al. [48] and the MIC of silymarin against the antibiotic-resistant strain was 500 mg/L, so it was much higher than in the present study. Such discrepancies may be the result of both different starting concentrations of silymarin and the use of bacteria that are less sensitive to the test compound.

Reports in the literature on the antifungal effect of silymarin confirm the effectiveness of the compound against both yeast and mold fungi [49]. Janeczko and Kochanowicz [50] studied the effect of silymarin against *Candida albicans*, *C. glabrata*, *C. parapsilosis*, *C. tropicalis*, and *C. krusei*, determining the MIC to be in the range from 30 to 300 µg/mL. In our study, silymarin was less active, its MIC value ranged from 20–255 mg·mL^{-1}. One yeast species, *Candida glabrata*, was not sensitive to even the highest concentration of the preparation.

The results of the present study indicate that the silymarin extract from organic milk thistle crops had a more effective antimicrobial effect than that from the conventional system. This was most clearly seen for *S. aureus*, whose development was inhibited 4 times more strongly by the organic product. The organic extract also turned out to be a more effective biocide against the MRSA strain. Two times higher anti-yeast activity of the organic extract was noticeable for *C. albicans* and *C. tropicalis*.

Table 4. Antimicrobial activity [mg·mL^{-1}] of silymarin extract from organic and conventional farming.

Microorganisms	Silymarin Extract from Organic Milk Thistle Crops		Silymarin Extract from Conventional Milk Thistle Crops	
	MIC	MBC/MFC	MIC	MBC/MFC
Staphylococcus aureus	62.5	125.0	250.0	250.0
MRSA	7.8	15.6	15.6	31.25
Escherichia coli	x	x	x	x
Pseudomonas aeruginosa	x	x	x	x
Candida albicans	125.0	125.0	125.0	250.0
Candida krusei	250.0	250.0	125.0	250.0
Candida glabrata	x	x	x	x
Candida tropicalis	125.0	125.0	250.0	250.0

MIC—Minimum inhibitory concentration; MBC/MFC—Minimum bactericidal/fungicidal concentration; x—no action.

Differences in the effectiveness of antimicrobial activity may result from the composition of silymarin. Among the raw materials from organic farming, three genotypes represented chemotype B with a predominance of silybinin and silychristin, while in the group of raw materials from conventional farming, all genotypes represented chemotype A with silybinin as the basic component. There are mentions in the literature of silydianin having a different action than silybinin [16]. According to Zielińska-Przyjemska and Wiktorowicz [51], silydianin is an effective inhibitor of free radicals production and release and this can partly explain the therapeutic use of silydianin as an anti-inflammatory drug. However, this issue requires further research. If the above hypothesis is confirmed, it will be justified to breed and register cultivars with a high content of silydianin in silymarin, adapted to the conditions of a temperate climate.

The present study brought the following new knowledge on the quality of milk thistle fruits:

- Herbal raw material from organic farming obtained in Central Europe may be characterized by a different composition of silymarin, and a high content of silychristin and silydianin in the composition of silymarin;
- Silymarin, especially genotypes with a high total content of silychristin and silydianin, is a promising raw material for production of ointments for skin infections;
- Despite the fact that milk thistle fruits from organic farming are more microbiologically contaminated than those from conventional crops, they meet the standards for food products, especially since *Escherichia coli*, *Salmonella* spp., and *Listeria monocytogenes* have not been found on them.

4. Conclusions

Milk thistle, due to its versatile use, is an interesting plant for organic farms. They can primarily offer food products such as whole or ground fruit or pressed oil. Raw material from some organic crops, compared to raw material from conventional crops, may differ in the composition of silymarin—a higher proportion of the sum of silydianin and silychristin than the sum of silybinins. Drying and storage processes carried out properly allow a raw material to be obtained in which the level of microbiological contamination (yeast, molds) is within the standards provided for food products. Importantly, this organic raw material was also found to be free of *Escherichia coli*, *Salmonella* spp., and *Listeria monocytogenes*. The antimicrobial properties of silymarin, especially the one rich in silydianin and silychristin, give an opportunity to expand the offer with certified ointments and creams used in cases of skin infection and damage. However, this still requires extended and in-depth research.

Author Contributions: Conception and design of the research, K.S., J.A. and A.L.; data acquisition K.S., J.A., J.K.-S. and M.H., methodology, K.S., J.A. and A.L., writing—original draft preparation, K.S., J.A., A.L., J.K.-S. and M.H., writing—review and editing, K.S. and J.A. All authors have read and agreed to the published version of the manuscript.

Funding: Bydgoszcz University of Science and Technology, Department of Agronomy, grant BN-WRiB-8/2022.

Informed Consent Statement: Not applicable.

Data Availability Statement: The data presented in this study are available on request from the corresponding author.

Acknowledgments: The authors would like to thank Poznańskie Zakłady Zielarskie Herbapol S.A. for a significant substantive contribution to the creation of this article.

Conflicts of Interest: The authors declare no conflict of interest.

References

1. European Commission. Communication from the Commission to the European Parliament, the Council, the European Economic and Social Committee and the Committee of the Regions. On an Action Plan for the Development of Organic Production. Brussels, 19 April 2021. Available online: https://eur-lex.europa.eu/resource.html?uri=cellar:13dc912c-a1a5-11eb-b85c-01aa75ed71a1.0003.02/DOC_1&format=PDF (accessed on 30 January 2023).
2. Eurostat. Eurostat Agricultural Data. 2021. Available online: https://ec.europa.eu/eurostat/databrowser/view/org_croppro/default/table?lang=en (accessed on 30 January 2023).
3. Andrzejewska, J.; Martinelli, T.; Sadowska, K. *Silybum marianum*: Non-medical exploitation of the species. *Ann. Appl. Biol.* **2015**, *167*, 285–297. [CrossRef]
4. Regulation (EU) 2018/848 of the European Parliament and of the Council of 30 May 2018 on Organic Production and Labelling of Organic Products and Repealing Council Regulation (EC) No 834/2007. Official Journal of the European Union, L 150/1. 14 June 2018. Available online: https://eur-lex.europa.eu/legal-content/EN/TXT/PDF/?uri=CELEX:32018R0848&from=PL (accessed on 30 January 2023).
5. Arunrat, N.; Kongsurakan, P.; Sereenonchai, S.; Hatano, R. Soil Organic Carbon in Sandy Paddy Fields of Northeast Thailand: A Review. *Agronomy* **2020**, *10*, 1061. [CrossRef]
6. Afshar, R.K.; Chaichi, M.R.; Asareh, M.H.; Hashemi, M.; Liaghat, A. Interactive efect of deficit irrigation and soil organic amendments on seed yield and favonolignan production of milk thistle (*Silybum marianum* L. Gaertn.). *Ind. Crops Prod.* **2014**, *58*, 166–172. [CrossRef]
7. Yadegari, H.; Khammari, I.; Fakheri, B.; Nouri, A.; Ebadi, M.-T. Flavonolignans of Milk Thistle (*Silybum marianum* L.) Seeds Affected by Fertilization Type and Plant Genotype. *IJHS* **2021**, *9*, 371–384.
8. Martinelli, T.; Potenza, E.; Moschella, A.; Zaccheria, F.; Benedettelli, S.; Andrzejewska, J. Phenotypic Evaluation of a Milk Thistle Germplasm Collection: Fruit Morphology and Chemical Composition. *Crop Sci.* **2016**, *56*, 3160–3172. [CrossRef]
9. Rosińska, A.; Dorna, H.; Szopińska, D.; Irzykowska, L.; Seidler-Łożykowska, K. Evaluation of milk thistle (*Silybum marianum* (L.) Gaertn.) seed germination in relation to seed health and seedling emergence. *Herba Pol.* **2018**, *64*, 1–10. [CrossRef]
10. Rosińska, A.; Dorna, H.; Szopińska, D.; Seidler-Łożykowska, K. The effect of colour grading of milk thistle (*Silybum marianum* (L.) Gaertn.) seeds on their quality for sowing. *Herba Pol.* **2017**, *63*, 7–19. [CrossRef]
11. Růžicková, G.; Fojtová, J.; Saŭcková, M. The yield and quality of milk thistle [*Silybum marianum* (L.) Gaertn.] seed oil from the perspective of environment and genotype—A pilot study. *Acta Fytotech. zootech.* **2011**, *1*, 9–12.
12. Andrzejewska, J.; Sadowska, K.; Mielcarek, S. Effect of sowing date and rate on the yield and flavonolignan content of the fruits of milk thistle (*Silybum marianum* L. Gaertn.) grown on light soil in a moderate climate. *Ind. Crops Prod.* **2011**, *33*, 462–468. [CrossRef]
13. Sadowska, K.; Andrzejewska, J.; Woropaj-Janczak, M. Effect of weather and agrotechnical conditions on the content of nutrients in the fruits of milk thistle (*Silybum marianum* L. Gaertn.). *Acta Sci. Pol. Hortorum Cultus* **2011**, *3*, 197–207.
14. Pickova, D.; Ostry, V.; Toman, J.; Malir, F. Food Supplements: A Review. Presence of Mycotoxins in Milk Thistle (*Silybum marianum*) Food Supplements: A Review. *Toxins* **2020**, *12*, 782. [CrossRef]
15. Wang, X.; Zhang, Z.; Wu, S.C. Health Benefits of *Silybum marianum*: Phytochemistry, Pharmacology, and Applications. *Agric. Food Chem.* **2020**, *68*, 11644. [CrossRef]
16. Abed, I.J.; Al-Moula, R.; Abdulhasan, G.A. Antibacterial Effect of Flavonoids Extracted from Seeds of *Silybum marianum* against Common Pathogenic Bacteria. *World J. Exp. Biosci.* **2015**, *3*, 36–39. [CrossRef]
17. Ali, B.S.M.; ElSayed, A.I.; Doheem, M.A.; Abo Eita, A.M.; Omar, A.A. Effect of Milk Thistle (*Silybum marianum* (L.) Gaertn) Seed Extract on Bacterial Activities and Growth of Human Liver Cancer Cells. *J. Biotech. Res.* **2020**, *6*, 27–33. [CrossRef]
18. Hu, X.; Qin, N.; Xue, J.; Li, S.; Huang, X.; Sun, J.; Xu, F.; Li, Z.; Li, D.; Hua, H. Dehydrodiconiferyl alcohol from *Silybum marianum* (L.) Gaertn accelerates wound healing via inactivating NF-kappaB pathways in macrophages. *J. Pharm. Pharmacol.* **2020**, *72*, 305–317. [CrossRef]

19. Miller, L.S.; Cho, J.S. Immunity against *Staphylococcus aureus* cutaneous infections. *Nat. Rev. Immunol.* **2011**, *11*, 505–518. [CrossRef]
20. Moran, G.J.; Amii, R.N.; Abrahamian, F.M.; Talan, D.A.D. Methicillin-resistant *Staphylococcus aureus* in community-acquired skin infections. *Emerg. Infect. Dis.* **2005**, *11*, 928–930. [CrossRef]
21. Eady, E.A.; Cove, J.H.J. Staphylococcal resistance revisited: Community-acquired methicillin resistant *Staphylococcus aureus*—an emerging problem for the management of skin and soft tissue infections. *Curr. Opin. Infect. Dis.* **2003**, *16*, 103–124. [CrossRef]
22. Ioannou, P.; Tsagkaraki, E.; Athanasaki, A.; Tsioutis, C.; Gikas, A. Gram-negative bacteria as emerging pathogens affecting mortality in skin and soft tissue infections. *Hippokratia* **2018**, *22*, 23.
23. Havlickova, B.; Czaika, V.A.; Friedrich, M. Epidemiological trends in skin mycoses worldwide. *Mycoses* **2008**, *51* (Suppl. S4), 2–15. [CrossRef]
24. Espinosa-Hernández, V.M.; Morales-Pineda, V.; Martínez-Herrera, E. Skin infections caused by emerging *Candida* species. *Curr. Fungal Infect.* **2020**, *14*, 99–105. [CrossRef]
25. Polskie Towarzystwo Farmaceutyczne. *Pharmacopea Polonica*, 7th ed.; Polskie Towarzystwo Farmaceutyczne: Warsaw, Poland, 2020; Volume II, pp. 1847–1848. (In Polish)
26. *PN-ISO 4833-2:2013-12*; Microbiology of the Food Chain. Horizontal Method for the Enumeration of Microorganisms—Part 2: Colony-Count at 30 Degrees C by the Surface Plating Technique. Polish Committee for Standardization: Warsaw, Poland, 2013.
27. *PN-ISO 21527-2:2009*; Microbiology of Food and Animal Feeding Stuffs. Horizontal Method for the Enumeration of Yeasts and moulds—Part 2: Colony Count Technique in Products with Water Activity Less Than or Equal to 0.95. Polish Committee for Standardization: Warsaw, Poland, 2009.
28. *PN-ISO 16649-2:2004*; Microbiology of Food and Animal Feeding Stuffs. Horizontal Method for the Enumeration of Beta-glucuronidase-positive Escherichia coli—Part 2: Colony-Count Technique at 44 C Using 5-Bromo-4-chloro-3-indolyl-D-glucuronide. Polish Committee for Standardization: Warsaw, Poland, 2004.
29. *PN-ISO 4832:2007*; Microbiology of Food and Animal Feeding Stuffs. Horizontal Method for the Enumeration of Coliforms—Colony-Count Technique. Polish Committee for Standardization: Warsaw, Poland, 2007.
30. *PN-EN ISO 11290-1:2017-07*; Food Chain Microbiology. Horizontal Method for the Detection and Enumeration of Listeria Monocytogenes and Other Listeria spp.—Part 1: Detection Method. Polish Committee for Standardization: Warsaw, Poland, 2017.
31. *PN-EN ISO 6579-1:2017-04*; Microbiology of the Food Chain—Horizontal Method for the Detection, Enumeration and Serotyping of Salmonella—Part 1: Detection of *Salmonella* spp. ISO: Geneva, Switzerland, 2017.
32. Food Safety Authority of Ireland. Commission Regulation (EC) No 2073/2005 of 15 November 2005 on microbiological criteria for foodstuffs. *Off. J. Eur. Union* **2005**, *50*, 1–26.
33. Regulation of the Minister of Health of 13 January 2003 on the maximum levels of chemical and biological contaminants that may be present in food, food ingredients, permitted additives, processing aids or on food. *J. Laws* **2003**, *37*, 326. (In Polish)
34. World Health Organization. *WHO Guidelines for Assessing Quality of Herbal Medicines with Reference to Contaminants and Residues*; World Health Organization: Geneva, Switzerland, 2007.
35. Fathi-Achachlouei, B.; Azadmard-Damirchi, S. Milk Thistle Seed Oil Constituents from Different Varieties Grown in Iran. *JAOCS* **2009**, *86*, 643–649. [CrossRef]
36. du Plessis-Stoman, D.; Downing, T.G.; Van de Venter, M.; Govender, S. Traditional herbal medicines: Potential degradation of sterols and sterolins by microbial contaminants. *S. Afr. J. Sci.* **2009**, *105*, 147–150. [CrossRef]
37. Ding, L.J.; Su, J.Q.; Sun, G.X.; Wu, J.S.; Wei, W.X. Increased microbial functional diversity under long-term organic and integrated fertilization in a paddy soil. *Appl. Microbiol. Biotechnol.* **2018**, *102*, 1969–1982. [CrossRef]
38. Alegbeleye, O.O.; Singleton, I.; Sant'Ana, A.S. Sources and contamination routes of microbial pathogens to fresh produce during field cultivation: A review. *Food Microbiol.* **2018**, *73*, 177–208. [CrossRef]
39. Gutierrez-Rodriguez, E.; Adhikari, A. Preharvest farming practices impacting fresh produce safety. *Microbiol. Spectrum* **2018**, *6*, PFS-0022-2018. [CrossRef]
40. Julien-Javaux, F.; Gérard, C.; Campagnoli, M.; Zuber, S. Strategies for the safety management of fresh produce from farm to fork. *Curr. Opin. Food Sci.* **2019**, *47*, 145–152. [CrossRef]
41. Schweiggert, U.; Carle, R.; Schieber, A. Conventional and alternative processes for spice production—A review. *Trends Food Sci Technol.* **2007**, *18*, 260–268. [CrossRef]
42. Ercole, C.; Del Gallo, M.; Mosiello, L.; Baccella, S.; Lepidi, A. *Escherichia coli* detection in vegetable food by a potentiometric biosensor. *Sens. Actuators B* **2003**, *91*, 163–168. [CrossRef]
43. Rosińska, A. The occurrence of fungi on the commercial dietary supplements based on the milk thistle (*Silybum marianum* (L.) Gaertn.) available on Polish market. *Herba Pol.* **2022**, *68*, 29–34. [CrossRef]
44. Tournas, V.H.; Sapp, C.; Trucksess, M.W. Occurrence of aflatoxins in milk thistle herbal supplements. *Food Addit. Contam. Part A* **2012**, *29*, 994–999. [CrossRef]
45. O'Riordan, M.J.; Wilkinson, M.G. A survey of the incidence and level of aflatoxin contamination in a range of imported spice preparations on the Irish retail market. *Food Chem.* **2008**, *107*, 1429–1435. [CrossRef]
46. Lee, D.G.; Kim, H.K.; Park, Y.; Park, S.C.; Woo, E.R.; Jeong, H.G.; Hahm, K.S. Gram-positive bacteria specific properties of silybin derived from *Silybum marianum*. *Arch. Pharm. Res.* **2003**, *26*, 597–600. [CrossRef]

47. Evren, E.; Yurtcu, E. In vitro effects on biofilm viability and antibacterial and antiadherent activities of silymarin. *Folia Microbiol.* **2015**, *60*, 351–356. [CrossRef]
48. Faezizadeh, Z.; Gharib, A.; Godarzee, M. In-vitro and in-vivo evaluation of silymarin nanoliposomes against isolated methicillin-resistant *Staphylococcus aureus*. *IJPR* **2015**, *14*, 627.
49. Yun, D.G.; Lee, D.G. Silymarin exerts antifungal effects via membrane-targeted mode of action by increasing permeability and inducing oxidative stress. *Biochim. Biophys. Acta Biomembr. BBA.* **2017**, *1859*, 467–474. [CrossRef]
50. Janeczko, M.; Kochanowicz, E. Silymarin, a popular dietary supplement shows anti–*Candida* activity. *Antibiotics* **2019**, *8*, 206. [CrossRef]
51. Zielińska-Przyjemska, M.; Wiktorowicz, K. An In vitro Study of the Protective Effect of the Flavonoid Silydianin against Reactive Oxygen Species. *Phytother. Res.* **2006**, *20*, 115–119. [CrossRef]

applied sciences

MDPI

Article

Strawberry (*Fragaria* × *ananassa*) and Kiwifruit (*Actinidia deliciosa*) Extracts as Potential Radioprotective Agents: Relation to Their Phytochemical Composition and Antioxidant Capacity

Margareth da Silva Ribeiro [1], Natividad Sebastià [2,3], Alegría Montoro [2,3] and Eva García-Martínez [1,*]

[1] Food Investigation and Innovation Group, Food Technology Department, Universitat Politècnica de València, Camino de Vera s/n, 46022 Valencia, Spain; margoribeiro2001@yahoo.com.br

[2] Servicio de Protección Radiológica, Hospital Universitario y Politécnico La Fe, Avenida Fernando Abril Martorell, 106, 46026 Valencia, Spain; natividad.sebastia@gmail.com (N.S.); montoro_ale@gva.es (A.M.)

[3] Biomedical Imaging Research Group GIBI230, Instituto de Investigación Sanitaria La Fe, Avenida Fernando Abril Martorell, 106, 46026 Valencia, Spain

* Correspondence: evgarmar@tal.upv.es

Abstract: Ionising radiation is an important form of treatment for human cancer; however, the side effects associated with oxidative damage caused by radiation compromise its effectiveness. This work aimed to quantify the major bioactive components of freeze-dried kiwifruit (KD) and strawberry (SD) extracts and assess their potential efficacy as radioprotective agents in human blood lymphocytes. Their possible genotoxic and cytotoxic effects were also evaluated. The study was conducted by pre-treating human lymphocytes with KD and SD (50, 400, and 800 μg/mL) before radiation at 2 Gy. The results showed that SD presented a higher antioxidant capacity (12.6 mmol Trolox equivalents/100 g db) and higher values of total phenolic compounds (2435 mg of gallic acid equivalents/100 g db), while KD had the highest vitamin C content (322 mg ascorbic acid/100 g db). Regarding phenolic compounds, pelargonidin-3-glucoside was the most abundant in SD (1439 mg/1000 g db) and quercetin-3-O-galactoside in KD (635 mg/1000 g db). None of the tested concentrations of both fruit extracts showed a genotoxic effect. SD (800 μg/mL) reduced the sister chromatid exchange frequency and mitotic index. The efficacy of KD (400 and 800 μg/mL) in lowering the dicentric chromosome frequency demonstrated its radioprotective activity.

Keywords: antioxidants; vitamin C; phenolic compounds; lymphocytes; genotoxicity; radioprotector

1. Introduction

The term ionising radiation refers to electromagnetic and particle radiation that is powerful enough to ionise atoms or molecules. For this reason, when passing through living tissues, a certain degree of damage may be produced. Both the well-known direct and indirect effects of ionising radiation cause this damage. The indirect impact is mainly responsible for the formation of reactive free radicals through the radiolysis of water, which react with macromolecules like DNA, proteins, and membranes so that they eventually induce cell damage or even dysfunction and death [1,2]. Since DNA is considered the primary target for ionising radiation damage, special attention has been paid to the effects on this molecule. If endogenous defence systems cannot effectively repair DNA lesions, this leads to genome instability and chromosome abnormalities. Radiation-induced genome aberrations play an important role in carcinogenesis caused by radiation.

Concerning radiation damage to humans, it is crucial to protect biological systems from radiation-induced damage [3]. To protect organisms from cellular and molecular damage during irradiation, radioprotectors agents act predominantly by enhancing antioxidant defence mechanisms by scavenging free radicals [2]. It has been suggested that administering radioprotective agents might lower radiation's harmful effects on cells. Antioxidants have the potential to function as radioprotectors by preventing the propagation of chain reactions initiated by free radicals and thus reducing certain DNA damage caused by ionising radiation [4].

Many natural and synthetic compounds have been examined in the recent past as radioprotector agents. However, some synthetic agents' inherent toxicity at radioprotective concentrations warrant further search for safer and more effective compounds [5]. Nowadays, attention is focused on natural compounds with a lower toxicity, more favourable administration routes, and improved pharmacokinetics than synthetic radioprotectors [6]. The genotoxic effects of natural compounds may be evaluated by using biomarkers. The sister chromatid exchange (SCE) test permits assessing the consequences of direct DNA damage. This assay also provides information on the genome's instability when exposed to potentially genotoxic substances. [7]. The mitotic index (MI) reflects the frequency of cell division, and is a reliable parameter to identify induced cytotoxicity [8]. For the analysis of proliferation kinetic index data (PI), the proportion of cells in the first, second, and third division in cell treatment cultures is commonly used [9].

As a result of their low toxicity and minimum adverse effects at the effective dosage, fruits or plants with antioxidant properties have been considered as an appropriate source of radioprotector agents [10]. The radioprotective effects of various phytochemicals such as resveratrol [11], chlorogenic and quinic acids [12], and phenolic compounds [2] have already been analysed. Additionally, some studies have assessed the radioprotective effects of extracts, for example, the ethanolic extract of propolis [13] and phenolic extracts of fruits [14].

The radioprotective efficiency of fruits is generally tested based on the reduction in ionising radiation damage on the peripheral blood lymphocytes as in vitro models. Enzymatic incorrect repair of ionising-radiation-induced DNA damage can result in large-scale rearrangements of the genome, such as translocations and dicentrics. Cell death, mutation, and neoplasia are just a few examples of the major phenotypic changes that these and other chromosomal exchange abnormalities may bring about. Radiation-induced chromosome aberrations (CAs) in lymphocytes are considered the most sensitive biological technique for measuring radiation exposure at doses over 0.1 Gy if an adequate number of metaphases are analysed [15].

Strawberries (*Fragaria × ananassa*) and kiwifruits (*Actinidia* sp.) have been thoroughly investigated for their health benefits [16]. Both fruits are an important source of micronutrients and bioactive compounds with antioxidant and health-promoting properties, including minerals, vitamins C (VC), and phenolic compounds [17,18]. In recent years, VC and polyphenols have been suggested to be the most important chemical species contributing to these fruits' functional properties, particularly those related to the antioxidant potential (AOP) [16,19]. The prevention of degenerative diseases such as various cancers, cardiovascular and neurological conditions, obesity, diabetes, and oxidative stress dysfunctions appears to be intimately linked to this antioxidant action [20]. Nevertheless, data concerning the radioprotective potential of kiwifruits and strawberries related to their antioxidant activity are limited.

With these considerations, the present work aimed to characterise the VC content, main phenolic compound content, and total polyphenol content (TPC) as foremost responsible for the AOP of kiwifruits and strawberries, and to evaluate the antioxidant effect of these fruits, as well as their genotoxic and radioprotective activities in human blood lymphocytes.

2. Materials and Methods

2.1. Chemicals and Reagents

Folin–Ciocalteau reagent, DPPH·(2,2-diphenyl-1-picryl-hydrazyl), ABTS (2,2-azinobis (3-ethylbenzothiaoline-6-sulfonic acid)), gallic acid, Trolox (6-hydroxy-2,5,7, 8-tetramethylchroman-2-carboxylic acid), iron (III) chloride ($FeCl_3 \cdot 6H_2O$), iron (II) sulfate ($FeSO_4 \cdot 7H_2O$), potassium persulfate, ellagic acid, and ferulic acid were purchased from Sigma-Aldrich (St. Louis, MO, USA). Pelargonidin-3-glucoside, Quercetin-3-glucoside, Quercetin-3-galactoside, and epicatechin were from Extrasynthese (Lyon, France). TPTZ

(2,4,6-tripyridyl-s-triazine) was purchased from Fluka Chemie (Buchs, Switzerland) and sodium carbonate was from J.T. Baker (Center Valley, PA, USA). DL-dithiothreitol (DTT), oxalic acid, and (+) ascorbic acid were from Scharlab S.L. (Barcelona, Spain). Unless otherwise stated, all chemicals employed were of analytical or superior quality. Methanol, ethanol, formic acid, and acetonitrile were of HPLC grade (Merck, Darmstadt, Germany). For cytogenetic analysis, Carnoy's lymphocyte fixative solution was prepared with methanol (Merck), acetic acid (Panreac, Barcelona, Spain) (3:1 *v*/*v*), bromodeoxyuridine (Sigma-Aldrich, Saint Louis, MO, USA), potassium chloride (Merck), antimitotic Colcemid (Gibco, Thermo Fisher Scientific, Waltham, MA, USA), bisbenzimideHoechst 33258 (Sigma), and Leishman dye (Merck).

2.2. Fruit Samples

Kiwifruit (*Actinidia deliciosa* cv.Hayward) and strawberry (*Fragaria* × *ananassa*) were bought in a local supermarket in Valencia, Spain. Fruit pieces were selected by visual appreciation based on a better physical appearance and similar colour, size, and firmness. The fruits were washed, and the navel and the peduncle parts were discarded. Then, they were cut into slices and triturated using a food processor (Thermomix TM 21 Vorwerk, Madrid, Spain). Kiwifruit was triturated with the peel. The pulp was homogenised and then freeze-dried. The obtained kiwifruit and strawberry puree were named K and S. They were characterised by the biochemical parameters and antioxidant activity, as described in Section 2.4, to be taken as control.

The purees were placed on aluminium pans (200 g in 0.5 cm thickness by pan) and frozen (CVF 545/86, Ing. Climas, Barcelona, Spain) at −40 °C for 48 h before freeze-drying (Telstar Lioalfa-6 Lyophyliser, Barcelona, Spain) at 10^{-2} Pa and −40 °C for 48 h. To obtain the powder, the freeze-dried fruit was ground in an electric mincer (Moulinette:320, Moulinex, Barcelona, Spain). The products thus obtained were vacuum packaged (Tecnotrip EVO86154, Barcelona, Spain) and stored in a desiccator with silica gel and silicone sealing protected from light until the preparation of the extracts (no more than 24 h). Freeze-dried kiwi and strawberry samples were designated as KD and SD, respectively.

2.3. Extracts Preparation

Fruit extracts used for quantifying bioactive compounds and for determining the AOP and genotoxic and radioprotective activities were obtained using a method adapted from Silva-Espinoza et al. (2022) [21]. Briefly, 0.5 g of sample mixed with a 9.5 mL mixture of methanol/water 70:30 (*v*/*v*) and homogenised at 400 rpm (Velp Scientifica, Usmate Velate, Italy) in the darkness for 20 min was centrifuged at 5867× *g* for 10 min at 10 °C (Selecta Medifriger-BL, Barcelona, Spain). The supernatants were collected, filtered (0.45 µm nylon filter, VWR, Radnor, PA, USA), and dried in a rotary evaporator (35 °C under vacuum, 20 min) to remove the organic solvent and then freeze-dried. Afterwards, the extract powder was collected and stored in hermetic opaque vials.

For analysis, dry extracts were reconstituted with the extraction solvent. For genotoxic and radioprotective activities, KD and SD extracts were screened at different concentrations based on previous studies [22]. In this case, dry extracts were reconstituted with methanol/water (70:30, *v*/*v*) to obtain the final concentrations in blood of 50, 400, and 800 µg/mL. For the rest of the analyses, reconstitution was until the initial volume before dehydration.

2.4. Chemical Determinations

Data are expressed as means and standard deviation of nine replicates (extractions in triplicate and measurements in triplicate) for chemical analysis. TP, individual phenolic compounds, VC, and AOP measured by the DPPH, FRAP, and ABTS methods were determined in kiwifruit samples (K and KD) and in strawberry samples (S and SD) before and after the freeze-drying process. Results were expressed per 100 g dry basis (db) for comparative purposes. Moreover, the changes in bioactive compounds caused by freeze-drying were calculated in reference to the fresh fruit's bioactive content (Equation (1), Table 1, and Figure 1).

$$\text{Changes in bioactive substances } (\%) = \frac{[BcF - BcE_x]}{[BcF]} \times 100 \quad (1)$$

where BcF is the bioactive compound content in the fresh fruit, K or S (mg/100 g db) and BcE_x is the bioactive compound content in each freeze-dried fruit extract (mg/100 g db).

2.4.1. Water Content

The water content (x_w) of fresh fruit was determined by vacuum drying (60 ± 1 °C, <100 mm Hg pressure, Vacioterm, J.P. Selecta, Barcelona, Spain) until constant weight [23]. For freeze-dried samples, x_w was evaluated using an automatic Karl Fischer titrator (C10S Mettler Toledo Compact Coulometric KF Titrator, Columbus, OH, USA).

2.4.2. VC

The process for determining VC involved converting dehydroascorbic acid to ascorbic acid (AA) using DTT as a reducing agent and ultra-high-performance liquid chromatography determination (UHPLC, Jasco equipment, Cremella, Italy) [21]. The conditions were Kromaphase 100-C18, 5 µm (4.6 × 250 mm) column (Scharlab S.L, Barcelona, Spain); mobile phase 0.1% oxalic acid, 20 µL volume injection, flow rate 1 mL/min, and detection at 243 nm and 25 °C (UV-visible detector MD-1510). An L (+) ascorbic acid calibration curve was used at 10–530 ppm intervals. The results were expressed in mg AA/100 g (db).

2.4.3. Phenolic Compounds

Three major phenolic compounds were identified and quantified using UHPLC (Jasco equipment, Cremella, Italy). Quercetin-3-O-galactoside, epicatechin, and ferulic acid were determined in kiwifruit and quercetin-3-glucoside, pelargonidin-3-glucoside, and ellagic acid were determined in strawberries.

Compound separations were achieved on a Synergi-C18, 4 µm, 150 × 4.6 mm, Hydro-RP column (Phenomenex, Madrid, Spain) (volume injected 10 µL, flow rate 1 mL/min) with 0.2% formic acid (A) and acetonitrile (B) used as mobile phases. The elution conditions were 10% B for 5 min and a linear gradient to 30% B in the subsequent 30 min [24].

Chromatograms were recorded at 280, 320, 360, and 510 nm (UV-visible detector MD-1510) at 25 °C. Anthocyanin (pelargonidin-3-glucoside) was quantified at 510 nm, flavonols (quercetins) at 360 nm, phenolic acids (ferulic acid and ellagic acid) at 320 nm, and flavan-3-ols (epicatechin) at 280 nm. Phenolics were identified by their retention time and they were quantified through the integration of the peak areas obtained from the corresponding standards chromatograms. The results were expressed as mg/1000 g db for each phenolic compound.

2.4.4. Total Polyphenols

TP content was determined spectrophotometrically according to the adapted Folin–Ciocalteau method [21]. The results were presented as mg gallic acid equivalent (GAE)/100 g db. All measurements were taken with a UV-visible spectrophotometer (Thermo Electron Corporation, Waltham, MA, USA).

2.4.5. AOP Evaluation

The AOP was determined by three complementary assays: ABTS, DPPH, and FRAP. Results of ABTS and DPPH were expressed as mmol Trolox Equivalent (TE)/100 g db using the corresponding Trolox calibration curves. For FRAP methanolic solutions of known Fe(II) concentrations, $FeSO_4 \cdot 7H_2O$ was used for calibration and results were expressed as mmol Fe^{2+}/100 g db. All measurements were taken with a UV-visible spectrophotometer (Thermo Electron Corporation, Waltham, MA, USA).

Firstly, the AOP was assessed by using the free radical scavenging activity of the stable radical 2,2-diphenyl-1-picryl-hydrazyl-hydrate (DPPH) [21]. Absorbance was recorded at 517 nm at the initial time ($A_{initial}$) and 15 min later (A_{sample}) once the reaction had achieved a steady state.

The percentage of DPPH· (%DPPH·) was calculated by Equation (2):

$$\%DPPH = \frac{\left(A_{initial} - A_{sample}\right)}{A_{initial}} \times 100 \quad (2)$$

For the ABTS assay, the procedure followed the spectrophotometric method of Re et al. (1999) [25] and the percentage inhibition of absorbance at 734 nm was calculated.

The FRAP assay was performed according to Silva-Espinoza et al. [21] with some modifications. The FRAP reagent was prepared by combining 25 mL of 0.3 M acetate buffer (pH 3.6), 2.5 mL 10 mM TPTZ (2,4,6-tripyridyl-s-triazine) solution in 40 mM HCl, and 2.5 mL of 20 mM $FeCl_3 \cdot 6H_2O$. The resulting mixture was heated at 37 °C before use. Fruit extracts (30 µL) were allowed to react with 900 µL of the FRAP solution. The absorbance was measured at 593 nm.

2.5. Biological Determinations

2.5.1. Culture Conditions

Culture conditions were implemented for the study of both genotoxicity and radioprotection biomarkers in accordance with Sebastià et al. (2013) [11] and Montoro et al. (2012) [13]. Heparinised human peripheral blood samples were used for both biomarkers of genotoxicity and radioprotection, and were treated with 250 µL of a methanol/water solution of KD and SD to obtain final concentrations in blood of 50, 400, and 800 µg/mL. The selection of the tested concentrations was made based on previous research work and other in vitro studies of the fruit extracts selected in this study [11,13,26–28]. For each treatment, separate cultures were established by mixing 0.75 mL of whole blood with 5 mL of PB-MaxTM Karyotiping medium and incubating for 48 and 72 h at 37 °C. To exclusively analyse first-division metaphases, a final concentration of 12 µg/mL of bromodeoxyuridine was present since the setting up of the cultures. An amount of 150 µL of Colcemid was added 2 h before harvesting the 48 h and 72 h cultures to stop the cell culture in the metaphase.

The study received approval from the Ethics Committee of the Hospital Universitario y Politécnico La Fe, and samples of human peripheral blood were collected following the consent exemption report.

2.5.2. Irradiation Conditions for Radioprotective Activity

Human peripheral blood samples containing lithium heparin as an anticoagulant were obtained using sterile Vacutainer tubes (Becton, Dickinson and Company, Franklin Lakes, NJ, USA). An amount of 250 µL of different concentrations of KD and SD was added to 12 mL of blood samples, which were collected 1 h prior to irradiation and kept at 37 °C in a water bath. The samples were then irradiated with a dose of 2 Gy using a linear accelerator (LINACC linaciX, Varian Medical System, Palo Alto, CA, USA) located at the Hospital Universitario y Politécnico La Fe (Valencia, Spain). To facilitate the repair mechanisms, incubation at 37 °C for 1 h immediately after irradiation was necessary. As controls, two tubes were irradiated with a dose of 2 Gy: one containing only peripheral blood and the other containing peripheral blood along with 250 µL of methanol/water (70:30, *v*/*v*) (solvent control). The irradiation conditions were established in accordance with the recommendations of the International Atomic Energy Agency [29].

2.5.3. Stain Technique

The stain technique was previously described by Sebastià et al. (2013) [11] and Montoro et al. (2012) [13]. Two- to three-day-old slides were stained with Fluorescence plus Giemsa stain technique. In summary, the old slides were treated with Hoechst 33,258 for 20 min at room temperature. Subsequently, the slides were washed with distilled water and covered with 2× saline-sodium citrate (SSC) before being exposed to UV light (300 W) for 2 min. After washing the slides with water and allowing them to dry for 30 min, they were stained with Leishman (Merck) for 5 min, enabling the identification of first-, second-, and third-division metaphases.

2.5.4. Cytogenetic Analysis

Sister Chromatid Exchange

The frequency of sister chromatid exchanges (SCE), as a genotoxicity biomarker, was determined by analysing 50 s division metaphases for each treatment (Figure S1, Supplementary Material). The assessment of SCE scores in lymphocytes involved recording the total number of exchanges in the analysed cells for each treatment to establish their frequency (Y_{SCE}). Additionally, we recorded the number of chromosomes with one (Y_{1SCE}), two (Y_{2SCE}), or three (Y_{3SCE}) chromatid exchanges in the total number of analysed cells for each treatment.

Evaluation of MI and PI

For each concentration tested, MI and PI as cytotoxicity biomarkers were analysed. According to Rojas et al. (1992) [30], the mitotic index was determined by exclusively considering the proportion of metaphases in 500 cells during the first-division metaphases.

Equation (3) was used to evaluate the relative mitotic index (RMI) [30]:

$$RMI = \frac{\text{MI treated}}{\text{MI control}} \times 100 \quad (3)$$

To evaluate the PI, the proportion of cells in the first (M1), second (M2), and third (M3) division was obtained from 100 consecutive metaphases for each concentration (Figure 1). The proliferation index (PI) was calculated using Equation (4), and the relative proliferation index (RPI) was evaluated using Equation (5) [30].

$$PI = \frac{M1 + 2M2 + 3M3}{100} \quad (4)$$

$$RPI = \frac{\text{PI treated}}{\text{PI control}} \times 100 \quad (5)$$

Chromosomal Aberrations Assays

The analysis of CAs for assessing radioprotective capacity was carried out via the dicentric chromosomes assay since this biomarker is the gold standard technique for this purpose [29].

Chromosomal analysis was performed on first-division metaphases, which consisted of 46 centromeres in a total of two hundred cells (Figure S1, Supplementary Material). Chromosomal aberrations were categorized into dicentric chromosomes (Dic) and rings, with the latter being considered only when an acentric fragment was present. Extra acentric fragments were classified as acentric fragments that were not associated with dicentric and ring chromosomes. Other abnormalities such as chromatid breaks and gaps were also taken into account.

2.6. Statistical Analysis

All the results were expressed as mean ± standard deviation values. An analysis of variance was used to calculate significant differences among samples (one way ANOVA). Differences of $p < 0.05$ were considered as significant. Means separation was performed according to a Student's *t*-test. Correlation was assessed using Spearman's rank correlation coefficient for genotoxicity and radioprotection analyses. All statistical analyses were performed using SPSS (Statistical Package for Social Sciences) version 10.0 for Windows.

The Poisson distribution was assessed using the dispersion index test statistic U (variance/mean). A value of U greater than 1.96 indicates over-dispersion at the 5% significance level.

3. Results and Discussion

3.1. Bioactive Compounds and Antioxidant Activity

The water content of the fruit batch used in this study was x_w 85.8 ± 0.3 g/100 g for K and 89.9 ± 1.3 g/100 g for S and decreased to around 2.5 ± 0.1 g/100 g for the corresponding freeze-dried samples. The concentration of each phytochemical component analysed and the AOP of kiwifruit and strawberry samples are shown in Table 1 and Figure 1. In general, the TP of kiwifruit samples was lower than in strawberries. The initial TP of fresh fruit samples varied from 349 to 2237 mg GAE/100 g db in K and S, respectively. K presented values similar to those reported by Du et al. (2009) [31], and S showed higher TP than those obtained in other studies with different strawberry varieties [32]. This variability can be attributed to the different fruit variety or states of maturity and the different solvents used in the extraction process.

Table 1. Total phenolic content (TP), vitamin C (VC), and antioxidant activity (measured by DPPH, ABTS, and FRAP methods) of fresh kiwifruit (K) and strawberry (S) samples and freeze-dried kiwifruit (KD) and strawberry (SD) samples.

Sample	TP (1)	VC (2)	DPPH (3)	ABTS (3)	FRAP (4)
K	349 ± 3 ^{a}	359 ± 10 ^{b}	0.84 ± 0.16 ^{a}	2.2 ± 0.5 ^{a}	4.41 ± 0.08 ^{a}
KD	371.6 ± 1.2 ^{b}	322 ± 5 ^{a}	0.90 ± 0.06 ^{a}	2.3 ± 0.2 ^{a}	6.24 ± 0.19 ^{b}
	−6.48%	10.31%	−7.24%	−4.55%	−41.50%
S	2237 ± 6 ^{y}	188.5 ± 1.3 ^{z}	5.8 ± 0.4 ^{z}	16.2 ± 0.9 ^{z}	17.2 ± 0.3 ^{z}
SD	2435 ± 11 ^{z}	176 ± 3 ^{y}	4.4 ± 0.2 ^{y}	12.6 ± 1.2 ^{y}	13.6 ± 0.5 ^{y}
	−8.85%	6.63%	24.14%	22.22%	20.43%

Values are expressed as means ± standard deviation. For each fruit, the same letters in a column indicate homogeneous groups according to a Student's *t*-test at $p > 0.05$ (ANOVA). Percentages indicate the changes caused by the freeze-drying process (Equation (1)). (1) mg of gallic acid equivalents/100 g dry basis (db), (2) mg ascorbic acid/100 g db, (3) mmol trolox equivalents/100 g db, (4) mmol Fe^{2+}/100 g db.

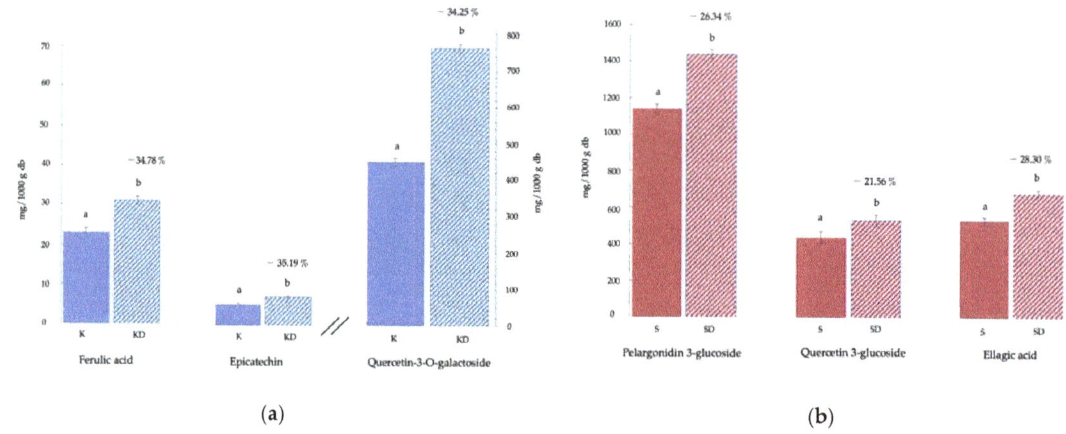

Figure 1. Polyphenol compound content (mg/100 g dried basis): (**a**) kiwifruit: fresh (K) and freeze-dried (KD); (**b**) strawberry: fresh (S) and freeze-dried (SD). Results are expressed as mg per 1000 g dry basis (db). For each fruit and each compound, different letters indicate different homogeneous groups according to a Student's *t*-test at $p > 0.05$ (ANOVA) between samples. Percentages indicate the changes in polyphenol compounds caused by the freeze-drying process (Equation (1)).

Generally, plant preparations are marketed in the form of liquid extracts or as powders resulting from the drying of plant material or the liquid extracts themselves. Solid forms (powders) of plants are being used increasingly because they offer numerous advantages over liquid extracts: ease of standardization; greater physical, chemical, and biological stability; lower transport and storage costs; and the possibility of achieving higher concentrations. For all these reasons, dehydration of samples is a crucial step in order to achieve a product that is suitable for industrial use. However, it is important to bear in mind that dehydration, to some extent, can affect the activity and stability of bioactive compounds due to chemical, enzymatic, and enzymatic degradation and losses through volatilisation and/or thermal degradation. The freeze-drying technique was chosen in this study because of its benefits over other dehydration techniques [21]. The comparison of K and S data with the corresponding KD and SD extracts showed that the freeze-drying process favoured ($p < 0.05$) the extraction of phenolic compounds. An increase in TP up to 6.48% and 8.85% was observed for kiwi and strawberry, respectively (Equation (1)). This rise could be explained based on the breaking of the remaining cellular structure of the fruit puree due to the ice crystals formed during the freezing step. In this way, the subsequent entry of the solvent is facilitated, and consequently, the extraction of the TP may be improved. The high porosity characteristic of freeze-dried products can also contribute in this sense. Other studies have also shown that the freeze-drying process increases the amount of phenolic compounds [33,34]. Freeze drying caused a reduction in the particle size of the sample, which favoured the contact surface with the solvent and could cause an increase in yield extraction. TP are considered important contributors to AOP in fruits. This activity may be related to several properties such as redox characteristics, which allow them to act as reducing agents, hydrogen donors, singlet oxygen quenchers, and even metal chelators [35].

Spectrophotometric techniques are helpful for estimating the TP and the antioxidant capacity. However, more selective techniques, such as chromatographic ones, must be utilised to determine the individual polyphenolic profile. Figure S2 (Supplementary Material) illustrates an example of an obtained chromatogram, and Figure 1 shows the content of each of the phenolic compounds identified in the fresh fruits and in the corresponding KD (Figure 1a) and SD (Figure 1b) methanolic extracts. The percentage of changes due to the freeze-drying process (Equation (1)) has also been included. The molecules were classified as different phenolic compounds: phenolic acids (ferulic and ellagic acid), flavonols (quercetin and epicatechin), and anthocyanins (pelargonidin-3-glucoside). Specifically, the most abundant phenolic compounds found in K were quercetin-3-O-galactoside (473 ± 7 mg/1000 g db), ferulic acid (23 ± 2 mg/1000 g db), and epicatechin (5.4 ± 0.9 mg/1000 g db) and in S, they were pelargonidin-3-glucoside (1139 ± 24 mg/1000 g db), ellagic acid (530 ± 17 mg/1000 g db), and quercetin-3-glucoside (436 ± 33 mg/1000 g db). Generally, the phenolic quantities obtained are in accordance with those described in other studies [24,36]. As described for TP, freeze-dried fruit extracts showed higher concentrations of these components than fresh fruits ($p < 0.05$). Particularly, phenolics increased by ≈35% for kiwifruit and 21–28% for strawberry in relation to fresh samples (Equation (1)). Concerning the kiwifruit phenolic composition, Jiao et al. (2018) [37] established the significance of certain phenolic compounds in kiwifruit methanolic extracts as antioxidants. According to their study, the sequence of the phenolic components that contribute to reducing power is as follows: epicatechin > vanillic acid > *p*-coumaric acid > ferulic acid > syringic acid > caffeic acid > catechin. The chemical antioxidant activity of kiwifruit measured in vitro is mainly due to compounds such as phenolic acids and flavonoids which can easily donate electrons by virtue of their aromatic structures [37]. Among the major phenolic components of kiwifruit, quercetin has been reported as a potent antioxidant with radioprotective abilities [38]. Additionally, ferulic acid has been shown to be a powerful antioxidant capable of scavenging hydroxyl and peroxyl radicals as well as nitric oxide radicals. It was revealed that ferulic acid pre-treatment significantly reduced the inflammatory response associated with ionising radiation in murine models [38]. On the other hand, strawberries are also considered a rich source of polyphenols, among them, anthocyanins are quantitatively the most significant ones [20].

Kiwifruits and strawberries may also be highlighted as important sources of VC [20,31]. Along with polyphenolic compounds, VC is considered the most significant water-soluble antioxidant and it intervenes in many physiological reactions. It is effective in scavenging superoxide radical anions, hydrogen peroxide, hydroxyl radicals, singlet oxygen, and reactive nitrogen oxide [35]. The physiological and biochemical effects of VC are a result of its role as an electron donor. Specifically, AA donates two electrons from a double bond between the second and third carbons of the 6-C molecule. In the present study, VC values of K were higher than those of S (Table 1). Some authors reported that the VC in kiwifruit is higher than other fruits such as oranges, strawberries, lemons, apples, peaches, or grapes [39]. In this case, freeze-drying caused significant losses ($p < 0.05$) of VC by 10.31% and 6.63% in KD and SD, respectively, compared to fresh fruit (Equation (1)). When evaluating the VC content, it is important to consider that the molecule is very labile, thermosensitive, and easily oxidisable, and thus it is easily degraded [40]. In this sense, the freeze-dried sample's increased porosity makes VC more exposed to oxygen, which could adversely affect its stability [34]. Amounts of VC in KD and SD found in this study were comparable to those described by Leontowicz et al. (2016) [41] and Nemzer et al. (2018) [42], respectively.

The AOP of fruits is crucial for determining their functional value. Over the last few decades, fruit antioxidants have attracted growing interest, suggesting that their consumption may help prevent and reduce oxidative reactions that adversely affect human health, contributing to the development of chronic diseases and cancers in various ways. Since the AOP is due to synergistic reactions between different compounds, using more than one method to measure this activity correctly is recommended. In our study, AOP was evaluated by the free radical scavenging capacity (DPPH and ABTS) and the ferric reducing antioxidant power (FRAP) assays (Table 1). With DPPH and ABTS methods, no significant differences ($p > 0.05$) were detected among the scavenging activity of kiwifruit samples, so neither the freeze-drying process nor the operations to obtain the dry extract affected this functional property. Our results agree with those reported in the literature [18]. However, the AOP of the kiwifruit samples determined using the FRAP assay showed that KD had significantly ($p < 0.05$) more ability to reduce Fe^{3+} to Fe^{2+} than K (gains of 41.5%). The FRAP reducing power appears to be correlated with the degree of hydroxylation and amount of conjugation in polyphenols [39]. The increase in KD antioxidant activity measured by FRAP may be caused by the formation of novel antioxidant substances, such as polymeric structures of polyphenols with greater antioxidant capacities. In any case, when phenol extracts contain small molecule reducing agents like ascorbic acid, interpretation becomes challenging. For strawberry, S showed the greater AOP ($p < 0.05$) for all three methods. Freeze drying caused losses ($p < 0.05$) of 20–24% depending on the analysis method (Equation (1)). Although a significant decrease in the AOP of freeze-dried strawberries was observed, these samples still retain appropriate quality concerning the original content. In general, our results were consistent with the observations reported by Floegel et al. (2011) [43].

3.2. *Genotoxicity and Cytotoxicity Biomarkers*

3.2.1. Sister Chromatid Exchanges (SCE)

Regarding the results of the SCE analysis, Y_{SCE}, Y_{1SCE}, Y_{2SCE}, and Y_{3SCE} values for each concentration of KD and SD are shown in Table 2. All tested concentrations of fruit extracts produced a decrease in the Y_{SCE} when the concentration–response data were analysed. In contrast, only the SD with the highest concentration (800 μg/mL) showed a significant reduction ($p < 0.05$) compared to the blood control and the treatment with methanol/water (70:30). The general reasons can be attributed to antioxidants, including VC and phenolic compounds, that can prevent genotoxic damage, as reported by Mrdanovic et al. (2012) [7].

The treatment with methanol/water (70:30) slightly increased Y_{SCE}; however, no significant differences ($p > 0.05$) were observed compared to the blood control.

Table 2. The frequency of sister chromatid exchange (Y_{SCE}) in human lymphocytes exposed to methanol/water (70:30 *v/v*) and different concentrations of freeze-dried kiwifruit (KD) and strawberry (SD) extracts (50, 400, and 800 μg/mL) and frequencies of chromosomes with one (Y_{1SCE}), two (Y_{2SCE}), or three (Y_{3SCE}) sister chromatid exchanges assessed in a total of 50 cells.

Conditions	Y_{SCE}	Y_{1SCE}	Y_{2SCE}	Y_{3SCE}
0 (control)	8.0 ± 0.5	7.2 ± 0.4	0.32 ± 0.08	0.06 ± 0.03
0 + methanol/water 70:30 (control + solvent)	8.2 ± 0.5	6.5 ± 0.4	0.86 ± 0.13	0.00 ± 0.00
KD 50	7.5 ± 0.4	6.0 ± 0.3	0.66 ± 0.11 *	0.06 ± 0.03 *
KD 400	7.8 ± 0.5	7.0 ± 0.4	0.32 ± 0.08 *	0.04 ± 0.03 *
KD 800	7.8 ± 0.5	6.3 ± 0.3	0.60 ± 0.11 *	0.10 ± 0.04 *
SD 50	6.4 ± 0.3	5.4 ±0.3 *	0.48 ± 0.10 *	0.02 ± 0.02 *
SD 400	6.4 ± 0.5	5.5 ± 0.5 *	0.48 ± 0.14 *	0.00 ± 0.00 *
SD 800	5.3 ± 0.3 *	4.8 ± 0.3 *	0.26 ± 0.08 *	0.00 ± 0.00 *

Results are expressed as means ± standard errors. * The difference was statistically significant ($p < 0.05$).

As shown in Table 2, regarding Y_{1SCE}, SD was the only extract that presented significant differences at any concentration tested ($p < 0.05$). Our data also revealed that all concentrations of SD and KD presented a statistically significant Y_{2SCE} ($p < 0.05$) when compared to the treatment with methanol/water (70:30). In our study, it was observed that SD at 800 μg/mL concentration was the only extract that presented statistical differences in chromosomes with one, two, or three exchanges, implicating the non-genotoxicity of this extract at high concentrations. In their study of the protective role of VC, Türkez and Aydın (2012) [44] concluded that ascorbic acid was not genotoxic and reduced the frequencies of SCEs. Mrdanovic et al. (2012) [7], in their study about the effects of orally administered antioxidants on the frequency of sister chromatid exchanges (SCEs) conducted on workers professionally exposed to antineoplastic agents, concluded that the combination of antioxidants could effectively prevent genotoxic damage induced by cytotoxic drugs. In this sense, the AOP of the studied extracts may have contributed to the decreased Y_{SCE}.

3.2.2. Mitotic Index and Proliferation Index

An increase or decrease in the MI can determine cytotoxicity levels induced by exposure to a wide range of compounds. An MI study of KD and SD on in vitro lymphocyte cultures was performed in a total of 500 cells for each experiment. Data concerning the MI are summarised in Table 3. Lymphocytes cultured in the presence of 50, 400, and 800 μg/mL of KD exhibited a similar MI to those recorded in lymphocytes without the fruit extracts (control) and with solvent (solvent control). However, a significant decrease ($p < 0.05$) in MI was observed in cultures treated with 800 μg/mL of SD. In addition, other natural extracts have demonstrated cytotoxic activity against human cells in vitro [11,45]. The observed decrease in mitotic activity could be attributed to various factors, such as the interruption of the mitotic cell cycle during interphase, inhibition of crucial nuclear protein synthesis required for normal mitotic progression, or suppression of DNA synthesis [46].

Data concerning the PI of KD and SD in lymphocytes are summarised in Table 3. Lymphocyte cultures treated only with the solvent or with the different concentrations of KD and SD showed PI values similar to the control value ($p > 0.05$). Therefore, it can be concluded that these extracts do not exhibit cytostatic effects on human peripheral blood lymphocytes in in vitro cultures and, consequently, are not genotoxic. Other natural extracts have also shown results similar to our in vitro study of PI [45].

Table 3. Mitotic (MI) and proliferation (PI) indices determined in human peripheral lymphocytes cultured for 72 h in the presence of methanol/water (70:30 v/v) solvent and various concentrations of freeze-dried kiwifruit (KD) and strawberry (SD) extracts (50, 400, and 800 μg/mL). The table also shows each concentration's relative mitotic index (RMI, %) and relative proliferation index (RPI, %).

Conditions	MI	RMI (%)	PI	RPI (%)
0 (control)	0.039 ± 0.009	100.00	1.58 ± 0.05	100.00
0 + methanol/water 70:30 (control + solvent)	0.039 ± 0.009	99.92	1.59 ± 0.04	100.63
KD 50	0.041 ± 0.009	103.47	1.57 ± 0.05	99.37
KD 400	0.039 ± 0.009	98.39	1.59 ± 0.05	100.63
KD 800	0.039 ± 0.009	98.39	1.59 ± 0.05	100.63
SD 50	0.041 ± 0.009	105.00	1.57 ± 0.05	99.37
SD 400	0.039 ± 0.009	98.39	1.57 ± 0.05	99.37
SD 800	0.035 ± 0.008 *	88.22	1.58 ± 0.05	100.58

MI and PI results are expressed as mean ± standard errors. * The difference was statistically significant ($p < 0.05$).

The relative indexes RMI and PRI were similar to the control ($p > 0.05$) (Table 3). Studies examining the relationship between SCE actions and various expressions of cytotoxicity have indicated that an increase in SCE frequency is often accompanied by decreased cell survival and alterations in cell cycle kinetics, resulting in lower values of the mitotic index (MI) [47]. In preclinical screening of natural products for potential antitumor activity, it is common to evaluate a single parameter, typically cell death. However, more comprehensive information can be obtained by assessing two biological endpoints, such as the MI and proliferation index (PI) assays [30]. For example, a reduction in the MI can indicate cell death or arrest of the cell cycle at any stage during interphase.

Our results have demonstrated that the highest concentration of SD reduced the MI, accompanied by a reduction in the Y_{SCE}; this might be because SD at 800 μg/mL was an antineoplastic and protective agent against SCE induction, therefore preventing the development, growth, or proliferation of malignant tumour cells.

3.2.3. Chromosomal Aberrations as Radioprotection Biomarkers

The cytogenetic results obtained from the analysis of human lymphocytes exposed to 2 Gy radiation in the presence of varying concentrations of KD and SD are presented in Table 4. After 2 Gy irradiation, the results revealed a decrease in the frequency of dicentric chromosomes (Y_{Dic}) with an increase in the concentration of KD and SD. The only concentrations for which the reduction in Y_{Dic} (and therefore reduction in damage) was statistically significant ($p < 0.05$) were KD at medium and high concentrations (400 μg/mL and 800 μg/mL) when compared to the control treated with methanol/water at 70:30. In our study, other types of CAs induced by the extracts were found (acentric fragments).

The distribution of cells treated and untreated with KD and SD containing a different number of dicentrics is shown in Table 5. After a uniform radiation exposure, the intercellular distribution of dicentrics followed a Poisson distribution in all cases, with or without extracts ($U = \pm 1.96$). A value of $U > 1.96$ indicates overdispersion at the 5% level of significance.

Table 4. Chromosomal aberrations in human lymphocytes exposed to a radiation dose of 2 Gy under different conditions: pre-treated with methanol/water (70:30 *v*/*v*) solvent and pre-treated with varying concentrations of kiwifruit (KD) and strawberry (SD) extracts (50, 400, and 800 μg/mL).

Conditions	N	Dic	Y_{Dic}
0	200	63	0.315 ± 0.037
0 + methanol/water 70:30	175	24	0.137 ± 0.027
KD 50	200	23	0.115 ± 0.024
KD 400	197	17	0.086 ± 0.020 *
KD 800	201	16	0.080 ± 0.019 *
SD 50	240	29	0.121 ± 0.022
SD 400	175	20	0.114 ± 0.025
SD 800	200	23	0.115 ± 0.023

N, number of analysed cells; Dic, dicentric chromosomes; Y_{Dic}, frequencies of dicentric chromosomes ± standard errors. * The difference in Y_{Dic} was statistically significant ($p < 0.05$).

Table 5. Distribution of dicentric cells (Dic), dispersion index (DI), and normalised unit of this index (U) for each condition treatment with 2 Gy radiation. Samples were pre-treated with methanol/water (70:30 *v*/*v*) solvent and pre-treated with different concentrations of freeze-dried kiwifruit (KD) and strawberry (SD) extracts (50, 400, and 800 μg/mL).

Conditions	N	Dic	0Dic (1)	1Dic (1)	2Dic (1)	3Dic (1)	DI	U
0	200	63	137	51	6	0	0.87	−1.30
0 + methanol/water 70:30	175	24	151	22	1	0	0.95	−0.47
KD 50	200	23	177	21	1	0	0.98	−0.24
KD 400	197	17	180	17	0	0	0.92	−0.83
KD 800	201	16	185	16	0	0	0.93	−0.77
SD 50	240	29	211	27	1	0	0.95	−0.54
SD 400	175	20	155	18	1	0	0.99	−0.09
SD 800	200	23	177	23	0	0	0.89	−1.13

N: number of analysed cells; (1) number of cells with zero, one, two, or three dicentric chromosomes.

Several studies have confirmed that the health protective effect associated with kiwifruit is mainly related to their phytochemicals, which may prevent oxidative damage to biological systems by lowering oxidative stress and altering signal transduction pathways involved in cell survival and proliferation [48]. A possible explanation for the irradiation protective effects of antioxidants is a model based on electron spin resonance studies. According to this model, radiation exposure leads to the formation of short-lived radicals like +OH, which are responsible for cell death, as well as long-lived radicals that have the potential to induce mutations and transformations [38]. Our data showed that kiwifruit extracts reduced the chromosomal aberration frequency after irradiation, probably due to its AOP. Regarding biochemicals contributing towards radioprotection, it is known that some phenolics and VC are able to initiate a wide range of physiological responses that reduce severe chromosomal damage [49]. In this sense, VC acted effectively in the sequestration of ROS and, consequently, in the significant reduction in Y_{Dic}. This antioxidant property of VC plays a significant role in safeguarding cellular macromolecules, particularly DNA, from oxidative damage [50]. Its protective effect has also been attributed to enhancing other antioxidant compounds' repair and recovery processes [51]. In this sense, the antioxidant and radioprotective properties of AA have been described by Jagetia et al. [52]. The radio protective activity of this component depends on its antioxidant and free radical scavenging property. Other studies also described that AA was radioprotective in reducing radiation-induced skin excisions in Swiss albino mice exposed to gamma radiation, and also increased their wound healing [53]. On the other hand, the radioprotective effect of phenolics may be mediated through several mechanisms besides free radical scavenging, such as improvements in the antioxidant state and anti-lipid peroxidation potential. These biological properties are conferred due to the presence of a variety of phenolic hydroxyl groups linked to the ring structure. Some studies attribute the protective effect of flavonoids to their direct hydroxyl radical scavenging potency and thus behave like an endogenous enzyme [54]. Polyphenols, especially flavonoid glycosides and their derivatives, contain ketone groups conjugated to aromatic rings activated by electron donor constituents. This prevents energy transfer, reduces oxidative stress, and stabilises redox processing in cells. Polyphenols may up-regulate mRNAs of antioxidant enzymes (catalase, superoxide dismutase, and glutathione peroxidase), thus mitigating the oxidative damage induced by ionising radiation. Up-regulation of DNA repair genes and inhibition of nitric oxide may also protect against radiation-induced DNA damage [55]. Although the mechanisms of radioprotection differ among groups, most compounds act primarily via scavenging of radiation-induced free radicals or by quenching of secondary biomolecular reactive species [53].

Regarding SD, it was observed that Y_{Dic} was reduced but did not present a significant difference ($p > 0.05$) compared to the methanol/water control. This means that strawberry extracts at all studied concentrations did not reduce the damage caused by the effect of ionising radiation. Despite the high AOP, phenolic compounds and TP, SD did not act effectively in reducing the Y_{Dic}. The different radioprotective responses of both fruits may indicate that radioprotection is more related to the type of phenolic compound than to the amount of them, and as it has already been commented, their biological action depends on their chemical structure.

Altogether, our findings encourage further studies to fully understand which molecules are effective and what mechanisms are involved in their radioprotective action. Research on human bioavailability of the phytochemicals involved is also advised.

4. Conclusions

Rapid technological development, radiation exposure from natural or artificial sources, radiotherapy, medical imaging, etc., cause unintended toxicity and contribute to severe radiation-related pathologies. It is crucial to safeguard the biological system from such a barrage of detrimental effects from radiation. Scientific research has led to the creation of several synthetic chemicals to combat radiation's potential dangers. However, the need for alternative natural sources is vital due to its toxicity and adverse effects. In this report, methanolic extracts were obtained from freeze-dried kiwifruit and strawberries. Comparing both fruits, kiwifruit stood out for its vitamin C, ferulic acid, epicatechin, and quercetin-3-O-galactoside content, and strawberry for its total phenol content, ellagic acid, pelargonidin-3-glucoside, quercetin-3-glucoside, and antioxidant activity. The freeze-drying process resulted in gains in the phenolic compounds of both samples and in the antioxidant activity of the kiwifruit. Then, the radioprotective effect of methanolic extracts from freeze-dried kiwifruit and strawberries against the damage induced by γ-irradiation in human peripheral blood lymphocytes was investigated. The results showed that kiwifruit methanolic extracts at concentrations of 400 and 800 μg/mL possessed radioprotective ability, as evidenced by the significant reduction in the chromosomal aberration frequency after irradiation at a dose of 2 Gy. However, freeze-dried strawberry methanolic extracts did not reduce the damage caused by the effect of ionising radiation. None of the samples were found to exhibit genotoxic or cytotoxic effects. Although the free radical scavenging capacity of the extracts represents a possible mechanism for their radioprotective effect, it cannot be concluded that a greater antioxidant activity implies radioprotective power. The different responses of both fruits may indicate that radioprotection may be related to the different antioxidants present, such as vitamin C, and to the type of phenolic compound rather than to the amount of them, as their biological action depends on their chemical structure. Additionally, the synergistic effect between antioxidant substances must also be taken into account. Further investigation on radioprotection mechanisms is worth conducting in the future. Although potentially protective agents against harmful radiation have been investigated for decades, there is currently no ideal radioprotectant available.

Therefore, it is very important that natural compounds with a potential radioprotective effect are studied for medical applications and in radiological and nuclear emergencies. The next step carried out by our research group will be in vivo studies.

Supplementary Materials: The following supporting information can be downloaded at: https://www.mdpi.com/article/10.3390/app13158996/s1, Figure S1: Metaphases with radioprotective and genotoxic biomarkers used in this study. Figure S2: Representative HPLC chromatogram of polyphenols of strawberries. (1) Pelargonidin-3-Glucoside, (2) quercetin-3-Glucoside, (3) ellagic acid.

Author Contributions: Conceptualisation, A.M. and E.G.-M.; methodology, N.S., A.M. and E.G.-M.; formal analysis, N.S., A.M. and E.G.-M.; data curation, M.d.S.R.; writing—original draft preparation, N.S., A.M. and E.G.-M.; writing—review and editing, E.G.-M.; supervision, A.M. and E.G.-M.; project administration, A.M. and E.G.-M. All authors have read and agreed to the published version of the manuscript.

Funding: This research was funded by Vicerrectorado de Investigación, Universitat Politècnica de València (UPV-FE-2013-56) (Spain).

Institutional Review Board Statement: The study was conducted in accordance with the Declaration of Helsinki and approved by the Ethics Committee of Hospital Universitario y Politécnico La Fe (date of approval: 4 April 2018) for studies involving humans.

Informed Consent Statement: Informed consent was obtained from all subjects involved in the study.

Data Availability Statement: Not applicable.

Acknowledgments: The authors thank Alberto Yuste del Carmen for the technical support. M.d.S.R. thanks CAPES for the financial support (no. 99999.013619/2013-06-DOC-PLENO; Coordenação de Aperfeiçoamento de Pessoal de Nível Superior/Coordination for the Improvement of Higher Education of Brazil).

Conflicts of Interest: The authors declare no conflict of interest.

References

1. Fuchs-Tarlovsky, V. Role of antioxidants in cancer therapy. *Nutrition* **2013**, *29*, 15–21. [CrossRef] [PubMed]
2. Szeijk, A.; Tomasz, P.; Czubatka-Bienkowska, A.; Olejnik, A.K.; Pawlaczyk-Graja, I.; Gancarz, R.; Zbikowska, H.M. A comparative study on the radioprotective potential of the polyphenolic glycoconjugates from medicinal plants of *Rosaceae* and *Asteraceae* families versus their aglycones. *J. Photochem. Photobiol.* **2017**, *171*, 50–57. [CrossRef] [PubMed]
3. Shahani, S.; Rostamnezhad, M.; Ghaffarirad, V.; Ghasemi, A.; Pourfallah, T.A.; Hosseinimehr, S.J. Radioprotective effect of *Achillea millefolium* L. against genotoxicity induced by ionising radiation in human normal lymphocytes. *Dose-Response* **2015**, *13*, 1559325815583761. [CrossRef] [PubMed]
4. Smith, T.A.; Kirkpatrick, D.R.; Smith, S.; Smith, T.K.; Pearson, T.; Kailasam, A.; Herrmann, K.Z.; Schubertv, J.; Agrawal, D.K. Radioprotective agents to prevent cellular damage due to ionising radiation. *J. Transl. Med.* **2017**, *15*, 232. [CrossRef]
5. Mettler, F.E.J.; Brenner, D.; Coleman, C.N.; Kaminski, J.M.; Kennedy, A.R. Can radiation risks to patients be reduced without reducing radiation exposure? The status of chemical radioprotectants. *Med. Phys. Inf. Clin. Perspect.* **2011**, *196*, 616–618. [CrossRef]
6. Zbikowska, H.M.; Szejk, M.; Saluk, J.; Gancarz, R.; Olejnik, A.K. Polyphenolic–polysaccharide conjugates from plants of *Rosaceae/Asteraceae* family as potential radioprotectors. *Int. J. Biol. Macromol.* **2016**, *86*, 329–337. [CrossRef]
7. Mrdanovic, J.; Jungic, S.; Šolajic, S.; Bogdanovic, V.; Jurišic, V. Effects of orally administered antioxidants on micronuclei and sister chromatid exchange frequency in workers professionally exposed to antineoplastic agents. *Food Chem. Toxicol.* **2012**, *50*, 2937–2944. [CrossRef]
8. Bianchi, J.; Fernandes, T.C.C.; Marin-Morales, M.A. Induction of mitotic and chromosomal abnormalities on *Allium cepa* cells by pesticides imidacloprid and sulfentrazone and the mixture of them. *Chemosphere* **2016**, *144*, 475–483. [CrossRef]
9. Rojas, E.; Herrera, L.A.; Sordo, M.; Gonsebatt, M.E.; Montero, R.; Rodríguez, R.; Ostrosky-Wegman, P. Mitotic index and cell proliferation kinetics for identification of antineoplastic activity. *Anti-Cancer Drugs* **1993**, *4*, 599–672. [CrossRef]
10. Kuntic, V.S.; Stankovic, M.B.; Vujic, Z.B.; Brboric, S.J.; Markovic, S. Radioprotectors—The Evergreen Topic. *Chem. Biodivers.* **2013**, *10*, 1791–1803. [CrossRef]
11. Sebastià, N.; Almonacid, M.; Villaescusa, J.I.; Cervera, J.; Such, E.; Silla, M.A.; Soriano, J.M.; Montoro, A. Radioprotective activity and cytogenetic effect of resveratrol in human lymphocytes: An in vitro evaluation. *Food Chem. Toxicol.* **2013**, *51*, 391–395. [CrossRef]

12. Cinkilic, N.; Cetintas, S.K.; Zorlu, T.; Vatan, O.; Yilmaz, D.; Cavas, T.; Tunc, S.; Ozkan, L.; Bilaloglu, R. Radioprotection by two phenolic compounds: Chlorogenic and quinic acid, on X-ray induced DNA damage in human blood lymphocytes in vitro. *Food Chem. Toxicol.* **2013**, *53*, 359–363. [CrossRef]
13. Montoro, A.; Soriano, J.M.; Barquinero, J.F.; Almonacid, M.; Verdú, G.; Sahuquillo, V.; Villaescusa, J.I.; Sebastià, N. Assessment in vitro of cytogenetic and genotoxic effects of propolis on human lymphocytes. *Food Chem. Toxicol.* **2012**, *50*, 216–221. [CrossRef] [PubMed]
14. Kamran, M.Z.; Ranjan, A.; Kaur, N.; Sur, S.; Tandon, V. Radioprotective Agents: Strategies and Translational Advances. *Med. Res. Rev.* **2016**, *36*, 461–493. [CrossRef]
15. Hlatky, L.; Sachs, R.K.; Vazquez, M.; Cornforth, M.N. Radiation-induced chromosome aberrations: Insights gained from biophysical modelling. *BioEssays* **2002**, *24*, 14–723. [CrossRef]
16. Mazzoni, L.; Perez-Lopez, P.; Giampieri, F.; Alvarez-Suarez, J.M.; Gasparrini, M.; Forbes-Hernandez, T.Y.; Quiles, J.L.; Mezzetti, B.; Battino, M. The genetic aspects of berries: From field to health. *J. Sci. Food Agric.* **2016**, *96*, 365–371. [CrossRef]
17. Forbes-Hernández, T.Y.; Gasparrini, M.; Afrin, S.; Bompadre, S.; Mezzetti, B.; Quiles, J.L.; Giampieri, F.; Battino, M. The healthy effects of strawberry polyphenols: Which strategy behind antioxidant capacity? *Crit. Rev. Food Sci. Nutr.* **2016**, *56*, S46–S59. [CrossRef]
18. Wojdyło, A.; Nowicka, P.; Oszmian, J.; Golis, T. Phytochemical compounds and biological effects of Actinidia fruits. *J. Funct. Foods* **2017**, *30*, 194–202. [CrossRef]
19. Burillo, P.S.; Oliveras, M.J.; Quesada, J.; Rufián-Henares, J.A.; Pastoriza, S. Relationship between composition and bioactivity of persimmon and kiwifruit. *Food Res. Int.* **2018**, *105*, 461–472. [CrossRef]
20. Amatori, S.; Mazzoni, L.; Alvarez-Suarez, J.M.; Giampieri, F.; Gasparrini, M.; Forbes-Hernandez, T.Y.; Afrin, S.; Errico Provenzano, A.; Persico, G.; Mezzetti, B.; et al. Polyphenol-rich strawberry extract shows in vitro and in vivo biological activity against invasive breast cancer cells. *Sci. Rep.* **2016**, *6*, 30917. [CrossRef]
21. Silva-Espinoza, M.A.; García-Martínez, E.; Martínez-Navarrete, N. Protective capacity of gum Arabic, maltodextrin, different starches, and fibers on the bioactive compounds and antioxidant activity of an orange puree (*Citrus sinensis* (L.) Osbeck) against freeze-drying and in vitro digestion. *Food Chem.* **2021**, *357*, 129724. [CrossRef] [PubMed]
22. Lee, J.; Sowndhararajan, K.; Kim, M.; Kim, J.; Kim, D.; Kim, S.; Kim, G.Y.; Kim, S.; Jhoo, J.W. Antioxidant, inhibition of a-glucosidase and suppression of nitric oxide production in LPS-induced murine macrophages by different fractions of *Actinidia arguta* stem. *Saudi J. Biol. Sci.* **2014**, *21*, 532–538. [CrossRef] [PubMed]
23. AOAC. *Official Methods of Analysis of AOAC International*, 21st ed.; AOAC International: Rockville, MD, USA, 2019.
24. Gramza-Michałowska, A.; Bueschke, M.; Kulczyński, B.; Gliszczyńska-Świgło, A.; Kmiecik, D.; Bilska, A.; Purłan, M.; Wałęsa, L.; Ostrowski, M.; Filipczuk, M.; et al. Phenolic compounds and multivariate analysis of antiradical properties of red fruits. *J. Food Meas. Charact.* **2019**, *13*, 1739–1747. [CrossRef]
25. Re, R.; Pellegrini, N.; Proteggente, A.; Pannala, A.; Yang, M.; Rice-Evans, C. Antioxidant activity applying an improved ABTS radical cation decolorisation assay. *Free. Radic. Biol. Med.* **1999**, *26*, 1231–1237. [CrossRef]
26. Jurica, K.; Brčić Karačonji, I.; Mikolić, A.; Milojković-Opsenica, D.; Benković, V.; Kopjar, N. In vitro safety assessment of the strawberry tree (*Arbutus unedo* L.) water leaf extract and arbutin in human peripheral blood lymphocytes. *Cytotechnology* **2018**, *70*, 1261–1278. [CrossRef]
27. Iwasawa, H.; Morita, E.; Yui, S.; Yamazaki, M. Anti-oxidant effects of kiwi fruit in vitro and in vivo. *Biol. Pharm. Bull.* **2011**, *34*, 128–134. [CrossRef]
28. Weaver, J.; Briscoe, T.; Hou, M.; Goodman, C.; Kata, S.; Ross, H.; McDougall, G.; Stewart, D.; Riches, A. Strawberry polyphenols are equally cytotoxic to tumourigenic and normal human breast and prostate cell lines. *Int. J. Oncol.* **2009**, *34*, 777–786. [CrossRef]
29. International Atomic Energy Agency (IAEA). *Biological Dosimetry: Chromosomal Aberration Analysis for Dose Assessment*; Technical Reports Series; IAEA: Viena, Austria, 2011; p. 260.
30. Rojas, E.; Montero, R.; Herrera, L.A.; Sordo, M.; Gonsebatt, M.E.; Rodriguez, R.; Ostrosky-Wegman, P. Are mitotic index and lymphocyte proliferation kinetics reproducible endpoints in genetic toxicology testing? *Mutat. Res. Lett.* **1992**, *282*, 283–286. [CrossRef]
31. Du, G.; Li, M.; Ma, F.; Liang, D. Antioxidant capacity and the relationship with polyphenol and vitamin C in Actinidia fruits. *Food Chem.* **2009**, *113*, 557–562. [CrossRef]
32. Tudor, V.; Manole, C.G.; Teodorescu, R.; Asanica, A.; Barbulescu, J.D. Analysis of some phenolic compounds and free radical scavenging activity of strawberry fruits during storage period. *Agric. Agric. Sci. Procedia* **2015**, *6*, 157–164. [CrossRef]
33. Spigno, G.; Tramelli, L.; Faveri, D.M. Effects of extraction time, temperature and solvent on concentration and antioxidant activity of grape marc phenolics. *J. Food Eng.* **2007**, *81*, 200–208. [CrossRef]
34. García-Martínez, E.; Camacho, M.; Nuria Martínez-Navarrete, N. In Vitro Bioaccessibility of Bioactive Compounds of Freeze-Dried Orange Juice Co-Product Formulated with Gum Arabic and Modified Starch. *Molecules* **2023**, *28*, 810. [CrossRef]
35. Carocho, M.; Ferreira, I.C.F.R. A review on antioxidants, prooxidants and related controversy: Natural and synthetic compounds, screening and analysis methodologies and future perspectives. *Food Chem. Toxicol.* **2013**, *51*, 15–25. [CrossRef]
36. Hettihewa, S.K.; Hemar, Y.; Rupasinghe. H.P. Flavonoid-Rich Extract of *Actinidia macrosperma* (AWild Kiwifruit) inhibits Angiotensin-Converting Enzyme In Vitro. *Foods* **2018**, *7*, 146. [CrossRef]

37. Jiao, Y.; Kilmartin, P.A.; Fan, M.; Quek, S.Y. Assessment of phenolic contributors to antioxidant activity of new kiwifruit cultivars using cyclic voltammetry combined with HPLC. *Food Chem.* **2018**, *268*, 77–85. [CrossRef]
38. Prasad, J.; Pattnaik, S.; Arya, R.; Dash, R.; Sahoo, S.; Pradhan, B.; Bhuyan, P.; Behera, P.; Jena, M.; Sharma, A.; et al. Phytochemicals: A potential next generation agent for radioprotection. *Phytomedicine* **2022**, *106*, 154188. [CrossRef]
39. Tavarini, S.; Degl'Innocenti, E.; Remorini, D.; Massai, R.; Guidi, L. Antioxidant capacity, ascorbic acid, total phenols and carotenoids changes during harvest and after storage of Hayward kiwifruit. *Food Chem.* **2008**, *107*, 282–288. [CrossRef]
40. Klimczak, I.; Malecka, M.; Miroslawa, S.; Gliszczynska-Swiglo, A. Effect of storage on the content of polyphenols, vitamin C and the antioxidant activity of orange juices. *J. Food Compos. Anal.* **2007**, *20*, 313–322. [CrossRef]
41. Leontowicz, H.; Leontowicz, M.; Latocha, P.; Jesion, I.; Park, Y.S.; Katrich, E.; Barasch, D.; Nemirovski, A.; Gorinstein, S. Bioactivity and nutritional properties of hardy kiwi fruit *Actinidia arguta* in comparison with *Actinidia deliciosa* 'Hayward' and *Actinidia eriantha* 'Bidan'. *Food Chem.* **2016**, *196*, 281–291. [CrossRef]
42. Nemzer, B.; Vargas, L.; Xiaa, X.; Sintara, M.; Feng, H. Phytochemical and physical properties of blueberries, tart cherries, strawberries, and cranberries as affected by different drying methods. *Food Chem.* **2018**, *262*, 242–250. [CrossRef]
43. Floegel, A.; Kim, D.; Chung, S.; Sung, I.K.; Chun, C.K. Comparison of ABTS/DPPH assays to measure antioxidant capacity in popular antioxidant-rich US foods. *J. Food Comps. Anal.* **2011**, *24*, 1043–1048. [CrossRef]
44. Türkez, H.; Aydın, E. The protective role of ascorbic acid on imazalil-induced genetic damage assessed by the cytogenetic tests. *Toxicol. Ind. Health* **2012**, *28*, 648–654. [CrossRef] [PubMed]
45. Montoro, A.; Almonacid, M.; Serrano, J.; Saiz, M.; Barquinero, J.F.; Barrios, L.; Verdu, G.; Pérez, J.; Villaescusa, J.I. Assessment by cytogenetic analysis of the radioprotection properties of propolis extract. *Radiat. Prot. Dosim.* **2005**, *115*, 461–464. [CrossRef] [PubMed]
46. Mohanty, S.; Das, A.B.; Das, P.; Mohanty, P. Effect of a low dose of aluminum on mitotic and meiotic activity, 4C DNA content, and pollen sterility in rice, *Oryza sativa* L. cv. Lalat. *Ecotoxicol. Environ. Saf.* **2004**, *59*, 70–75. [CrossRef] [PubMed]
47. Mourelatos, D. Chromosomes study as predictor of chemo response of tumours. *Cancer J.* **1996**, *9*, 136–141.
48. Brevik, A.; Gaivão, I.; Medin, T.; Jørgenesen, A.; Piasek, A.; Elilasson, J.; Karlsen, A.; Blomhoff, R.; Veggan, T.; Duttaroy, A.K.; et al. Supplementation of a western diet with golden kiwifruits (*Actinidia chinensis* var 'Hort 16A') effects on biomarkers of oxidation damage and antioxidant protection. *Nutr. J.* **2011**, *10*, 54. [CrossRef]
49. Badjatia, N.; Satyam, A.; Singh, P.; Seth, A.; Sharma, A. Altered antioxidant status and lipid peroxidation in Indian patients with urothelial bladder carcinoma. *Urol. Oncol. Semin. Orig. Investig.* **2009**, *28*, 360–367. [CrossRef]
50. Montecinos, V.; Guzman, P.; Barra, V.; Villadgran, M.; Munoz-Motesino, C.; Sotomayor, K.; Escobar, E. Vitamin C is an essential antioxidant that enhances survival of oxidatively stressed human vascular endothelial cells in the presence of a vast molar excess of glutathione. *J. Biol. Chem.* **2007**, *282*, 15506–15515. [CrossRef]
51. Traber, M.G.; Stevens, J.F. Vitamins C and E: Beneficial effects from a mechanistic perspective. *Free. Radic. Biol. Med.* **2011**, *51*, 1000–1013. [CrossRef]
52. Jagetia, G.C.; Rajanikant, G.; Rao, S.K. Evaluation of the effect of ascorbic acid treatment on wound healing in mice exposed to different doses of fractionated gamma radiation. *Radiat. Res.* **2003**, *159*, 371–380. [CrossRef]
53. Prasad, N.R.; Srinivasan, M.; Pugalendi, K.; Menon, V.P. Protective effect of ferulic acid on γ-radiation-induced micronuclei, dicentric aberration and lipid peroxidation in human lymphocytes. *Mutat. Res.* **2006**, *603*, 129–134. [CrossRef]
54. Hosseinimehr, S.J. Trends in the development of radioprotective agents. *Drug Discov. Today* **2007**, *12*, 794–805. [CrossRef]
55. Kemertelidze, E.P.; Tsitsishvili, V.G.; Alaniya, M.D.; Sagareishvili, T.G. Structure-function analysis of the radioprotective and antioxidant activity of flavonoids. *Chem. Nat. Compd.* **2000**, *36*, 54–59. [CrossRef]

applied sciences

MDPI

Article

Effectiveness of Immature Asian Pear Extract on Pulmonary Injury Caused by Particulate Matter in Mice

Mi-Ran Kim [1,†], Jin-Hwa Lee [1,†], Mo-Un Ku [2], Ki-Young Kim [3], Su Shin [3], Eun-Jin Hong [3], Sae-Kwang Ku [4,*] and Jae-Suk Choi [1,*]

1 Department of Seafood Science and Technology, The Institute of Marine Industry, Gyeongsang National University, Tongyeong 53064, Republic of Korea; 210217@gnu.ac.kr (M.-R.K.); evolution_5237@gnu.ac.kr (J.-H.L.)
2 Department of Biomedical Laboratory Science, College of Health and Welfare, Kyungwoon University, Gumi 39160, Republic of Korea; 202121018@ikw.ac.kr
3 Research Institute, Bio Port Korea Inc., Busan 46034, Republic of Korea; kyk1967@bioportkorea.com (K.-Y.K.); sshin@bioportkorea.com (S.S.); ejhong@bioportkorea.com (E.-J.H.)
4 Department of Anatomy and Histology, College of Korean Medicine, Daegu Haany University, Gyeongsan 38610, Republic of Korea
* Correspondence: gucci200@dhu.ac.kr (S.-K.K.); jsc1008@gnu.ac.kr (J.-S.C.); Tel.: +82-53-819-1549 (S.-K.K.); +82-55-772-9142 (J.-S.C.)
† These authors contributed equally to this work.

Abstract: The use of natural products in developing respiratory-function-protective pharmaceuticals is actively progressing. However, in this context, the improvement effects of young Asian pear (*Pyrus pyrifolia* Nakai) extracts have not been evaluated yet. Thus, this study investigated the anti-inflammatory and lung damage improvement effects of immature Asian pear extract (IAP; 400, 200, and 100 mg/kg) using a particulate matter 2.5 μm ($PM_{2.5}$)-induced sub-acute lung injury mouse model. The experimental results were compared with dexamethasone (0.75 mg/kg), used as a control drug. After two intranasal instillations of $PM_{2.5}$ and ten doses of IAP extract for eight days, changes in macroscopic lung autopsy, leukocyte fractionation from bronchoalveolar lavage fluid, lung antioxidant defense system, lung histopathology, and mRNA expression in lung tissue were confirmed. Stress-induced inflammatory lung damage through the increased expression of $PM_{2.5}$-induced PI3K/Akt and p38 MAPK mRNA was significantly suppressed via the administration of IAP extract (400–100 mg/kg). Furthermore, IAP extract administration promoted serous fluid production in lung tissue, increased substance P and ACh levels, and decreased mucus-production-related expression of MUC5AC and MUC5B mRNA. Interestingly, the observed effects showed a dose-dependent manner without serious hepatotoxicity. The results of this study indicate that a proper oral administration of IAP extract could be helpful in protecting against lung diseases, positioning IAP extract as a potential candidate for an alternative agent to safeguard the respiratory system.

Keywords: Asian pear; *Pyrus pyrifolia*; lung injury; $PM_{2.5}$; mice; anti-inflammatory activity

1. Introduction

Due to the increase in particulate matter (PM) caused by air pollution, various respiratory diseases are rapidly emerging in East Asia, including China, Korea, and Japan. Beijing, the capital city of China, stands out among these regions, having a notably elevated prevalence of respiratory diseases attributed to PM exposure [1–4]. Efforts to curtail air pollution were intensified when Beijing was spotlighted as one of the most severely impacted cities by PM air pollution during the Beijing Olympics [5]. PM has a diameter of around 2.5 μm ($PM_{2.5}$) and contains organic contaminants, including polycyclic aromatic hydrocarbons, mineral dust and inorganic contaminants, such as sulfate, nitrate, elemental carbon, and ammonium [6,7].

Inhaling PM can lead to severe lung damage and, subsequently, harm the heart. Asthma is one of the most representative non-communicable respiratory tract diseases caused by air pollution [3,8]. According to the World Health Organization report in 2017 [9], the global prevalence of asthma was estimated to be approximately 235 million people. PM inhalation is recognized as a major cause of allergic inflammation, representing a significant and severe etiological factor in asthma [10]. Over 70% of inhaled PM settles in the lower trachea, while approximately 22% makes its way to the alveoli [11]. The buildup of PM induces oxidative stress in airway and lung tissue's epithelial cells, leading to localized tissue damage, and ultimately triggering inflammatory responses [5,12]. Key factors contributing to oxidative stress in airway epithelial cells include reactive glutathione peroxidase (GPX), reactive oxygen species (ROS), superoxide dismutase (SOD), heme-oxygenase-1 (HO-1), catalase (CAT), and glutathione (GSH) [13]. The initiation of an inflammatory response can easily be measured by an increase in proinflammatory cytokines and inflammatory mediators [14]. Inflammation of the lungs following PM inhalation is caused by an increase in ROS and cytokines (interleukin (IL)-6 and tumor necrosis factor (TNF)-α) in bronchoalveolar lavage fluid (BALF) and the migration of monocytes [1–3,15].

With the increase in respiratory disorders caused by PM, advancements have been made in the development of drugs aimed at treating or preventing respiratory damage [2,3,16]. In particular, natural products produce physiologically active substances with relatively minimal side effects and excellent antioxidant and anti-inflammatory activities [17,18]; thus, the use of $PM_{2.5}$-induced lung injury models to develop respiratory protection drugs from natural sources is actively progressing [1–3,16,19].

Pears (*Pyrus* spp.) are one of the most purchased fruits globally [20]. Given that pears do not have a distinct color or scent, the study of chemical substances or biological activities present in pears is relatively limited or sparse [21,22]. Based on the existing study on the chemical compounds found in *Pyrus pyrifolia* Nakai [23], an oriental pear most commonly cultivated in Korea, phenylpropanoid malate derivatives [23], caffeoyl triterpenes [24], and flavonoids have been isolated and identified as the major active compounds [25–28]. The main active ingredients in pears, including flavonoids, arbutin, caffeic acid, malaxinic acid, chlorogenic acid, and phenolic compounds, as well as antioxidant activity, decrease with maturation [29], whereas immature pears possess high levels of phenolic compounds and antioxidant activity compared to those of mature pears [22]. Therefore, there is a high opportunity to utilize the immature pears as a source of active components for functional food materials [22]. In order to gather high-quality mature fruits on orchard farms, immature pear fruits are discarded immediately after flowering [30]; through this process, just one cluster out of seven or eight is retained, with the rest being discarded [31]. Considering the annual production of mature pears in Korea [32], the amount of discarded immature pears is estimated to be approximately 15,000 tons per year [22], most of which are discarded [22]. To the best of our knowledge, the use of immature pear extracts to improve $PM_{2.5}$-induced lung disease has not been evaluated yet.

Dexamethasone (DEXA) is a representative synthetic adrenocortical steroid that exhibits approximately 20–30 times stronger anti-inflammatory effects than in vivo hydrocortisone. In addition, its effect is approximately 4–5 times better than that of prednisolone, another commonly used synthetic corticosteroid [33]. The anti-inflammatory effects of DEXA are well-known in various inflammatory respiratory diseases [16,34,35], including $PM_{2.5}$-induced lung damage [19]. Therefore, DEXA was selected as the positive control drug in this study.

During this study, the anti-inflammatory and lung damage improvement effects of immature Asian pear extract (IAP) were evaluated using a $PM_{2.5}$-induced sub-acute lung injury mouse model [1–3,16].

2. Materials and Methods

2.1. Test Material

The immature Asian pear extract (IAP; *P. pyrifolia* Nakai) used in the present study was supplied by Bioport Korea Inc., Yangsan, Republic of Korea. The immature Asian pears (2500 g) were mixed with 12,500 mL of purified water and extracted for 4 h at 100 °C. The solution was concentrated under reduced pressure using Rotavapor® R-220 (BÜCHI Labortechnik AG, Flawil, Switzerland) and spray-dried using a spray-dryer (Einsystem Co., Anyang, Republic of Korea) to obtain the powdered extract. The IAP powder, light brown in color, was dissolved in distilled water (D.W.) to prepare a stock solution at a concentration of 40 mg/mL and stored at −20 °C for further steps. The IAP powder was catalogued in the herbarium of the Medical Research Center for Herbal Convergence on Liver Disease at Daegu Haany University, Gyeongsan, Republic of Korea, with a sample number–IAP2022BPK01.

2.2. High-Performance Liquid Chromatography (HPLC)

An Agilent HPLC 1200 series (Agilent Technologies, Inc., Santa Clara, CA, USA) was employed to assess the amount of arbutin in the IAP extract using Capcell Pak C18 UG120 column (Osaka Soda Co., Ltd., Osaka, Japan) and a wavelength detector (G1314B; Agilent Technologies, Inc.). IAP extract and arbutin were dissolved in the mobile phase solvent (a mixture of acetonitrile and 10 mM KH_2PO_4 aqueous solution (1.1:98.9)) and filtered through 0.45-μm membrane filters before injection. During the analysis, the column was kept at a temperature of 30 °C, and arbutin was examined at a wavelength of 280 nm. The arbutin standard was procured from Alfa Aesar, MA, USA. For quantification, each sample was injected at a volume of 10 μL and a flow rate of 0.8 mL/min.

2.3. Animals

Seventy-two specific virus antibody-free/pathogen-free inbred male BALB/c mice, sourced from OrientBio in Seongnam, Republic of Korea, were acclimated for a period of seven days. After acclimatization, ten mice per group were allocated to six groups based on their body weight (average body weight of the normal vehicle control group: 21.23 ± 0.69 g; average body weight of the $PM_{2.5}$ lung injury-induced experimental group: 21.19 ± 0.82 g) measured one day before the first injection of $PM_{2.5}$ and test substance administration. The animal experiment was approved by the Animal Experiment Ethics Committee of Daegu Haany University under the approval number–DHU2022-098. Prior to test substance administration, a fasting period of 18 h was imposed on all experimental animals. On the final day leading up to the necropsy, there were no restrictions on water intake.

The experimental groups were as follows (six groups consisting of ten mice in each group):

1. Intact (vehicle) control = D.W. (10 mL/kg) administered and intranasal saline instillation (0.1 mL/kg) mice.
2. $PM_{2.5}$ control = D.W. (10 mL/kg) administered and intranasal $PM_{2.5}$ instillation (1 mg/kg) mice.
3. DEXA = DEXA (0.75 mg/kg) administered and intranasal $PM_{2.5}$ instillation (1 mg/kg) mice.
4. IAP400 = IAP (400 mg/kg) administered and intranasal $PM_{2.5}$ instillation (1 mg/kg) mice.
5. IAP200 = IAP (200 mg/kg) administered and intranasal $PM_{2.5}$ instillation (1 mg/kg) mice.
6. IAP100 = IAP (100 mg/kg) administered and intranasal $PM_{2.5}$ instillation (1 mg/kg) mice.

2.4. Induction of Lung Damage

To induce sub-acute lung injury, $PM_{2.5}$ at a dose of 0.1 mL/kg (equivalent to 1 mg/kg B.W.) was twice intranasal injected into the mice, one hour before test substance administration and with a 48 h gap between (Day 0 and Day 2). To prevent excessive aggregation of $PM_{2.5}$ molecules in the suspension, sonication was performed for 30 min using an ultrasonicator (Model 5210, Branson, St. Louis, MO, USA) before the intranasal injection.

2.5. Administration of Test Material

IAP concentrations of 40, 20, and 10 mg/mL were prepared in sterile D.W. IAP solution was orally administered to the corresponding animal once daily by utilizing a sonde attached to a 1-mL syringe for 10 days. The dose of IAP solution was set as 10 mL/kg (400, 200, and 100 mg/kg). DEXA (Sigma-Aldrich, St. Louis, MO, USA) was dissolved in sterile D.W. to prepare a 0.075 mg/mL solution and orally administered once daily for 10 days at 10 mL/kg (0.75 mg/kg) to the responding animals. To implement the similar handling, restraint, and administration stress for the normal vehicle and the $PM_{2.5}$ control group, the same volume of vehicle and sterile D.W. was supplied orally in the same way rather than the test substance or DEXA. The DEXA dose of 0.75 mg/kg used in this experiment has been previously reported [16,19,34,35]. The highest dose of IAP selected as the test substance was 400 mg/kg, and medium and low doses were established at 200 mg/kg and 100 mg/kg, respectively, with a common ratio of two.

2.6. Removal of Lung Tissue

Lung tissue was weighed using an automatic scale (XB320M; Precisa Instrument, Dietikon, Switzerland). NB 324 3-0 sterilized nylon thread (AILEE, Busan, Republic of Korea) was used for ligating right lower secondary lobe and left secondary bronchus. The lung left lobe was used for RT-PCR (real-time reverse transcription polymerase chain reaction) analysis, macroscopic, and histopathological observations. BALF collection was performed using the right upper and middle lobes of the lungs. The long right lower lobe was utilized to analyze cytokines (IL-6, TNF-α, CXCL-1, and -2; chemokine ligand 1 and 2), the levels of matrix metalloproteinases (MMP), substance P, acetylcholine (ACh), ROS, lipid peroxidation, and antioxidant defense system.

2.7. Observation Items

B.W. and weight gain, lung weight and gross autopsy findings, aspartate transaminase (AST) and alanine transaminase (ALT) levels in serum, total cell count in BALF, total white blood cell count, leukocyte count, TNF-α, IL-6, CXCL1, CXCL2, MMP-9, MMP-12, ACh, and substance P levels in lung tissue, antioxidant defense system in lung tissue, histopathological changes in lung tissue, mucus production in lung tissue, oxidative stress and inflammation, and changes in mRNA expression associated with apoptosis were observed in this study.

2.8. Lung Gross Necropsy Findings

The percentage of congested areas in the lung was measured using the method of Lee et al. [36]. After taking a picture of the lung left lobe with a digital camera, the percentage of congestion was calculated using an automated image analyzer (iSolution FL ver 9.1; IMT i-solution Inc., Burnaby, BC, Canada).

2.9. BALF Leukocyte Fractionation

The calculation of white blood cells, neutrophils, lymphocytes, eosinophil, and monocytes was done using automatic hemocytometer following the method of Shu et al. [37]. BALF was collected from blood following the method of Glynos et al. [38].

2.10. Lung Antioxidant Defense System

Lipid peroxidation (malondialdehyde [MDA] level), ROS and GSH levels, and CAT and SOD activity were analyzed according to the methods of Glynos et al. [38], Abei [39], and Bannister [40].

2.11. Histopathological Changes in Lung Tissue

The histopathological changes, such as mean alveolar surface area (ASA; $\%/mm^2$), the number of inflammatory cells infiltrating the alveoli ($cells/mm^2$), number of secondary bronchial mucosal periodic acid Schiff (PAS)-positive mucus-producing cells ($cells/mm^2$), alveolar septum thickness (μm), and secondary bronchial mucosal thickness (μm) were observed in the lung tissue following the methods of Lebargy et al. [41] and Shu et al. [37].

2.12. mRNA Expression in Lung Tissue

The mRNA expression of mucus-production-related mucin 5AC (MUC5AC) and MUC5B, oxidative stress- and inflammation-related nuclear factor (NF)-κB, p38 mitogen-activated protein kinase (MAPK), protein kinase B (Akt), phosphatase and tensin homolog (PTEN), phosphoinositide 3-kinase (PI3K), and apoptosis-related Bax and Bcl-2 genes, in lung tissue were analyzed using the methods of Deng et al. [42] and Duong et al. [43] (Table S1). Briefly, TRIzol reagent (Invitrogen, Carlsbad, CA, USA) was utilized to extract the RNA and reverse transcription of RNA was conducted using the cDNA Reverse Transcription kit (Thermo Fisher Scientific Inc., Rockford, IL, USA) following the manufacturer's instructions. RT-PCR was performed on CFX96^{TM} Real-Time System (Bio-Rad, Hercules, CA, USA), and data were normalized using the comparative threshold cycle method [44] in comparison with the β-actin (Actb) mRNA expression, used as a control.

2.13. Statistical Analysis

All numerical data are represented as mean values ± standard deviation, based on a sample size of ten mice in each group. Significance between groups was examined using SPSS 18.0 (SPSS Inc., Chicago, IL, USA). When there was no significance in variance observed in the Levene test, a one-way analysis of variance (ANOVA) test was employed. When the variance was significantly different, Dunnett's T3 test was performed to analyze the significance ($p < 0.05$) [45–47]. The change rate of the test substance or reference-drug-administered group compared with the $PM_{2.5}$ control or vehicle control group, respectively, was analyzed using the following equations:

$$\text{Percentage change compared to the intact vehicle control (\%)} = ((\text{data of } PM_{2.5} \text{ control} - \text{data of intact vehicle control mice})/\text{data of intact vehicle control mice}) \times 100 \quad (1)$$

$$\text{Percentage change compared to } PM_{2.5} \text{ control (\%)} = ((\text{data of test material- or reference-treated mice} - \text{data of } PM_{2.5} \text{ control mice})/\text{data of } PM_{2.5} \text{ control mice}) \times 100 \quad (2)$$

3. Results

3.1. Level of Arbutin in IAP Extract

The HPLC analysis of IAP extract showed arbutin level of 8.17 mg/g (Figure 1).

3.2. Changes on the Body Weight and Gains

In the DEXA-administered group, significant weight loss ($p < 0.05$ or $p < 0.01$) was noted in comparison with the normal vehicle control group 7 days after test substance administration, and 10 days weight gain also significantly decreased ($p < 0.05$ or $p < 0.01$) in comparison with the control group. In comparison with the normal vehicle control group, no substantial differences were observed in B.W. or weight gains in any of the experimental groups. Significant ($p < 0.05$ or $p < 0.01$) changes in weight loss were observed in the DEXA-administered group 8 days after administration in comparison with the $PM_{2.5}$ control group. The changes in weight gain over 10 days were also significantly reduced ($p < 0.05$ or $p < 0.01$) in comparison with the $PM_{2.5}$ control group. A significant difference in B.W. and weight gains was not observed in any of the three IAP-administered groups (IAP 400, 200, and 100; Figure 2).

3.3. Findings from the Macroscopic Examination of the Lungs and Changes in Weight

The $PM_{2.5}$ control group exhibited lung enlargement with significant local blockage and a significant increase ($p < 0.01$) in the blockage area and relative and absolute lung weight was observed in comparison with the normal vehicle control. Nevertheless, compared to the $PM_{2.5}$ control, the three IAP groups showed a significant and dose-dependent decrease ($p < 0.01$) in gross congestion in the lungs and both absolute and relative weights. In particular, in the IAP400 group, $PM_{2.5}$-induced pulmonary congestion and lung enlargement and suppression of absolute and relative lung weight increases were comparable to those in the DEXA group (Figure 3; Table 1).

Table 1. Lung weights and gross inspections in vehicle or $PM_{2.5}$-treated pulmonary-injured mice.

Items (Unit) Groups	Lung Weights		Congestional Regions (%)–Gross Findings
	Absolute (g)	Relative (%)	
Controls			
vehicle	0.125 ± 0.005	0.640 ± 0.033	1.64 ± 1.58
$PM_{2.5}$	0.185 ± 0.008 [a]	0.947 ± 0.061 [a]	68.07 ± 12.46 [a]
DEXA	0.136 ± 0.007 [bc]	0.735 ± 0.036 [ac]	8.72 ± 2.66 [ac]
Test substance–IAP			
400 mg/kg	0.142 ± 0.015 [c]	0.722 ± 0.080 [c]	9.02 ± 2.57 [ac]
200 mg/kg	0.153 ± 0.009 [ac]	0.771 ± 0.046 [ac]	31.68 ± 11.46 [ac]
100 mg/kg	0.161 ± 0.010 [ac]	0.829 ± 0.063 [ac]	42.49 ± 11.09 [ac]

Values are expressed as mean ± SD. DT3 = Dunnett's T3. [a] $p < 0.01$ and [b] $p < 0.05$ versus vehicle control via DT3 test. [c] $p < 0.01$ versus $PM_{2.5}$ control via DT3 test.

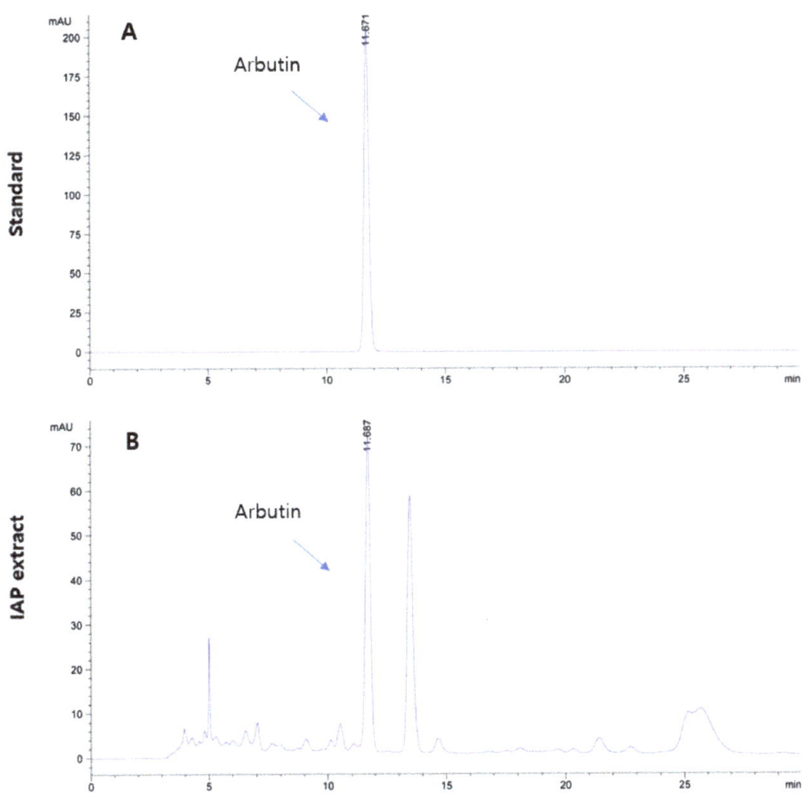

Figure 1. Profile of arbutin: (**A**) standard and (**B**) immature Asian pear extract.

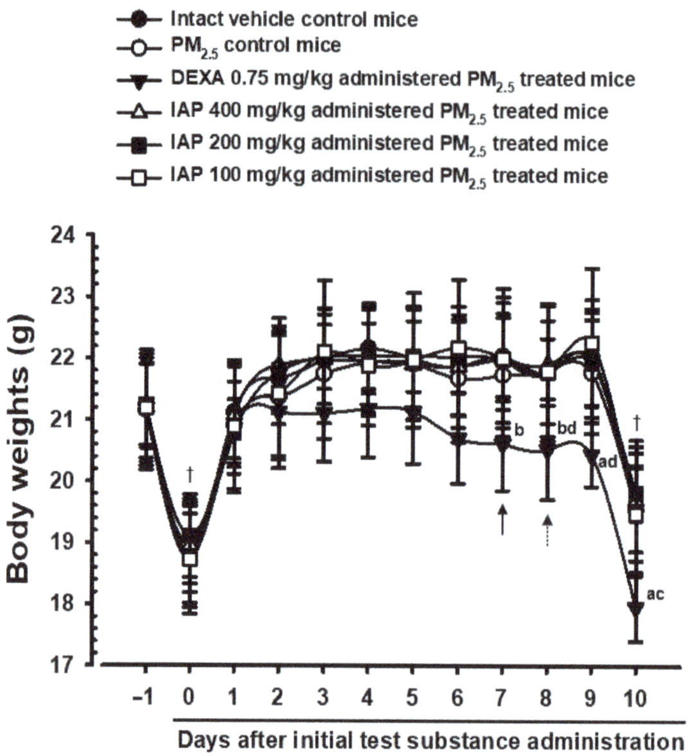

Figure 2. B.W. changes in vehicle or $PM_{2.5}$-treated pulmonary injured mice. THSD = Tukey's Honest Significant Difference. Day-1 signifies one day prior to the initial administration of the test substance. Day 10 represents the day of sacrifice, which is 24 h after the last (10th) administration of the test substance. All animals underwent overnight fasting prior to both the initial administration of the test article and sacrifice (†). ^{a} $p < 0.01$ and ^{b} $p < 0.05$ versus vehicle control by THSD test. ^{c} $p < 0.01$ and ^{d} $p < 0.05$ versus $PM_{2.5}$ control by THSD test.

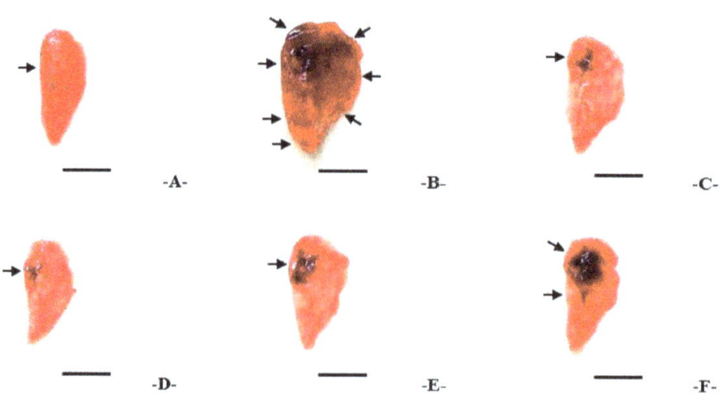

Figure 3. Representative gross left lung images, taken from vehicle or $PM_{2.5}$-treated pulmonary injured mice. (**A**) = vehicle control (D.W. PO (per oral) with saline). (**B**) = $PM_{2.5}$ control (D.W. PO with $PM_{2.5}$). (**C**) = DEXA (0.75 mg/kg of DEXA PO with $PM_{2.5}$). (**D**) = IAP400 (400 mg/kg of IAP PO mice with $PM_{2.5}$). (**E**) = IAP200 (200 mg/kg of IAP PO with $PM_{2.5}$). (**F**) = IAP100 (100 mg/kg of IAP PO with $PM_{2.5}$). Scale bars = 6.00 mm. Arrows indicated congestion regions.

3.4. Changes in Total Cell Count, Total Leukocyte Count, and Leukocyte Differential Count in BALF

In the $PM_{2.5}$ control, the total number of cells, neutrophils, monocytes, total leukocytes, lymphocytes, and acidophilus leukocytes in the BALF were significantly increased ($p < 0.01$) in comparison with the normal vehicle control. In contrast, the lymphocytes, total number of cells, neutrophils, total leukocytes, eosinophil, and monocytes in the BALF in all three IAP-treated groups decreased significantly ($p < 0.01$) in a dose-dependent manner. In particular, in the IAP400 group, the inhibition of an increase in the total amount of leukocytes and cells in the BALF was comparable to that in the DEXA group (Table 2).

Table 2. Cytology of BALF in vehicle or $PM_{2.5}$-treated pulmonary injured mice.

Items Groups	Total Cells	Total Leukocytes	Differential Counts			
			Lymphocytes	Neutrophils	Eosinophils	Monocytes
Controls						
vehicle	9.90 ± 2.02	6.70 ± 1.83	3.50 ± 1.35	1.25 ± 0.49	0.02 ± 0.02	1.28 ± 0.41
$PM_{2.5}$	79.70 ± 9.27 [a]	60.30 ± 6.88 [a]	35.70 ± 6.38 [a]	13.72 ± 1.51 [a]	1.85 ± 0.20 [a]	7.45 ± 0.87 [a]
DEXA	20.70 ± 3.86 [ac]	13.30 ± 3.33 [ac]	7.30 ± 3.27 [c]	2.46 ± 0.81 [bc]	0.28 ± 0.22 [bc]	2.20 ± 0.40 [ac]
Test substance–IAP						
400 mg/kg	21.40 ± 4.33 [ac]	13.80 ± 2.82 [ac]	7.40 ± 2.22 [ac]	2.64 ± 0.86 [ac]	0.32 ± 0.19 [ac]	2.32 ± 0.75 [bc]
200 mg/kg	43.30 ± 4.88 [ac]	32.20 ± 4.94 [ac]	19.50 ± 3.87 [ac]	6.77 ± 1.60 [ac]	0.93 ± 0.36 [ac]	3.70 ± 1.04 [ac]
100 mg/kg	54.40 ± 7.26 [ac]	40.00 ± 5.03 [ac]	23.50 ± 4.95 [ac]	8.74 ± 1.48 [ac]	1.21 ± 0.25 [ac]	4.79 ± 1.03 [ac]

Values are expressed as Mean ± SD, $\times 10^4$ cells/mL. BALF = Bronchoalveolar lavage fluid; DT3 = Dunnett's T3. [a] $p < 0.01$ and [b] $p < 0.05$ versus vehicle control via DT3 test. [c] $p < 0.01$ versus $PM_{2.5}$ control via DT3 test.

3.5. Changes in ALT and AST Levels in Serum

In contrast to the normal vehicle, all $PM_{2.5}$-induced lung injury experimental groups did not display significant changes in the serum AST and ALT levels. In comparison with the $PM_{2.5}$ control group, significant changes in serum ALT and AST levels were not found in the DEXA, IAP400, IAP200, and IAP100 groups (Figure 4).

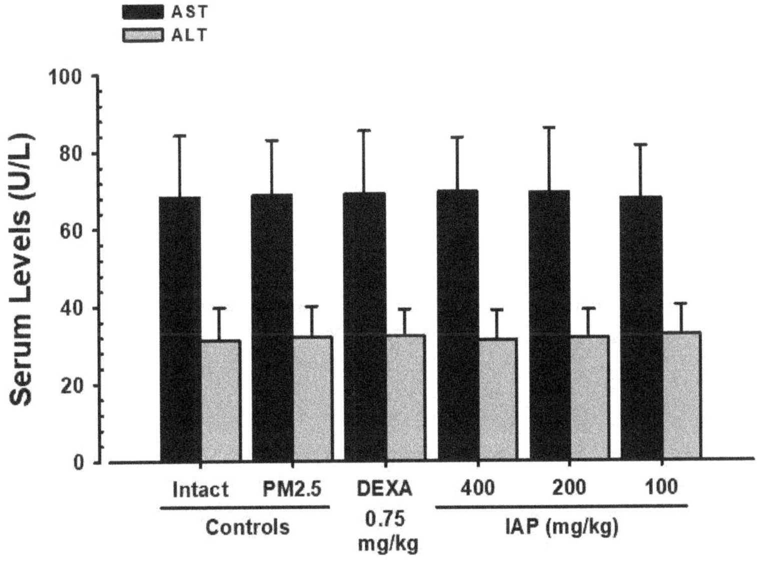

Figure 4. Serum AST and ALT levels in vehicle or $PM_{2.5}$-treated pulmonary-injured mice. Values are expressed mean ± SD. AST = Aspartate aminotransferase; ALT = Alanine aminotransferase.

3.6. Changes in the Levels of Cytokines IL-6, TNF-α, CXCL1, and CXCL2 in Lung Tissue

In the $PM_{2.5}$ control group, the amount of cytokines IL-6, TNF-α, CXCL1, and CXCL2 increased significantly ($p < 0.01$) in lung tissue in comparison with the normal vehicle control group. On the other hand, compared to the $PM_{2.5}$ control group, the levels of IL-6, TNF-α, CXCL2, and CXCL1 significantly decreased ($p < 0.01$) in groups supplied with all three doses of IAP extract. In addition, these changes were in a dose-dependent manner. In particular, in the IAP400 group, the suppression of the increase in IL-6, TNF-α, CXCL2, and CXCL1 in $PM_{2.5}$-induced lung tissue was comparable to that in the DEXA group (Table 3).

Table 3. Lung Cytokine-TNF-α, IL-6, CXCL1 and CXCL2 contents in vehicle or $PM_{2.5}$-treated pulmonary-injured mice.

Items (Unit)	Lung Contents (pg/mL)			
Groups	TNF-α	IL-6	CXCL1	CXCL2
Controls				
vehicle	30.55 ± 10.65	30.32 ± 12.46	32.54 ± 10.30	17.92 ± 5.80
$PM_{2.5}$	250.19 ± 43.99 [a]	453.02 ± 86.77 [a]	352.12 ± 80.52 [a]	198.62 ± 25.01 [a]
DEXA	68.92 ± 11.80 [ab]	119.94 ± 47.36 [ab]	97.70 ± 21.55 [ab]	54.88 ± 16.88 [ab]
Test substance–IAP				
400 mg/kg	69.65 ± 12.87 [ab]	122.59 ± 28.96 [ab]	101.86 ± 29.95 [ab]	56.88 ± 19.16 [ab]
200 mg/kg	115.42 ± 28.16 [ab]	212.47 ± 53.70 [ab]	162.32 ± 37.62 [ab]	94.80 ± 32.39 [ab]
100 mg/kg	156.02 ± 26.86 [ab]	289.37 ± 56.65 [ab]	215.30 ± 27.37 [ab]	124.52 ± 20.14 [ab]

Values are expressed mean ± SD. TNF = Tumor necrosis factor; IL = interleukin; CXCL = The chemokine (C-X-C motif) ligand; DT3 = Dunnett's T3. [a] $p < 0.01$ versus vehicle control via DT3 test. [b] $p < 0.01$ versus $PM_{2.5}$ control via DT3 test.

3.7. Changes in MMP-9 and MMP-12 Levels in Lung Tissue

In the $PM_{2.5}$ control, the levels of MMP-9 and -12 increased significantly ($p < 0.01$) in the lung tissue when compared to the normal vehicle control group. In contrast, in a dose-dependent manner, the levels of MMP-9 and -12 decreased significantly ($p < 0.01$) in all IAP-extract-administered groups. In particular, in the IAP400 group, the suppression of the increase in MMP-9 and MMP-12 levels in $PM_{2.5}$-induced lung tissue was comparable to that in the DEXA group (Figure 5).

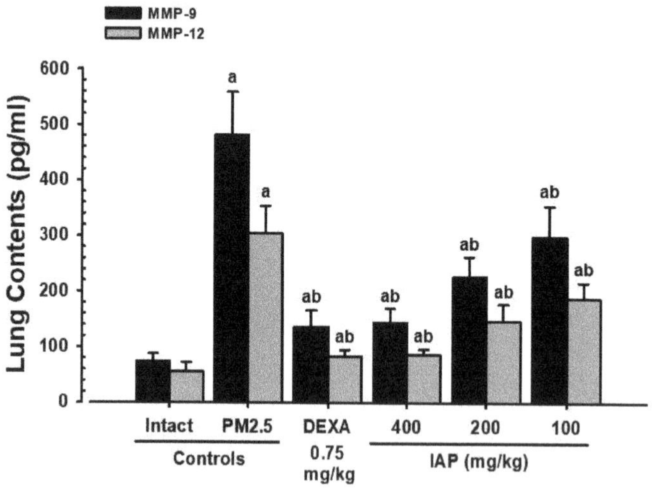

Figure 5. Lung MMP-9 and MMP-12 contents in vehicle or $PM_{2.5}$-treated pulmonary injured mice. Values are expressed mean ± SD. MMP = Matrix metalloproteinase; DT3 = Dunnett's T3. [a] $p < 0.01$ versus vehicle control via DT3 test. [b] $p < 0.01$ versus $PM_{2.5}$ control via DT3 test.

3.8. Changes in Substance P and ACh Levels in Lung Tissue

In the $PM_{2.5}$ control, substance P and ACh levels significantly increased ($p < 0.01$) in the lung tissue compared with the normal vehicle control group. In contrast, substance P and ACh levels decrease significantly ($p < 0.01$) in the DEXA group. In all three IAP-extract-administered groups, substance P and ACh levels in the lung tissue increased significantly ($p < 0.01$) in a dose-dependent manner compared to the $PM_{2.5}$ control (Figure 6).

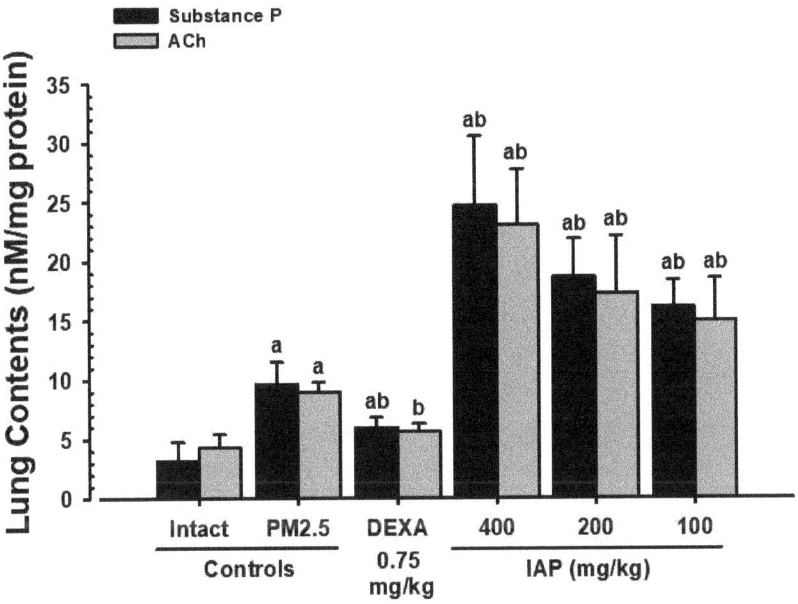

Figure 6. Lung substance P and ACh contents in vehicle or $PM_{2.5}$-treated pulmonary injured mice. Values are expressed mean ± SD. ACh = Acetylcholine; DT3 = Dunnett's T3. [a] $p < 0.01$ versus vehicle control via DT3 test. [b] $p < 0.01$ versus $PM_{2.5}$ control via DT3 test.

3.9. Changes in Lipid Peroxidation and Antioxidant Defense System in Lung Tissue

In the $PM_{2.5}$ control group, lipid peroxidation (MDA level) and ROS increased significantly in the lung tissue ($p < 0.01$), along with a decrease in the GSH level and CAT and SOD activity when compared with the normal vehicle control group. On the other hand, a significant reduction ($p < 0.01$) in lipid peroxidation and ROS and increases in the CAT, GSH, and SOD activity were observed in the lung tissue of mice treated with all three IAP extract doses. In addition, these effects were in a dose-dependent manner. Particularly, the IAP400 group exhibited comparable antioxidant activity to the DEXA-administered group (Table 4).

3.10. Alterations in the mRNA Expression of MUC5B and MUC5AC, Which Are Related to Mucus Production in Lung Tissue

In the $PM_{2.5}$ control group, the expression of mucus-production-related MUC5AC and MUC5B mRNA in the lung tissue significantly increased ($p < 0.01$) when compared with the normal vehicle control group. On the other hand, compared to the $PM_{2.5}$ control group in the lung tissue, MUC5AC and MUC5B mRNA expression were reduced significantly ($p < 0.01$) in all three IAP groups in a dose-dependent manner. In particular, MUC5AC and MUC5B mRNA expression was comparable between the IAP400 and DEXA groups (Table 5).

Table 4. Lung lipid peroxidation (MDA Contents), GSH Contents, and CAT and SOD activities in vehicle or $PM_{2.5}$-treated pulmonary injured mice.

Items (Unit) Groups	Lung Contents (nM/mg Protein)			Lung Enzyme Activity (U/mg Protein)	
	MDA	ROS	GSH	SOD	CAT
Controls					
vehicle	3.39 ± 0.97	23.29 ± 10.28	42.70 ± 10.63	337.60 ± 51.59	75.30 ± 14.03
$PM_{2.5}$	21.87 ± 4.56 ^{c}	96.66 ± 11.71 ^{a}	6.36 ± 2.05 ^{c}	60.00 ± 18.68 ^{cd}	9.40 ± 2.01 ^{c}
DEXA	8.58 ± 1.44 ^{cd}	43.15 ± 8.08 ^{ab}	18.84 ± 4.56 ^{cd}	189.60 ± 39.04 ^{cd}	30.00 ± 7.32 ^{cd}
Test substance–IAP					
400 mg/kg	8.91 ± 2.46 ^{cd}	44.30 ± 9.58 ^{ab}	20.28 ± 3.86 ^{cd}	197.40 ± 34.58 ^{cd}	30.80 ± 12.45 ^{cd}
200 mg/kg	12.65 ± 2.97 ^{cd}	57.72 ± 13.11 ^{ab}	16.44 ± 3.03 ^{cd}	157.00 ± 23.97 ^{cd}	24.90 ± 7.17 ^{cd}
100 mg/kg	14.82 ± 1.69 ^{cd}	66.53 ± 13.73 ^{ab}	12.74 ± 3.28 ^{cd}	129.30 ± 14.26 ^{cd}	19.60 ± 3.10 ^{cd}

Values are expressed mean ± SD. MDA = Malondialdehyde; ROS = Reactive oxygen species; GSH = Glutathione; CAT = Catalase; SOD = Superoxide dismutase; THSD = Tukey's Honest Significant Difference; DT3 = Dunnett's T3. ^{a} $p < 0.01$ versus vehicle control by THSD test. ^{b} $p < 0.01$ versus $PM_{2.5}$ control by THSD test. ^{c} $p < 0.01$ versus vehicle control via DT3 test. ^{d} $p < 0.01$ versus $PM_{2.5}$ control via DT3 test.

Table 5. Changes in the lung tissue mRNA expressions in vehicle or $PM_{2.5}$-treated pulmonary injured mice.

Groups Items	Controls			Test Substance–IAP		
	Vehicle	$PM_{2.5}$	DEXA	400 mg/kg	200 mg/kg	100 mg/kg
MUC5AC	1.00 ± 0.07	5.15 ± 0.57 ^{c}	2.05 ± 0.43 ^{ce}	2.10 ± 0.43 ^{ce}	2.91 ± 0.46 ^{ce}	3.61 ± 0.76 ^{ce}
MUC5B	1.00 ± 0.06	2.95 ± 0.46 ^{c}	1.61 ± 0.25 ^{ce}	1.64 ± 0.22 ^{ce}	1.93 ± 0.12 ^{ce}	2.08 ± 0.19 ^{ce}
NF-κB	1.00 ± 0.06	9.20 ± 1.59 ^{c}	2.32 ± 0.83 ^{ce}	2.57 ± 1.16 ^{de}	4.08 ± 0.91 ^{ce}	5.62 ± 1.03 ^{ce}
p38 MAPK	1.00 ± 0.06	8.31 ± 0.88 ^{c}	2.33 ± 0.53 ^{ce}	2.39 ± 0.58 ^{ce}	3.89 ± 1.17 ^{ce}	5.38 ± 1.01 ^{ce}
PTEN	1.00 ± 0.07	0.28 ± 0.10 ^{a}	0.71 ± 0.12 ^{ab}	0.70 ± 0.08 ^{ab}	0.60 ± 0.10 ^{ab}	0.52 ± 0.05 ^{ab}
PI3K	1.00 ± 0.07	6.36 ± 0.72 ^{c}	1.93 ± 0.36 ^{ce}	1.99 ± 0.24 ^{ce}	3.09 ± 0.81 ^{ce}	4.33 ± 0.88 ^{ce}
Akt	1.00 ± 0.06	5.45 ± 1.43 ^{c}	1.83 ± 0.34 ^{ce}	1.85 ± 0.19 ^{ce}	2.65 ± 0.43 ^{ce}	3.55 ± 0.47 ^{cf}
Bcl-2	1.00 ± 0.06	0.35 ± 0.07 ^{a}	0.72 ± 0.11 ^{ab}	0.71 ± 0.11 ^{ab}	0.60 ± 0.10 ^{ab}	0.52 ± 0.06 ^{ab}
Bax	1.00 ± 0.06	7.18 ± 1.41 ^{c}	2.70 ± 0.84 ^{ce}	2.72 ± 1.23 ^{de}	3.85 ± 0.82 ^{ce}	4.75 ± 0.60 ^{ce}

Values are expressed as mean ± SD, relative expressions/β-actin mRNA. RT-PCR = Reverse transcription polymerase chain reaction; NF-κB = Nuclear factor kappa-light-chain-enhancer of activated B cells; MAPK = Mitogen-activated protein kinases; PTEN = Phosphatase and tensin homolog, PI3K = Phosphoinositide 3-kinase; Akt = Protein kinase B; Bcl-2 = B-cell lymphoma 2; Bax = Bcl-2-associated X protein; THSD = Tukey's Honest Significant Difference; DT3 = Dunnett's T3. ^{a} $p < 0.01$ versus vehicle control by DT3 test. ^{b} $p < 0.01$ versus $PM_{2.5}$ control via DT3 test. ^{c} $p < 0.01$ and ^{d} $p < 0.05$ versus vehicle control by DT3 test. ^{e} $p < 0.01$ and ^{f} $p < 0.05$ versus $PM_{2.5}$ control via DT3 test.

3.11. Alterations in the mRNA Expression of NF-κB, p38 MAPK, Akt, PI3K, and PTEN, Which Are Related to Oxidative Stress and Inflammation in Lung Tissue

In $PM_{2.5}$ control, PI3K, NF-κB, Akt, and p38 MAPK mRNA expression increased significantly ($p < 0.01$), and PTEN mRNA expression reduced significantly ($p < 0.01$) compared to the normal vehicle control. In contrast, NF-κB, p38 MAPK, PI3K, and Akt mRNA expression reduced significantly ($p < 0.01$ or $p < 0.05$) and PTEN mRNA expression increased significantly ($p < 0.01$ or $p < 0.05$) in the lung tissue in all IAP groups in a dose-dependent manner. In particular, in the IAP400 group, p38 MAPK, inflammation-related NF-κB, Akt, PI3K, and PTEN mRNA expression and $PM_{2.5}$-induced oxidative stress were comparable to those in the DEXA supplied group (Table 5).

3.12. Changes in Apoptosis-Related Bcl-2 and Bax mRNA Expression in Lung Tissue

In lung tissue of the $PM_{2.5}$ control group, apoptosis-related Bcl-2 mRNA expression significantly decreased ($p < 0.01$), while Bax mRNA expression increased in comparison with the normal vehicle control group. In contrast, Bcl-2 mRNA expression in all three IAP extract groups increased significantly ($p < 0.01$) and Bax mRNA expression decreased in a dose-dependent manner. Particularly, in the IAP400 group, the changes in $PM_{2.5}$-induced apoptosis-related Bax and Bcl-2 mRNA expression in lung tissue were comparable to those in the DEXA-treated group (Table 5).

3.13. Histopathological Alterations Observed in the Lungs

In the $PM_{2.5}$ control group, significant alveolar septal thickening was observed, attributed to inflammatory cell infiltration, accompanied by elevated secondary bronchial mucosal thickening and an increased abundance of PAS-positive mucus-producing cells. Furthermore, there was a noteworthy increase in the average alveolar septal and secondary bronchial mucosal thicknesses ($p < 0.01$), as well as an augmented number of infiltrating inflammatory cells surrounding the alveoli and an increased count of PAS-positive mucus-producing cells in the secondary bronchial mucosa. Additionally, there was an associated decrease in ASA in the $PM_{2.5}$ control group when compared with the normal vehicle control group. Nevertheless, in all three IAP-extract-administered groups, significant increases ($p < 0.01$) in ASA and average alveolar septum thickness were observed in a dose-dependent manner. Additionally, there was a reduction in the quantity of infiltrating inflammatory cells surrounding the alveoli. Notably, in the IAP400 group, the inhibitory activity against $PM_{2.5}$-induced alveolar septum thickening, inflammatory cell infiltration, and ASA reduction was comparable to that observed in the DEXA-administered group. Furthermore, in line with the expectorant activity [16], all three IAP-extract-administered groups demonstrated significant increases ($p < 0.05$ or $p < 0.01$) in mean thickness of the secondary bronchial mucosa and the quantity of PAS-positive mucus-producing cells in a dose-dependent manner. Contrastingly, in the DEXA-administered group, no significant changes were observed in the average thickness of the secondary bronchial mucosa and the quantity of PAS-positive mucus-producing cells when compared to the $PM_{2.5}$ control group (Table 6; Figure 7).

Table 6. Histomorphometric analysis of the left lobe tissue of the lungs in mice treated with the vehicle or $PM_{2.5}$ to assess pulmonary injury.

Items Groups	Mean ASA (%/mm^2)	Mean Alveolar Septal Thickness (μm)	Mean Thickness of SB (μm)	Mean IF Cell Numbers Infiltrated in AR (×10 cells/mm^2)	PAS-Positive Cells on the SB (Cells/mm^2)
Controls					
vehicle	81.34 ± 8.03	7.52 ± 1.30	15.68 ± 1.81	62.00 ± 13.70	23.00 ± 3.16
$PM_{2.5}$	35.24 ± 7.72 ^{a}	7.25 ± 10.07 ^{a}	21.39 ± 1.34 ^{a}	732.40 ± 140.45 ^{a}	36.80 ± 4.54 ^{a}
DEXA	71.16 ± 7.40 ^{b}	21.77 ± 4.66 ^{ab}	21.05 ± 2.04 ^{a}	180.20 ± 64.34 ^{ab}	34.00 ± 12.07
Test substance–IAP					
400 mg/kg	57.25 ± 7.79 ^{ab}	21.95 ± 5.39 ^{ab}	31.60 ± 2.77 ^{ab}	184.80 ± 30.01 ^{ab}	86.20 ± 14.19 ^{ab}
200 mg/kg	60.39 ± 7.96 ^{ab}	26.37 ± 5.00 ^{ab}	28.82 ± 2.25 ^{ab}	305.00 ± 59.75 ^{ab}	69.60 ± 14.57 ^{ab}
100 mg/kg	50.54 ± 2.87 ^{ab}	30.76 ± 4.50 ^{ab}	26.59 ± 1.63 ^{ab}	399.60 ± 90.98 ^{ab}	53.60 ± 10.00 ^{ac}

Values are expressed as mean ± SD. ASA = Alveolar surface area; AR = Alveolar region; SB = Secondary bronchus mucosa; IF = Inflammatory; PAS = Periodic acid Schiff; DT3 = Dunnett's T3. ^{a} $p < 0.01$ versus vehicle control via DT3 test. ^{b} $p < 0.01$ and ^{c} $p < 0.05$ versus $PM_{2.5}$ control via DT3 test.

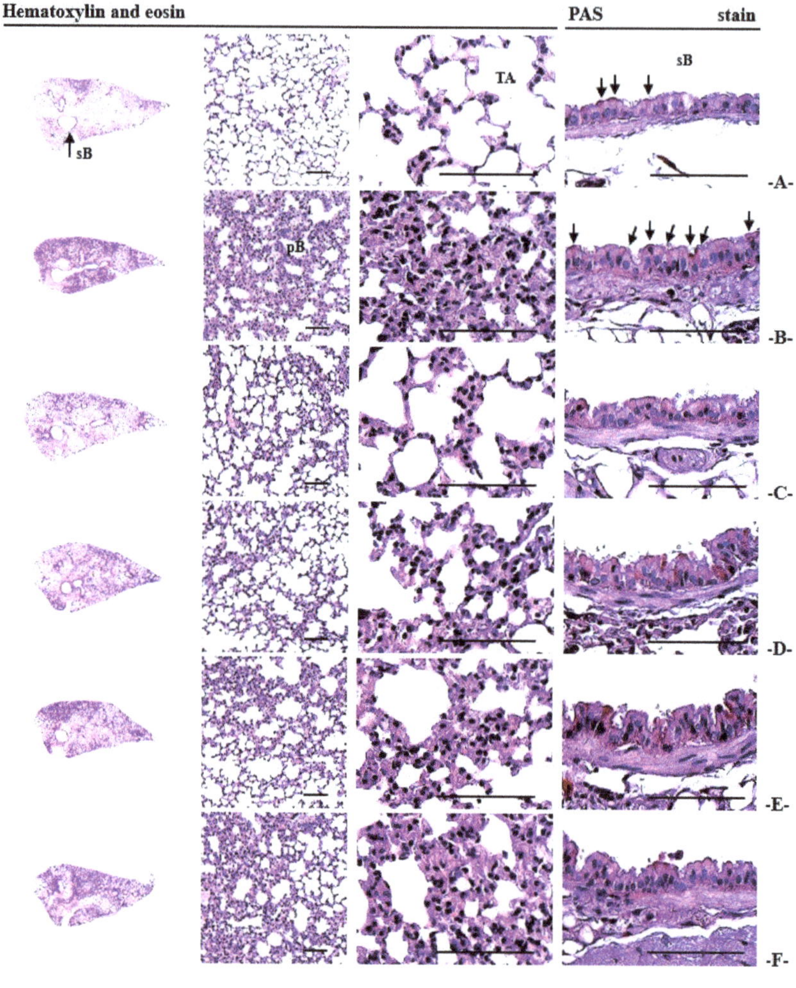

Figure 7. Representative and general histopathological profiles of the lung tissues taken from vehicle or $PM_{2.5}$-treated pulmonary-injured mice. ASA = Alveolar surface area; PAS = Periodic acid Schiff. sB = Secondary bronchus; pB = Primary bronchiole; TA = Terminal respiratory bronchiole-alveoli. (**A**) = vehicle control (DW PO (per oral) with saline). (**B**) = $PM_{2.5}$ control (DW PO with $PM_{2.5}$). (**C**) = DEXA (0.75 mg/kg of DEXA PO with $PM_{2.5}$). (**D**) = IAP400 (400 mg/kg of IAP PO mice with $PM_{2.5}$). (**E**) = IAP200 (200 mg/kg of IAP PO with $PM_{2.5}$). (**F**) = IAP100 (100 mg/kg of IAP PO with $PM_{2.5}$). Scale bars = 200 μm; arrows indicated PAS-positive mucus-producing cells.

4. Discussion

Due to the increase in the incidence of PM-induced respiratory diseases, an alternative method for treating or preventing respiratory damage caused by PM has become necessary. In this context, the $PM_{2.5}$-induced lung injury model using BALB/c mouse has shown pathologies similar to human PM-induced respiratory diseases [1–3,16]. Therefore, in the present study, the dose-dependent lung damage improvement effects of IAP extract were evaluated using a $PM_{2.5}$-induced mouse lung injury model to find new candidates for natural drugs and functional foods that can improve lung health [1–3,16].

Significant weight loss is a well-known side effect of DEXA [16,34,48,49]. In this study, significant weight loss in the DEXA-supplied group at 7 and 8 days was noticed in comparison with the normal vehicle and $PM_{2.5}$ control groups, respectively. Furthermore, compared to the normal vehicle and $PM_{2.5}$ control groups during the 10-day administration period, there was a significant decrease in B.W. and weight gain. On the other hand, in comparison with the $PM_{2.5}$ and vehicle control groups, the weight gain and B.W. were not significantly changed in any of the three IAP-extract-administered groups, and the animal weight was similar to that of BALB/c mice of the identical age [50,51].

AST and ALT have been used as representative blood biochemical indicators to determine liver damage [52]. No significant difference in the amount of serum AST and ALT levels was noticed between the vehicle control group and the $PM_{2.5}$-induced lung injury experimental groups. Moreover, no significant differences in serum AST and ALT levels were noticed in DEXA, IAP400, IAP200, and IAP100 groups. All experimental mice, including the $PM_{2.5}$ control group, showed changes in serum AST and ALT levels within the normal range for male BALB/c mice of the same age [50–52]. Hence, the current findings indicate that two intranasal injections of $PM_{2.5}$ (1 mg/kg) 48 h apart or repeated oral administration of DEXA (0.75 mg/kg) or IAP (400, 200, or 100 mg/kg) one time each day during 10 days does not cause serious liver toxicity in mice.

Lung weight is commonly used as an indicator for observing pulmonary edema as a result of vascular leakage [16,34,53]. Similar to the ovalbumin-induced asthma model [54,55] in $PM_{2.5}$-induced mice, as part of the inflammatory response, the vascular endothelial cells' binding force can weaken, resulting in pulmonary edema as an effect of vascular leakage increase [2,3,16]. Following intranasal injection of $PM_{2.5}$, lung enlargement with pronounced local congestion, indicating pulmonary edema, was noted, and significant increases in gross congestion and absolute and relative lung weights were observed compared to the normal vehicle control group. Nevertheless, in all IAP-extract-administered groups, significant reductions in pulmonary congestion (visibly observed) and lung weights (relative and absolute) were observed in a dose-dependent manner. Particularly, $PM_{2.5}$-induced lung blockage, weight gain (relative and absolute), and lung enlargement were comparable between the IAP400 and DEXA groups. These results provide evidence that the oral administration of IAP extract significantly and dose-dependently suppresses $PM_{2.5}$-induced pulmonary edema.

Regarding inflammatory lung damage caused by intranasal injection of $PM_{2.5}$, there were significant increases in lymphocytes, neutrophils, amount of total leukocytes, eosinophil, monocytes, and total cell number in BALF. In contrast, in all three IAP-extract-administered groups, the number of cells, white blood cells, lymphocytes, monocytes, neutrophils, and eosinophils in BALF significantly decreased in a dose-dependent manner. In particular, the amount of leukocytes and cells in BALF of the IAP400 group were comparable to those of the DEXA group, indicating that the oral administration of IAP extract significantly suppresses lung damage in $PM_{2.5}$-induced inflammation.

Mucus, present on the surface of the respiratory system, serves as the primary defense against various foreign substances and infectious agents. Mucin, a glycoprotein, is a crucial component of mucus [56,57]. The cleansing action of the microcilia in the respiratory system is directly influenced by the quantity and viscosity of mucin. In other words, impaired cleansing action of the microcilia is indicated by the presence of a large amount of mucin with high viscosity [58]. MUC5AC and MUC5B are the predominant polymeric substances secreted by the respiratory system [56,57], and alterations in the expression of MUC5AC and MUC5B are induced in various respiratory diseases [59–61]. Particularly, $PM_{2.5}$ is known to promote mucin production by increasing the expression of MUC5AC and MUC5B mRNA, resulting in the formation of high-viscosity sputum [16,56,57].

In the respiratory system, secretion is primarily regulated through the nervous system [62]. Stimulation or excitation of the vagus nerve leads to the local secretion of ACh and significant glandular secretion [63]. Indeed, Substance P is a well-known representative factor that promotes serous secretion in the respiratory system [64–66]. Hence, serous secretion in the respiratory system is primarily regulated by ACh and Substance P [62,63]. Furthermore, Substance P also functions as a neurogenic inflammatory factor in the respiratory system [67]. The inhalation of $PM_{2.5}$ leads to a noteworthy increase in Substance P levels [68,69]. Moreover, $PM_{2.5}$ exposure escalates vascular permeability, contributing to the inflammatory response [2,3], as a result causing an increase in ACh levels, which acts as a vasorelaxant factor [16,70]. However, in $PM_{2.5}$-induced lung injury mice, the vascular reactivity to ACh notably diminishes [71]. In the current experiment, the $PM_{2.5}$ control group exhibited heightened production of high-viscosity mucus, driven by increased levels of ACh and substance P [16]. Additionally, there was an increase in mRNA expression of MUC5AC and MUC5B [16,64,65], contributing to a pronounced neurogenic inflammatory response within lung tissue. On the contrary, as part of the mucolytic expectorant activity, significant increases in the secretion-promoting factors Substance P and ACh [16,62,63] were observed. Additionally, a reduction in mRNA expression of mucus-production-related genes MUC5AC and MUC5B [16,56,57] in lung tissue was observed in a dose-dependent manner in all three IAP-extract-administered groups. These results verify that the oral administration of IAP extract (400, 200, or 100 mg/kg) results in mucolytic expectorant activity through mucin-reducing effects by decreasing MUC5AC and MUC5B mRNA expression, as well as promoting serous secretion by increasing substance P and ACh production. In contrast, a significant reduction in MUC5AC and MUC5B mRNA expression was observed in the DEXA group. These expression levels were comparable to those observed in the IAP400 group. However, the levels of substance P and ACh were significantly decreased in the DEXA group when compared to the $PM_{2.5}$ control group.

The inhalation of $PM_{2.5}$ triggers an oxidative stress-induced inflammatory response characterized by increased lipid peroxidation and a depletion of the antioxidant defense system, leading to decreased GSH, CAT, and SOD activity [1–3,13,16,72,73]. As a consequence of the inflammatory response, there is an upsurge in the secretion of cytokines, such as TNF-α, IL-6, CXCL1, and CXCL2 [1–3,13,16,72,73]. Indeed, lipid peroxidation is an autocatalytic process that results in the oxidative degradation of cell membranes [74,75]. Destruction of the cell membrane by lipid peroxidation promotes the formation of toxic reactive aldehyde metabolites and cell death represented by MDA, i.e., free radicals [76,77]. Oxidative damage can affect a range of biological macromolecules, including but not limited to proteins, DNA, and lipids, as a result of exposure to various forms of ROS [77,78]. During the lipid peroxidation process, the amount of MDA produced is used to determine the lipid peroxidation index [77]. As an endogenous antioxidant, it serves as a representative molecule that effectively eliminates ROS even at low concentrations within cells, thereby controlling tissue damage [79]. SOD is also an endogenous antioxidant enzyme that acts as part of a cell's enzymatic defense system, and CAT is a representative endogenous antioxidant enzyme that converts toxic hydrogen peroxide (H_2O_2) to harmless water (H_2O) [80]. The decrease in GSH, SOD and CAT activity indicates the failure of compensatory action due to oxidative stress induced by inhalation of $PM_{2.5}$ [1–3,13,16].

$PM_{2.5}$ causes an increase in the level of chemotactic cytokines CXCL1 and CXCL2 in blood and respiratory tissues as part of an inflammatory response [16,72,73]. NF-κB is recognized as having an important role in inflammation associated with some adhesion molecules, ROS, and binds to promoters encoded in IL-6, IL-1, and TNF-α to trigger their transcription [81,82]. It also regulates various cytokine expressions, including IL-6, IL-10, CXCL1, and CXCL2, which are directly related to the inflammatory responses [83]. The amount of TNF-α, IL-6, CXCL1, and CXCL2 increases with an increase in NF-κB expression even during $PM_{2.5}$-induced lung injury [16,72]. Furthermore, p38 MAPK, PI3K, and Akt signaling pathways are involved in ROS-induced oxidative stress caused by $PM_{2.5}$-induced lung injury [42,84]. In particular, Akt activity triggers an inflammatory response by increasing the secretion of inflammatory chemotactic factor CXCL1 through an increase in NF-κB expression [85]. PTEN is a representative tumor suppressor gene, and abnormalities in PTEN cause the pathogenesis of several malignant tumors [86–88]. The MAPK signaling pathway regulates the expression of PTEN [89]. In other words, the inhibition of the MAPK signaling pathway maintains PTEN expression and significantly reduces the occurrence of malignant tumors [88,90]. The possibility of malignant tumor formation increases in $PM_{2.5}$-treated lung tissues, as there is a significant decrease in the expression of the tumor suppressor gene PTEN [91,92]. In the current study, the intranasal injection of $PM_{2.5}$ (1 mg/kg) showed significant lipid peroxidation in lung tissue, increased MDA and ROS levels, decreased GSH, CAT, and SOD activity, increased inflammatory cytokines TNF-α, IL-6, CXCL-1, and -2, together with inflammation-related NF-κB and oxidative stress, increased PI3K, Akt, and p38 MAPK expression, and decreased PTEN expression. Meanwhile, oxidative stress and inflammatory lung damage through $PM_{2.5}$-induced NF-κB, p38 MAPK, and PI3K/Akt expression decreased in a dose-dependent manner in all three IAP-extract-administered groups. The IAP400 group showed anti-inflammatory and antioxidant activity comparable to that in the DEXA group, suppressing $PM_{2.5}$-induced sub-acute lung injury. The results of this study indicate that the IAP extract administration dose-dependently inhibits the $PM_{2.5}$-induced oxidative stress and inflammatory lung damage via antioxidant and anti-inflammatory activities by the reduction of p38 MAPK, NF-κB, and PI3K/Akt expression.

MMPs are a group of structurally similar endopeptidases [93]. MMPs play an important role in tumor invasion, tissue morphogenesis, skin aging, angiogenesis, arthritis, and tissue repair, as they cause the degradation of various extracellular matrices [94,95]. MMPs can be subdivided into collagen-degrading enzymes, gelatinases, matrix-lysing enzymes, membranous MMPs, and various other types according to their structural characteristics and substrates [96,97]. $PM_{2.5}$ increases the expression of various MMPs, especially MMP-9 and MMP-12, as part of an inflammatory response, destroying the parenchyma of the surrounding respiratory system and causing respiratory distress through airway damage [16,93,98,99]. In this experiment, there were significant increases in MMP-9 and -12 levels in the lung tissue of the $PM_{2.5}$ control group compared to the normal vehicle control group. However, in contrast, the levels of MMP-9 and -12 in the lung tissue of all IAP-extract-administered groups were significantly reduced in a dose-dependent manner. Notably, the IAP400 group exhibited similar levels of MMP-9 and MMP-12 in $PM_{2.5}$-induced lung tissue compared to the DEXA group. These findings indicate that oral administration of IAP extract can effectively mitigate $PM_{2.5}$-induced airway damage in a dose-dependent manner by reducing MMP-9 and MMP-12 expression.

Indeed, excessive ROS imposes oxidative stress on proteins and lipids within the mitochondrial membrane, leading to severe damage. This damage may include loss of membrane potential, increased membrane permeability, and release of cytochrome c [100], activating caspase-3 and mitochondria-dependent apoptosis [101]. Oxidative stress caused by $PM_{2.5}$ induces mitochondria-dependent apoptosis [102]. Exactly, in the context of mitochondria-dependent apoptosis, Bax and Bcl-2 serve as representative pro-apoptotic and anti-apoptotic proteins, respectively. Bax plays a critical role in this process by inducing the loss of mitochondrial membrane potential, leading to the activation of caspase-3 and the release of cytochrome c, ultimately triggering apoptosis [103]. Exposure to $PM_{2.5}$ affects a rise in Bax expression and a decrease in Bcl-2 expression [16,100,101]. In this experiment, the expression of apoptosis-related Bcl-2 mRNA was significantly decreased, and the expression of Bax mRNA was significantly increased in the lung tissue of the $PM_{2.5}$ control group compared to the normal vehicle control group. On the other hand, the expression of Bcl-2 mRNA was increased, and the expression of Bax mRNA was decreased in the lung tissue of all three IAP-extract-administered mice, in a dose-dependent manner. In particular, Bcl-2 and Bax mRNA expression in $PM_{2.5}$-induced lung tissue was comparable between the IAP400 and DEXA groups. These results show that the oral administration of IAP extract can dose-dependently suppress lung damage caused by $PM_{2.5}$-induced apoptosis by regulating Bax and Bcl-2 mRNA expression.

The air space area (ASA; %/mm^2) could act as an indirect indicator to measure the gas exchange capacity histomorphometrically. A reduction in gas exchange surface area decreases ASA, indicating a decrease in lung function. In several lung diseases, a decrease in ASA has been noticed histopathologically [16,34,49,104]. PAS staining is a classic histochemical technique utilized to selectively stain mucus-secreting cells. An increase in PAS staining indicates heightened activity of mucus-producing cells [16,105,106]. Along with bronchial mucosal thickening, an increase in the number of PAS-positive mucus-producing cells in the bronchial mucosa is histopathologically considered as an indicator of an expectorant effect [16,106,107]. Consistent with the findings of previous drug efficacy experiments using a $PM_{2.5}$-induced lung injury model [2,3,13,16], the present study also demonstrated that intranasal injection of $PM_{2.5}$ (1 mg/kg) results in histopathologically significant lesions, including alveolar septum thickening due to inflammatory cell infiltration, and secondary bronchial mucosal thickening. Additionally, there was an increase in PAS-positive mucus-producing cells. Compared to the vehicle control group, the $PM_{2.5}$-induced lung injury mouse model exhibited significant increases in average alveolar septum and secondary bronchial mucosal thickness, as well as in the number of infiltrating inflammatory cells around the alveoli and amount of PAS-positive mucus-producing cells in the secondary bronchial mucosa. Concurrently, there were associated reductions in ASA in lung tissues. In contrast to the $PM_{2.5}$ control group, all IAP-extract-administered groups demonstrated notable increases in ASA and average alveolar septum thickness, along with a decrease in the number of infiltrating inflammatory cells surrounding the alveoli. The observed changes were in a dose-dependent manner. In particular, in the IAP400 group, the inhibition of $PM_{2.5}$-induced alveolar septum thickening and inflammatory cell infiltration and related ASA reduction were comparable to those in the DEXA group. Also, in a part of expectorant activity [16,106,107], significant increases in the amount of PAS-positive mucus-producing cells and the mean thickness of secondary bronchial mucosa were found in a dose-dependent manner in all IAP-extract-administered groups. However, in the DEXA group, compared to the $PM_{2.5}$ control group, significant changes in the amount of PAS-positive mucus-producing cells and average thickness of secondary bronchial mucosa were not observed. These findings show that the administration of IAP extract significantly activates not only the anti-inflammatory activity but also the mucolytic expectorant effect through the stimulation of gland fluid production in $PM_{2.5}$-induced sub-acute lung injury in a dose-dependent manner.

In addition, IAP extract showed expectorant effects on $PM_{2.5}$-induced sub-acute pulmonary injury in mice [108] and an alleviation effect on ovalbumin-induced asthma [109]. Based on the findings of this study, IAP extract shows promise as a potential natural drug or health-functional food for enhancing respiratory function and mitigating the impact of respiratory disorders caused by $PM_{2.5}$-induced lung injury. In this study, during HPLC analysis of IAP extract, substances corresponding to the remaining peaks, excluding arbutin, were not chemically identified. While the protective effect of IAP extract against lung injury was confirmed in this study, the protective effect of each HPLC peak could be studied in the future.

5. Conclusions

Stress-induced inflammatory lung damage through the increased expression of $PM_{2.5}$-induced PI3K/Akt and p38 MAPK mRNA was significantly suppressed via the administration of IAP extract (400–100 mg/kg) in a dose-dependent manner. As part of the mucolytic expectorant activity of IAP extract, significant increases in the amount of PAS-positive mucus-producing cells and mean thickness of the secondary bronchial mucosa were observed as compared to the $PM_{2.5}$ control group. Furthermore, IAP extract administration promoted serous fluid production in lung tissue, increased substance P and ACh levels, and decreased mucus production-related expression of MUC5AC and MUC5B mRNA in a dose-dependent manner. In particular, the administration of 400 mg/kg of IAP extract showed anti-inflammatory and antioxidative activity similar to that observed in the DEXA (0.75 mg/kg) administered group and significantly suppressed the $PM_{2.5}$-induced sub-acute lung injury. Unlike DEXA, IAP extract exhibited mucolytic expectorant activity by promoting meaningful serous fluid production (increase in substance P and ACh production and decrease in MUC5AC and MUC5B mRNA expression); therefore, it has great potential as a new pharmaceutical or health-functional food materials for improving effective respiratory function.

Supplementary Materials: The following supporting information can be downloaded at https://www.mdpi.com/article/10.3390/app13179578/s1, Table S1: Oligonucleotides for quantitative RT-PCR analysis.

Author Contributions: Conceptualization, K.-Y.K., S.-K.K. and J.-S.C.; methodology, S.S., E.-J.H. and S.-K.K.; software, J.-H.L. and M.-U.K.; validation, K.-Y.K. and S.S.; formal analysis, J.-H.L., M.-U.K. and S.-K.K.; investigation, M.-U.K. and J.-H.L.; resources, K.-Y.K.; data curation, M.-U.K. and S.-K.K.; writing, original draft preparation, M.-R.K., J.-H.L. and S.-K.K.; writing, review and editing, S.-K.K. and J.-S.C.; visualization, M.-R.K., J.-H.L. and M.-U.K.; supervision, S.-K.K. and J.-S.C.; project administration, K.-Y.K. and J.-S.C.; funding acquisition, K.-Y.K. and J.-S.C. All authors have read and agreed to the published version of the manuscript.

Funding: This study was supported by the Korea Institute of Planning and Evaluation for Technology in Food, Agriculture, Forestry (IPET) through the High Value-added Food Technology Development Program, funded by the Ministry of Agriculture, Food and Rural Affairs (MAFRA) (grant no. 122027-2).

Institutional Review Board Statement: The animal experiment was approved by the Animal Experiment Ethics Committee of Daegu Haany University with approval number DHU2022-098 (21 October 2022).

Informed Consent Statement: Not applicable.

Data Availability Statement: The data used to support the findings of this study are available from the corresponding author upon request.

Conflicts of Interest: The authors declare no conflict of interest.

References

1. Fernando, I.P.S.; Jayawardena, T.U.; Kim, H.S.; Lee, W.W.; Vaas, A.P.J.P.; De Silva, H.I.C.; Abayaweera, G.S.; Nanayakkara, C.M.; Abeytunga, D.T.U.; Lee, D.S.; et al. Beijing urban particulate matter-induced injury and inflammation in human lung epithelial cells and the protective effects of fucosterol from *Sargassum binderi* (Sonder ex J. Agardh). *Environ. Res.* **2019**, *172*, 150–158. [CrossRef]
2. Lee, W.; Ku, S.K.; Kim, J.E.; Cho, S.H.; Song, G.Y.; Bae, J.S. Inhibitory effects of protopanaxatriol type ginsenoside fraction (Rgx365) on particulate matter-induced pulmonary injury. *J. Toxicol. Environ. Health A* **2019**, *82*, 338–350. [CrossRef] [PubMed]
3. Lee, W.; Ku, S.K.; Kim, J.E.; Cho, S.H.; Song, G.Y.; Bae, J.S. Inhibitory effects of black ginseng on particulate matter-induced pulmonary injury. *Am. J. Chin. Med.* **2019**, *47*, 1237–1251. [CrossRef] [PubMed]
4. Zhuang, G.; Guo, J.; Yuan, H.; Zhao, C. The compositions, sources, and size distribution of the dust storm from China in spring of 2000 and its impact on the global environment. *Chin. Sci. Bull.* **2001**, *46*, 895–900. [CrossRef]
5. Wang, W.; Primbs, T.; Tao, S.; Simonich, S.L. Atmospheric particulate matter pollution during the 2008 Beijing Olympics. *Environ. Sci. Technol.* **2009**, *43*, 5314–5320. [CrossRef] [PubMed]
6. Huang, X.F.; He, L.Y.; Hu, M.; Zhang, Y.H. Annual variation of particulate organic compounds in $PM_{2.5}$ in the urban atmosphere of Beijing. *Atmos.Environ.* **2006**, *40*, 2449–2458. [CrossRef]
7. Lv, B.; Zhang, B.; Bai, Y. A systematic analysis of $PM_{2.5}$ in Beijing and its sources from 2000 to 2012. *Atmos. Environ.* **2016**, *124*, 98–108. [CrossRef]
8. Chiu, H.F.; Tsai, S.S.; Yang, C.Y. Short-term effects of fine particulate air pollution on hospital admissions for hypertension: A time-stratified case-crossover study in Taipei. *J. Toxicol. Environ. Health A* **2017**, *80*, 258–265. [CrossRef]
9. WHO. Air Pollution. Available online: https://www.who.int/health-topics/air-pollution#tab=tab_1 (accessed on 15 December 2022).
10. Nunes, C.; Pereira, A.M.; Morais-Almeida, M. Asthma costs and social impact. *Asthma Res. Pract.* **2017**, *3*, 1. [CrossRef]
11. Wang, G.; Huang, L.; Gao, S.; Gao, S.; Wang, L. Measurements of PM_{10} and $PM_{2.5}$ in urban area of Nanjing, China and the assessment of pulmonary deposition of particle mass. *Chemosphere* **2002**, *48*, 689–695. [CrossRef]
12. Schaumann, F.; Borm, P.J.; Herbrich, A.; Knoch, J.; Pitz, M.; Schins, R.P.; Luettig, B.; Hohlfeld, J.M.; Heinrich, J.; Krug, N. Metal-rich ambient particles (particulate matter 2.5) cause airway inflammation in healthy subjects. *Am. J. Respir. Crit. Care Med.* **2004**, *170*, 898–903. [CrossRef] [PubMed]
13. Wang, F.; Wang, R.; Liu, H. The acute pulmonary toxicity in mice induced by *Staphylococcus aureus*, particulate matter, and their combination. *Exp. Anim.* **2019**, *68*, 159–168. [CrossRef] [PubMed]
14. Pozzi, R.; De Berardis, B.; Paoletti, L.; Guastadisegni, C. Inflammatory mediators induced by coarse ($PM_{2.5-10}$) and fine ($PM_{2.5}$) urban air particles in RAW 264.7 cells. *Toxicology* **2003**, *183*, 243–254. [CrossRef]

15. Valavanidis, A.; Vlachogianni, T.; Fiotakis, K.; Loridas, S. Pulmonary oxidative stress, inflammation and cancer: Respirable particulate matter, fibrous dusts and ozone as major causes of lung carcinogenesis through reactive oxygen species mechanisms. *Int. J. Environ. Res. Public Health* **2013**, *10*, 3886–3907. [CrossRef] [PubMed]
16. Park, S.M.; Jung, C.J.; Lee, D.G.; Choi, B.R.; Ku, T.H.; La, I.J.; Cho, I.J.; Ku, S.K. *Adenophora stricta* root extract protects lung injury from exposure to particulate matter 2.5 in mice. *Antioxidants* **2022**, *11*, 1376. [CrossRef] [PubMed]
17. Devipriya, D.; Gowri, S.; Nideesh, T.R. Hepatoprotective effect of *Pterocarpus marsupium* against carbon tetrachloride induced damage in albino rats. *Anc. Sci. Life* **2007**, *27*, 19–25. [PubMed]
18. Kim, H.S.; Park, S.I.; Choi, S.H.; Song, C.H.; Park, S.J.; Shin, Y.K.; Han, C.H.; Lee, Y.J.; Ku, S.K. Single oral dose toxicity test of blue honeysuckle concentrate in mice. *Toxicol. Res.* **2015**, *31*, 61–68. [CrossRef]
19. Wang, P.; Liu, H.; Fan, X.; Zhu, Z.; Zhu, Y. Effect of San'ao decoction on aggravated asthma mice model induced by $PM_{2.5}$ and TRPA1/TRPV1 expressions. *J. Ethnopharmacol.* **2019**, *236*, 82–90. [CrossRef]
20. Lin, L.Z.; Harnly, J.M. Phenolic compounds and chromatographic profiles of pear skins (*Pyrus* spp.). *J. Agric. Food Chem.* **2008**, *56*, 9094–9101. [CrossRef]
21. Tanrioven, D.; Eksi, A. Phenolic compounds in pear juice from different cultivars. *Food Chem.* **2005**, *93*, 89–93. [CrossRef]
22. Lee, S.W.; Cho, J.Y.; Jeong, H.Y.; Na, T.W.; Lee, S.H.; Moon, J.H. Enhancement of antioxidative and antimicrobial activities of immature pear (*Pyrus pyrifolia* cv. Niitaka) fruits by fermentation with *Leuconostoc mesenteroides*. *Food Sci. Biotechnol.* **2016**, *25*, 1719–1726. [CrossRef] [PubMed]
23. Lee, K.H.; Cho, J.Y.; Lee, H.J.; Ma, Y.K.; Kwon, J.; Park, S.H.; Lee, S.H.; Cho, J.A.; Kim, W.S.; Park, K.H.; et al. Hydroxycinnamoylmalic acids and their methyl esters from pear (*Pyrus pyrifolia* Nakai) fruit peel. *J. Agric. Food Chem.* **2011**, *59*, 10124–10128. [CrossRef] [PubMed]
24. Cho, J.Y.; Kim, C.H.; Lee, H.J.; Lee, S.H.; Cho, J.A.; Kim, W.S.; Park, K.H.; Moon, J.H. Caffeoyl triterpenes from pear (*Pyrus pyrifolia* Nakai) fruit peels and their antioxidative activities against oxidation of rat blood plasma. *J. Agric. Food Chem.* **2013**, *61*, 4563–4569. [CrossRef] [PubMed]
25. Lee, K.H.; Cho, J.Y.; Lee, H.J.; Park, K.Y.; Ma, Y.K.; Lee, S.H.; Cho, J.A.; Kim, W.S.; Park, K.H.; Moon, J.H. Isolation and identification of phenolic compounds from an Asian pear (*Pyrus pyrifolia* Nakai) fruit peel. *Food Sci. Biotechnol.* **2011**, *20*, 1539–1545. [CrossRef]
26. Lee, H.S.; Isse, T.; Kawamoto, T.; Baik, H.W.; Park, J.Y.; Yang, M. Effect of Korean pear (*Pyrus pyrifolia* cv. Shingo) juice on hangover severity following alcohol consumption. *Food Chem. Toxicol.* **2013**, *58*, 101–106. [CrossRef] [PubMed]
27. Lee, Y.G.; Cho, J.Y.; Kim, C.M.; Jeong, H.Y.; Lee, D.I.; Kim, S.R.; Lee, S.H.; Kim, W.S.; Park, K.H.; Moon, J.H. Isolation and identification of 3 low-molecular compounds from pear (*Pyrus pyrifolia* Nakai cv. Chuhwangbae) fruit peel. *Korean J. Food Sci. Technol.* **2013**, *45*, 174–179. [CrossRef]
28. Cho, J.Y.; Lee, Y.G.; Lee, S.H.; Kim, W.S.; Park, K.H.; Moon, J.H. An ether and three ester derivatives of phenylpropanoid from pear (*Pyrus pyrifolia* Nakai cv. Chuhwangbae) fruit and their radical-scavenging activity. *Food Sci. Biotechnol.* **2014**, *23*, 253–259. [CrossRef]
29. Cho, J.Y.; Kim, E.H.; Yun, H.R.; Jeong, H.Y.; Lee, Y.G.; Kim, W.S.; Moon, J.H. Change in chemical constituents and free radical-scavenging activity during pear (*Pyrus pyrifolia*) cultivar fruit development. *Biosci. Biotechnol. Biochem.* **2015**, *79*, 260–270. [CrossRef]
30. Lee, Y.G.; Cho, J.Y.; Park, J.; Lee, S.H.; Kim, W.S.; Park, K.H.; Moon, J.H. Large-scale isolation of highly pure malaxinic acid from immature pear (*Pyrus pyrifolia* Nakai) fruit. *Food Sci. Biotechnol.* **2013**, *22*, 1539–1545. [CrossRef]
31. Lee, S.W.; Lee, Y.G.; Cho, J.Y.; Kim, Y.C.; Lee, S.H.; Kim, W.S.; Moon, J.H. Establishment of a simple method for purification of high purity chlorogenic acid from immature fruit of pear (*Pyrus pyrifolia* Nakai). *Appl. Biol. Chem.* **2015**, *58*, 335–341. [CrossRef]
32. Cho, J.Y.; Park, K.Y.; Lee, K.H.; Lee, H.J.; Lee, S.H.; Cho, J.A.; Kim, W.S.; Shin, S.C.; Park, K.H.; Moon, J.H. Recovery of arbutin in high purity from fruit peels of pear (*Pyrus pyrifolia* Nakai). *Food Sci. Biotechnol.* **2011**, *20*, 801–807. [CrossRef]
33. Hollander, J.L. Clinical use of dexamethasone: Role in treatment of patients with arthritis. *J. Am. Med. Assoc.* **1960**, *172*, 306–310. [CrossRef] [PubMed]
34. Min, B.G.; Park, S.M.; Choi, Y.W.; Ku, S.K.; Cho, I.J.; Kim, Y.W.; Byun, S.H.; Park, C.A.; Park, S.J.; Na, M.; et al. Effects of *Pelargonium sidoides* and Coptis Rhizoma 2 : 1 mixed formula (PS + CR) on ovalbumin-induced asthma in mice. *Evid. Based Complement. Altern. Med.* **2020**, *2020*, 9135637. [CrossRef] [PubMed]
35. Piao, C.H.; Fan, Y.J.; Nguyen, T.V.; Song, C.H.; Chai, O.H. Mangiferin alleviates ovalbumin-induced allergic rhinitis via Nrf2/HO-1/NF-κB signaling pathways. *Int. J. Mol. Sci.* **2020**, *21*, 3415. [CrossRef] [PubMed]
36. Lee, Y.S.; Cho, I.J.; Kim, J.W.; Lee, M.K.; Ku, S.K.; Choi, J.S.; Lee, H.J. Hepatoprotective effects of blue honeysuckle on CCl_4-induced acute liver damaged mice. *Food Sci. Nutr.* **2019**, *7*, 322–338. [CrossRef] [PubMed]
37. Shu, J.; Li, D.; Ouyang, H.; Huang, J.; Long, Z.; Liang, Z.; Chen, Y.; Chen, Y.; Zheng, Q.; Kuang, M.; et al. Comparison and evaluation of two different methods to establish the cigarette smoke exposure mouse model of COPD. *Sci. Rep.* **2017**, *7*, 15454. [CrossRef] [PubMed]
38. Glynos, C.; Bibli, S.I.; Katsaounou, P.; Pavlidou, A.; Magkou, C.; Karavana, V.; Topouzis, S.; Kalomenidis, I.; Zakynthinos, S.; Papapetropoulos, A. Comparison of the effects of e-cigarette vapor with cigarette smoke on lung function and inflammation in mice. *Am. J. Physiol. Lung Cell Mol. Physiol.* **2018**, *315*, L662–L672. [CrossRef] [PubMed]
39. Aebi, H. Catalase. In *Methods in Enzymatic Analysis*; Bergmeyer, H.U., Ed.; Academic Press: New York, NY, USA, 1974; pp. 673–686.
40. Bannister, J.V.; Calabrese, L. Assays for superoxide dismutase. In *Methods of Biochemical Analysis*; Glick, D., Ed.; John Wiley & Sons, Inc.: Hoboken, NJ, USA, 1987; Volume 32, pp. 279–312.

41. Lebargy, F.; Lenormand, E.; Pariente, R.; Fournier, M. Morphological changes in rat tracheal mucosa immediately after antigen challenge. *Bull. Eur. Physiopathol. Respir.* **1987**, *23*, 417–421.
42. Deng, X.; Rui, W.; Zhang, F.; Ding, W. $PM_{2.5}$ induces Nrf2-mediated defense mechanisms against oxidative stress by activating PIK3/AKT signaling pathway in human lung alveolar epithelial A549 cells. *Cell Biol. Toxicol.* **2013**, *29*, 143–157. [CrossRef]
43. Duong, C.; Seow, H.J.; Bozinovski, S.; Crack, P.J.; Anderson, G.P.; Vlahos, R. Glutathione peroxidase-1 protects against cigarette smoke-induced lung inflammation in mice. *Am. J. Physiol.-Lung Cell. Mol. Physiol.* **2010**, *299*, L425–L433. [CrossRef]
44. Schmittgen, T.D.; Livak, K.J. Analyzing real-time PCR data by the comparative CT method. *Nat. Protoc.* **2008**, *3*, 1101–1108. [CrossRef] [PubMed]
45. Ludbrook, J. Update: Microcomputer statistics packages. A personal view. *Clin. Exp. Pharmacol. Physiol.* **1997**, *24*, 294–296. [CrossRef] [PubMed]
46. Lee, S.; Lee, D.K. What is the proper way to apply the multiple comparison test? *Korean J. Anesthesiol.* **2018**, *71*, 353–360. [CrossRef]
47. Sauder, D.C.; DeMars, C.E. An updated recommendation for multiple comparisons. *Adv. Methods Pract. Psychol. Sci.* **2019**, *2*, 26–44. [CrossRef]
48. Piper, S.L.; Laron, D.; Manzano, G.; Pattnaik, T.; Liu, X.; Kim, H.T.; Feeley, B.T. A comparison of lidocaine, ropivacaine and dexamethasone toxicity on bovine tenocytes in culture. *J. Bone Jt. Surg. Br.* **2012**, *94*, 856–862. [CrossRef] [PubMed]
49. Ku, S.K.; Kim, J.W.; Cho, H.R.; Kim, K.Y.; Min, Y.H.; Park, J.H.; Kim, J.S.; Park, J.H.; Seo, B.I.; Roh, S.S. Effect of β-glucan originated from *Aureobasidium pullulans* on asthma induced by ovalbumin in mouse. *Arch. Pharm. Res.* **2012**, *35*, 1073–1081. [CrossRef]
50. Fox, J.G.; Cohen, B.J.; Loew, F.M. *Laboratory Animal Medicine*; Academic Press. Inc.: Orlando, FL, USA, 1984.
51. Tajima, Y. *Biological Reference Data Book on Experimental Animals*; Soft Science Inc.: Tokyo, Japan, 1989.
52. Sodikoff, C.H. *Laboratory Profiles of Small Animal Diseases, A Guide to Laboratory Diagnosis*; Mosby: St. Louise, MO, USA, 1995; pp. 1–36.
53. Tumes, D.J.; Cormie, J.; Calvert, M.G.; Stewart, K.; Nassenstein, C.; Braun, A.; Foster, P.S.; Dent, L.A. Strain-dependent resistance to allergen-induced lung pathophysiology in mice correlates with rate of apoptosis of lung-derived eosinophils. *J. Leukoc. Biol.* **2007**, *81*, 1362–1373. [CrossRef]
54. Okada, S.; Kita, H.; George, T.J.; Gleich, G.J.; Leiferman, K.M. Migration of eosinophils through basement membrane components in vitro: Role of matrix metalloproteinase-9. *Am. J. Respir. Cell Mol. Biol.* **1997**, *17*, 519–528. [CrossRef]
55. Tillie-Leblond, I.; Gosset, P.; Le Berre, R.; Janin, A.; Prangère, T.; Tonnel, A.B.; Guery, B.P. Keratinocyte growth factor improves alterations of lung permeability and bronchial epithelium in allergic rats. *Eur. Respir. J.* **2007**, *30*, 31–39. [CrossRef]
56. Na, H.G.; Kim, Y.D.; Choi, Y.S.; Bae, C.H.; Song, S.Y. Diesel exhaust particles elevate MUC5AC and MUC5B expression via the TLR4-mediated activation of ERK1/2, p38 MAPK, and NF-κB signaling pathways in human airway epithelial cells. *Biochem. Biophys. Res. Commun.* **2019**, *512*, 53–59. [CrossRef]
57. Wang, J.; Li, Y.; Zhao, P.; Tian, Y.; Liu, X.; He, H.; Jia, R.; Oliver, B.G.; Li, J. Exposure to air pollution exacerbates inflammation in rats with preexisting COPD. *Mediators Inflamm.* **2020**, *2020*, 4260204. [CrossRef] [PubMed]
58. Rose, M.C.; Nickola, T.J.; Voynow, J.A. Airway mucus obstruction: Mucin glycoproteins, MUC gene regulation and goblet cell hyperplasia. *Am. J. Respir. Cell Mol. Biol.* **2001**, *25*, 533–537. [CrossRef]
59. Groneberg, D.A.; Eynott, P.R.; Oates, T.; Lim, S.; Wu, R.; Carlstedt, I.; Nicholson, A.G.; Chung, K.F. Expression of MUC5AC and MUC5B mucins in normal and cystic fibrosis lung. *Respir. Med.* **2002**, *96*, 81–86. [CrossRef] [PubMed]
60. Henke, M.O.; Renner, A.; Huber, R.M.; Seeds, M.C.; Rubin, B.K. MUC5AC and MUC5B mucins are decreased in cystic fibrosis airway secretions. *Am. J. Respir. Cell Mol. Biol.* **2004**, *31*, 86–91. [CrossRef] [PubMed]
61. Kim, D.H.; Chu, H.S.; Lee, J.Y.; Hwang, S.J.; Lee, S.H.; Lee, H.M. Up-regulation of MUC5AC and MUC5B mucin genes in chronic rhinosinusitis. *Arch. Otolaryngol. Head Neck Surg.* **2004**, *30*, 747–752. [CrossRef] [PubMed]
62. Ballard, S.T.; Spadafora, D. Fluid secretion by submucosal glands of the tracheobronchial airways. *Respir. Physiol. Neurobiol.* **2007**, *159*, 271–277. [CrossRef] [PubMed]
63. Ueki, I.; German, V.F.; Nadel, J.A. Micropipette measurement of airway submucosal gland secretion. Autonomic effects. *Am. Rev. Respir. Dis.* **1980**, *121*, 351–357.
64. Haxhiu, M.A.; Haxhiu-Poskurica, B.; Moracic, V.; Carlo, W.A.; Martin, R.J. Reflex and chemical responses of trachea; submucosal glands in piglets. *Respir. Physiol.* **1990**, *82*, 267–278. [CrossRef]
65. Trout, L.; Corboz, M.R.; Ballard, S.T. Mechanism of substance P-induced liquid secretion across porcine bronchial epithelium. *Am. J. Physiol.* **2001**, *281*, L639–L645.
66. Phillips, J.E.; Hey, J.A.; Corboz, M.R. Tachykinin NK_3 and NK_1 receptor activation elicits secretion from porcine airway submucosal glands. *Br. J. Pharmacol.* **2003**, *138*, 254–260. [CrossRef]
67. Matsumoto, K.; Hosoya, T.; Tashima, K.; Namiki, T.; Murayama, T.; Horie, S. Distribution of trans ient receptor potential vanilloid 1 channel expressing nerve fibers in mouse rectal and colonic enteric nervous system: Relationship to peptidergic and nitrergic neurons. *Neuroscience* **2011**, *172*, 518–534. [CrossRef] [PubMed]
68. Liu, H.; Fan, X.; Wang, N.; Zhang, Y.; Yu, J. Exacerbating effects of $PM_{2.5}$ in OVA-sensitized and challenged mice and the expression of TRPA1 and TRPV1 proteins in lungs. *J. Asthma* **2017**, *54*, 807–817. [CrossRef] [PubMed]
69. Ji, Z.; Wang, Z.; Chen, Z.; Jin, H.; Chen, C.; Chai, S.; Lv, H.; Yang, L.; Hu, Y.; Dong, R.; et al. Melatonin attenuates chronic cough mediated by oxidative stress via transient receptor potential melastatin-2 in guinea pigs exposed to particulate matter 2.5. *Physiol. Res.* **2018**, *67*, 293–305. [CrossRef] [PubMed]

70. Fan, R.; Ren, Q.; Zhou, T.; Shang, L.; Ma, M.; Wang, B.; Xiao, C. Determination of endogenous substance change in $PM_{2.5}$-induced rat plasma and lung samples by UPLC-MS/MS method to identify potential markers for lung impairment. *Environ. Sci. Pollut. Res. Int.* **2019**, *26*, 22040–22050. [CrossRef] [PubMed]
71. Weldy, C.S.; Luttrell, I.P.; White, C.C.; Morgan-Stevenson, V.; Cox, D.P.; Carosino, C.M.; Larson, T.V.; Stewart, J.A.; Kaufman, J.D.; Kim, F.; et al. Glutathione (GSH) and the GSH synthesis gene Gclm modulate plasma redox and vascular responses to acute diesel exhaust inhalation in mice. *Inhal. Toxicol.* **2013**, *25*, 444–454. [CrossRef] [PubMed]
72. Jin, Y.; Wu, W.; Zhang, W.; Zhao, Y.; Wu, Y.; Ge, G.; Ba, Y.; Guo, Q.; Gao, T.; Chi, X.; et al. Involvement of EGF receptor signaling and NLRP12 inflammasome in fine particulate matter-induced lung inflammation in mice. *Environ. Toxicol.* **2017**, *32*, 1121–1134. [CrossRef] [PubMed]
73. Park, S.K.; Kang, J.Y.; Kim, J.M.; Kim, H.J.; Heo, H.J. *Ecklonia cava* attenuates $PM_{2.5}$-induced cognitive decline through mitochondrial activation and anti-inflammatory effect. *Mar. Drugs* **2021**, *19*, 131. [CrossRef] [PubMed]
74. Videla, L.A. Energy metabolism, thyroid calorigenesis, and oxidative stress: Functional and cytotoxic consequences. *Redox. Rep.* **2000**, *5*, 265–275. [CrossRef]
75. Subudhi, U.; Das, K.; Paital, B.; Bhanja, S.; Chainy, G.B. Alleviation of enhanced oxidative stress and oxygen consumption of L-thyroxine induced hyperthyroid rat liver mitochondria by vitamin E and curcumin. *Chem. Biol. Interact.* **2008**, *173*, 105–114. [CrossRef]
76. Venditti, P.; Di Meo, S. Thyroid hormone-induced oxidative stress. *Cell Mol. Life Sci.* **2006**, *63*, 414–434. [CrossRef]
77. Messarah, M.; Boumendjel, A.; Chouabia, A.; Klibet, F.; Abdennour, C.; Boulakoud, M.S.; Feki, A.E. Influence of thyroid dysfunction on liver lipid peroxidation and antioxidant status in experimental rats. *Exp. Toxicol. Pathol.* **2010**, *62*, 301–310. [CrossRef] [PubMed]
78. Das, K.; Chainy, G.B. Modulation of rat liver mitochondrial antioxidant defence system by thyroid hormone. *Biochim. Biophys. Acta* **2001**, *1537*, 1–13. [CrossRef] [PubMed]
79. Odabasoglu, F.; Cakir, A.; Suleyman, H.; Aslan, A.; Bayir, Y.; Halici, M.; Kazaz, C. Gastroprotective and antioxidant effects of usnic acid on indomethacin-induced gastric ulcer in rats. *J. Ethnopharmacol.* **2006**, *103*, 59–65. [CrossRef] [PubMed]
80. Cheeseman, K.H.; Slater, T.F. An introduction to free radical biochemistry. *Br. Med. Bull.* **1993**, *49*, 481–493. [CrossRef] [PubMed]
81. Ribbons, K.A.; Thompson, J.H.; Liu, X.; Pennline, K.; Clark, D.A.; Miller, M.J. Anti-inflammatory properties of interleukin-10 administration in hapten-induced colitis. *Eur. J. Pharmacol.* **1997**, *323*, 245–254. [CrossRef] [PubMed]
82. Carini, M.; Aldini, G.; Piccone, M.; Facino, R.M. Fluorescent probes as markers of oxidative stress in keratinocyte cell lines following UVB exposure. *Farmaco* **2000**, *55*, 526–534. [CrossRef] [PubMed]
83. Zaki, M.H.; Man, S.M.; Vogel, P.; Lamkanfi, M.; Kanneganti, T.D. Salmonella exploits NLRP12-dependent innate immune signaling to suppress host defenses during infection. *Proc. Natl. Acad. Sci. USA* **2014**, *111*, 385–390. [CrossRef]
84. Ryter, S.W.; Alam, J.; Choi, A.M. Heme oxygenase-1/carbon monoxide: From basic science to therapeutic applications. *Physiol. Rev.* **2006**, *86*, 583–650. [CrossRef]
85. Dong, Y.L.; Kabir, S.M.; Lee, E.S.; Son, D.S. CXCR2-driven ovarian cancer progression involves upregulation of proinflammatory chemokines by potentiating NF-κB activation via EGFR-transactivated Akt signaling. *PLoS ONE* **2013**, *8*, e83789. [CrossRef]
86. Zheng, H.; Ying, H.; Yan, H.; Kimmelman, A.C.; Hiller, D.J.; Chen, A.J.; Perry, S.R.; Tonon, G.; Chu, G.C.; Ding, Z.; et al. p53 and Pten control neural and glioma stem/progenitor cell renewal and differentiation. *Nature* **2008**, *455*, 1129–1133. [CrossRef]
87. Qi, Q.; Ling, Y.; Zhu, M.; Zhou, L.; Wan, M.; Bao, Y.; Liu, Y. Promoter region methylation and loss of protein expression of PTEN and significance in cervical cancer. *Biomed. Rep.* **2014**, *2*, 653–658. [CrossRef] [PubMed]
88. Li, Z.H.; Li, L.; Kang, L.P.; Wang, Y. MicroRNA-92a promotes tumor growth and suppresses immune function through activation of MAPK/ERK signaling pathway by inhibiting PTEN in mice bearing U14 cervical cancer. *Cancer Med.* **2018**, *7*, 3118–3131. [CrossRef]
89. Jiang, X.; Li, H. Overexpression of LRIG1 regulates PTEN via MAPK/MEK signaling pathway in esophageal squamous cell carcinoma. *Exp. Ther. Med.* **2016**, *12*, 2045–2052. [CrossRef] [PubMed]
90. Ebbesen, S.H.; Scaltriti, M.; Bialucha, C.U.; Morse, N.; Kastenhuber, E.R.; Wen, H.Y.; Dow, L.E.; Baselga, J.; Lowe, S.W. Pten loss promotes MAPK pathway dependency in HER2/neu breast carcinomas. *Proc. Natl. Acad. Sci. USA* **2016**, *113*, 3030–3035. [CrossRef] [PubMed]
91. Chen, S.; Li, D.; Zhang, H.; Yu, D.; Chen, R.; Zhang, B.; Tan, Y.; Niu, Y.; Duan, H.; Mai, B.; et al. The development of a cell-based model for the assessment of carcinogenic potential upon long-term $PM_{2.5}$ exposure. *Environ. Int.* **2019**, *131*, 104943. [CrossRef] [PubMed]
92. Chao, X.; Yi, L.; Lan, L.L.; Wei, H.Y.; Wei, D. Long-term $PM_{2.5}$ exposure increases the risk of non-small cell lung cancer (NSCLC) progression by enhancing interleukin-17a (IL-17a)-regulated proliferation and metastasis. *Aging* **2020**, *12*, 11579–11602. [CrossRef] [PubMed]
93. Jin, X.J.; Kim, E.J.; Oh, I.K.; Kim, Y.K.; Park, C.H.; Chung, J.H. Prevention of UV-induced skin damages by 11,14,17-eicosatrienoic acid in hairless mice in vivo. *J. Korean Med. Sci.* **2010**, *25*, 930–937. [CrossRef] [PubMed]
94. Kahari, V.M.; Saarialho-Kere, U. Matrix metalloproteinases in skin. *Exp. Dermatol.* **1997**, *6*, 199–213. [CrossRef]
95. Woessner, J.F., Jr. Role of matrix proteases in processing enamel proteins. *Connect. Tissue Res.* **1998**, *39*, 69–73. [CrossRef]
96. Nagase, H.; Woessne, J.F., Jr. Matrix metalloproteinases. *J. Biol. Chem.* **1999**, *274*, 21491–21494. [CrossRef]
97. Sternlicht, M.D.; Werb, Z. How matrix metalloproteinases regulate cell behavior. *Annu. Rev. Cell Dev. Biol.* **2001**, *17*, 463–516. [CrossRef] [PubMed]
98. Yang, B.; Guo, J.; Xiao, C. Effect of $PM_{2.5}$ environmental pollution on rat lung. *Environ. Sci. Pollut. Res. Int.* **2018**, *25*, 36136–36146. [CrossRef] [PubMed]

99. Jiang, Y.; Zhao, Y.; Wang, Q.; Chen, H.; Zhou, X. Fine particulate matter exposure promotes M2 macrophage polarization through inhibiting histone deacetylase 2 in the pathogenesis of chronic obstructive pulmonary disease. *Ann. Transl. Med.* **2020**, *8*, 1303. [CrossRef] [PubMed]
100. Monian, P.; Jiang, X. Clearing the final hurdles to mitochondrial apoptosis: Regulation post cytochrome C release. *Exp. Oncol.* **2012**, *34*, 185–191. [PubMed]
101. Wang, A.S.; Xu, Y.; Zhang, Z.W.; Lu, B.B.; Yin, X.; Yao, A.J.; Han, L.Y.; Zou, Z.Q.; Li, Z.; Zhang, X.H. Sulforaphane protects MLE-12 lung epithelial cells against oxidative damage caused by ambient air particulate matter. *Food Funct.* **2017**, *8*, 4555–4562. [CrossRef] [PubMed]
102. Li, X.; Ding, Z.; Zhang, C.; Zhang, X.; Meng, Q.; Wu, S.; Wang, S.; Yin, L.; Pu, Y.; Chen, R. MicroRNA-1228(*) inhibit apoptosis in A549 cells exposed to fine particulate matter. *Environ. Sci. Pollut. Res. Int.* **2016**, *23*, 10103–10113. [CrossRef] [PubMed]
103. Xiong, S.; Mu, T.; Wang, G.; Jiang, X. Mitochondria-mediated apoptosis in mammals. *Protein Cell.* **2014**, *5*, 737–749. [CrossRef] [PubMed]
104. Choi, H.Y.; Jung, T.Y.; Ku, S.K.; Yang, H.B.; Lee, H.S. Toxico-pathological study of p,p-DDE after experimental aerosol exposed to ICR Mouse. *Toxicol. Res.* **2005**, *21*, 151–160.
105. Honda, H.; Fujimoto, M.; Miyamoto, S.; Ishikawa, N.; Serada, S.; Hattori, N.; Nomura, S.; Kohno, N.; Yokoyama, A.; Naka, T. Sputum Leucine-rich alpha-2 glycoprotein as a marker of airway inflammation in asthma. *PLoS ONE* **2016**, *11*, e0162672. [CrossRef]
106. Hu, J.R.; Jung, C.J.; Ku, S.M.; Jung, D.H.; Ku, S.K.; Choi, J.S. Antitussive, expectorant, and anti-inflammatory effects of Adenophorae Radix powder in ICR mice. *J. Ethnopharmacol.* **2019**, *239*, 111915. [CrossRef]
107. Wang, D.; Wang, S.; Chen, X.; Xu, X.; Zhu, J.; Nie, L.; Long, X. Antitussive, expectorant and anti-inflammatory activities of four alkaloids isolated from Bulbus of *Fritillaria wabuensis*. *J. Ethnopharmacol.* **2012**, *139*, 189–193. [CrossRef]
108. Negara, B.F.S.P.; Kim, J.W.; Bashir, K.M.I.; Lee, J.H.; Ku, M.U.; Kim, K.Y.; Shin, S.; Hong, E.-J.; Ku, S.K.; Choi, J.-S. Expectorant effects of immature Asian pear extract on $PM_{2.5}$-induced subacute pulmonary injury in mice. *J. Food Biochem.* **2023**, *2023*, 5671679. [CrossRef]
109. Kim, M.R.; Bashir, K.M.I.; Lee, J.H.; Ku, M.U.; Kim, J.W.; Kim, K.Y.; Shin, S.; Hong, E.-J.; Ku, S.K.; Choi, J.-S. Efficacy confirm test of immature Asian pear extracts on the ovalbumin-induced asthma in mice. *Appl. Sci.* **2023**, *13*, 9342. [CrossRef]

applied sciences

MDPI

Article

The Inhibition Activities of the Fruit Extract of *Plinia cauliflora* against Melanoma Cells and the Single-Stranded DNA-Binding Protein (SSB) from *Klebsiella pneumoniae*

En-Shyh Lin [1] and Cheng-Yang Huang [2,3,*]

1 Department of Beauty Science, National Taichung University of Science and Technology, Taichung City 403, Taiwan
2 Department of Biomedical Sciences, Chung Shan Medical University, Taichung City 402, Taiwan
3 Department of Medical Research, Chung Shan Medical University Hospital, Taichung City 402, Taiwan
* Correspondence: cyhuang@csmu.edu.tw

Abstract: *Plinia cauliflora* has been associated with numerous ethnobotanical applications. In this study, we uncovered that the fruit extract of *P. cauliflora,* obtained using 50% ethanol, possesses inhibition activity against the *Klebsiella pneumoniae* single-stranded DNA-binding protein (KpSSB). SSB plays a critical role in cell survival, making it an attractive target for the development of anti-infective drugs. The inhibition activity against KpSSB by the *P. cauliflora* extract demonstrated an IC_{50} value of 73 ± 8 μg/mL. By using gas chromatography–mass spectrometry, the chemical content of this extract was tentatively determined. The top 15 compounds (>0.7%) were as follows: 5-hydroxymethylfurfural, 2,3-dihydro-3,5-dihydroxy-6-methyl-4h-pyran-4-one, 2,5-diformylfuran, furfural, ace-tic acid, citraconic anhydride, formic acid, ethyl 4-hydroxy-3-methylbut-2-enoate, furfuryl alcohol, furyl hydroxymethyl ketone, 3-acetyl-3-hydroxyoxolane-2-one, 2,3-dihydro-5-hydroxy-2-(hydroxymethyl)-4H-pyran-4-one, 2(3H)-furanone, dihy-dro-4-hydroxy-, hydroxyacetone, and 1-hydroxybut-3-en-2-one. To analyze the possible binding modes, the three most abundant compounds were then subjected to docking analysis. We also investigated whether the *P. cauliflora* extract exhibited any cytotoxic and antiproliferative effects on the survival of B16F10 melanoma cells. Additionally, we found that the extract of *P. cauliflora* could inhibit the migration and induce apoptosis of B16F10 cells. The results of this study collectively suggest that *P. cauliflora* holds potential pharmacological benefits, warranting further exploration for therapeutic applications.

Keywords: *Plinia cauliflora*; SSB; single-stranded DNA; antipathogen; B16F10 melanoma; GC–MS analysis; *Klebsiella pneumoniae*; 5-hydroxymethylfurfural; anticancer; molecular docking

1. Introduction

Plants have been characterized as a rich source of pharmaceutical products with therapeutic or prophylactic properties, which are associated with antioxidant, anti-inflammatory, antimicrobial, and antiproliferative activities [1–5]. The strategy of exploring medicinal plants as a source of therapeutic compounds has demonstrated its utility and productivity for potential pharmacological uses [4]. One notable advantage of utilizing natural extracts in therapeutic applications lies in their ability to exert multi-targeted modes of action [6]. Several active constituents derived from plant extracts, including vincristine, vinblastine, and paclitaxel, have already shown promise as effective anticancer drugs [7,8]. Thus, it is worth determining new beneficial properties of plant extracts for various therapeutic purposes.

Jaboticaba is the edible fruit of the jabuticabeira (*Plinia cauliflora*; also known as *Myrciaria cauliflora* previously) [9]. The fruit of *P. cauliflora* (jaboticaba) grows directly on the trunk of the tree (Figure 1A). The *P. cauliflora* fruit is a sweet, thick, and sticky white pulp covered in a dark purple-black outer skin [9,10]. The fruits of *P. cauliflora* are used as food and contain carbohydrates, minerals, amino acids, and vitamins. The fruits of *P. cauliflora* are eaten raw or used to make jellies, jams, juices, or wine. This superfruit [11] also contains polyphenols, flavonoids, anthocyanins, carotenoids, and gallotannins [12,13]. *P. cauliflora* has long been employed for the treatment of diarrhea, labyrinthitis, skin irritations, flu, genitourinary problems, and asthma [10]. Recently, the potential anticarcinogenic and antiproliferative effects of this fruit on several cancer cell lines such as breast [14] and lung cancer cells [15] are reported. Antibacterial activities of the fruit extracts of *P. cauliflora* were also demonstrated [16,17]. Therefore, determining the molecular targets and understanding the reaction mechanisms are of considerable interest for further biotechnological and pharmacological uses. In this study, the fruit extract of *P. cauliflora* was used to test for the suppression of B16F10 melanoma cells and the inhibition of the DNA-binding activity of the recombinant single-stranded DNA-binding protein (SSB), an essential DNA replication protein in bacteria [18].

Figure 1. Preparation and analysis of the fruit extract from *Plinia cauliflora*. (**A**) *P. cauliflora*, commonly used in traditional Brazilian medicine, boasts various ethnobotanical applications. The fruit, also known as jaboticaba, grows directly on the tree trunk. (**B**) Jaboticaba is the edible fruit of *P. cauliflora*. (**C**) GC chromatogram. The content was analyzed by GC–MS. By matching the generated spectra with mass spectral libraries, compounds were tentatively identified. The most abundant compound detected in this extract was 5-hydroxymethylfurfural, with a retention time of 19.13 min.

SSB is essential for all aspects of DNA metabolic processes, including replication, repair, recombination, and replication restart in bacteria [19–24]. SSB binds tightly and cooperatively to single-stranded DNA (ssDNA), regardless of sequence [18]. SSB is known to form interactions with numerous DNA metabolism proteins, creating what is referred to as the SSB interactome [25]. Given its crucial roles, SSB presents itself as a potential target for antibacterial drug development [26–28]. In eukaryotes, replication protein A (RPA) serves as the equivalent of bacterial SSB [29]. Although RPA shares similar functions with SSB, it differs in terms of structure and numerous other functions [30]. While SSB functions as a homotetramer(s) [24], human RPA (RPA1, RPA2, and RPA3) operates as a heterotrimer [30]. Additionally, RPA binds to its partner proteins through mechanisms that are different from those of SSB [20,30]. Accordingly, pharmacologically inhibiting SSB could be a suitable approach for targeting bacterial pathogens.

As recognized by the Infectious Disease Society of America (IDSA), a faction of antibiotic-resistant bacteria, namely *Enterococcus faecium*, *Staphylococcus aureus*, *Klebsiella pneumoniae*, *Acinetobacter baumannii*, *Pseudomonas aeruginosa*, and *Enterobacter* spp., acronymically dubbed the ESKAPE pathogens, can effectively "escape" the biocidal action of antibiotics [31]. These pathogens cause a number of lethal diseases and antimicrobial resistance (AMR) creates significant challenges in treating bacterial infections in both animals and humans [32,33]. As described by IDSA, these opportunistic ESKAPE pathogens are significantly associated with a high degree of morbidity, mortality, and increased costs. The rise and spread of ESKAPE pathogens with AMR pose a serious threat to global health. *K. pneumoniae* are one of the dangerous ESKAPE organisms and are a major cause of hospital- and community-acquired infections, including pneumonia, liver abscesses, and sepsis [34,35]. Given that the antibacterial activities of the fruit extracts of *P. cauliflora* were established [36,37], the molecular targets in *K. pneumonia* inhibited by extracts of *P. cauliflora* should be explored. Thus, we assessed here whether the fruit extract of *P. cauliflora* had activity against *K. pneumonia* SSB (KpSSB). The pharmacological inhibition of *P. cauliflora* extract against KpSSB may be useful in suppressing the DNA replication and decreasing the virulence of *K. pneumonia*.

Considering that DNA replication is an essential process for cell survival [38], DNA replication proteins should be suitable targets for drug development. In this study, we identified that KpSSB, an essential DNA replication protein in *K. pneumonia*, could be inhibited by the fruit extract of *P. cauliflora* obtained using 50% ethanol (Figure 1B). Given that human RPA [30] and SSB [20] are different in structure and function, bacterial SSB should be a prime target in antibiotic development. By using gas chromatography–mass spectrometry (GC–MS), the chemical content of this extract was determined. We have recently solved the crystal structure of KpSSB, and thus, the three most abundant compounds in the extract of *P. cauliflora* were further selected for docking analysis of KpSSB to realize the possible inhibition modes. Given that many natural products have anti-skin-cancer activities [39,40], we also investigated whether the extract of *P. cauliflora* exhibited any antiproliferative effects on the growth of B16F10 melanoma cells. Additionally, we found that the extract of *P. cauliflora* could induce apoptosis and inhibit the migration of B16F10 cells. The results of this study collectively suggest that *P. cauliflora* holds potential pharmacological benefits, warranting further exploration for therapeutic applications.

2. Materials and Methods

2.1. Materials

All solvents and chemicals used were of the highest grade, commercially obtained from Sigma-Aldrich (Saint Louis, MO, USA). The *E. coli* strain BL21(DE3) pLysS (Novagen, Worcestershire, UK) was used for protein expression. The cell line B16F10 melanoma, obtained from the Food Industry Research and Development Institute, Hsinchu, Taiwan, was maintained as a monolayer culture in Dulbecco's modified Eagle medium (GibcoTM; Thermo Fisher Scientific, Waltham, MA, USA) supplemented with 10% fetal bovine serum (FBS), 100 unit/mL penicillin, and 100 μg/mL streptomycin. Cells were incubated at 37 °C in a 95% air and 5% CO_2 incubator.

2.2. Expression and Isolation of the Recombinant Protein

The gene encoding KpSSB was cloned into pET21b, which incorporates a 6xHis tag at the protein's N-terminus. The resulting plasmid, referred to as pET21b-KpSSB, was introduced into *E. coli* BL21 (DE3) cells. These transformed cells were cultivated to an OD_{600} of 0.9 in LB medium supplemented with 250 μg/mL ampicillin, under vigorous shaking at 37 °C. Recombinant KpSSB was then induced by the addition of 1 mM isopropyl thiogalactopyranoside for 9 h at 25 °C. The induced cells were harvested and resuspended in Buffer A (20 mM Tris–HCl, 5 mM imidazole, and 0.5 M NaCl, pH 7.9), followed by disruption through sonication. The recombinant protein was purified from the soluble supernatant using Ni^{2+}-affinity chromatography (HisTrap HP; GE Healthcare Bio-Sciences, Piscataway, NJ, USA). Elution was performed with Buffer B (20 mM Tris–HCl, 250 mM imidazole, and 0.5 M NaCl, pH 7.9), and the purified protein was subsequently dialyzed against a dialysis buffer (20 mM HEPES and 100 mM NaCl, pH 7.0; Buffer C). Protein purity was assessed by electrophoresis on a 12% SDS-PAGE gel.

2.3. Electrophoretic Mobility Shift Analysis (EMSA)

For analysis of KpSSB binding, a biotinylated deoxythymidine (dT) homopolymer dT30 was employed as a standard assay. The ssDNA dT30 was biotinylated at its 5′ terminal. Incubation of this labeled ssDNA (30 fmol/μL) was with purified KpSSB at various concentrations (0, 20, 39, 78, 156, 312, 625, 1250, 2500, and 5000 nM). The electrophoretic mobility shift analysis (EMSA) was conducted using the LightShift Chemiluminescent EMSA Kit. Briefly, KpSSB and the DNA substrate were incubated together for 60 min at 37 °C in 40 mM Tris–HCl (pH 7.5) and 50 mM NaCl. After adding a dye mixture, an 8% native polyacrylamide gel was used. The reaction samples were analyzed at 100 V for 1 h in TBE running buffer. The protein–DNA complexes were electroblotted onto a positively charged nylon membrane (GE, Boston, MA, USA). A UV-light instrument equipped with 312 nm bulbs was used for a cross-linking reaction. After a 10-min exposure, the transferred DNA was then cross-linked to the nylon membrane. The streptavidin–horseradish peroxidase conjugate and chemiluminescent substrate (Pierce Biotechnology, Waltham, MA, USA) were used for detecting the DNA. The $[Protein]_{50}$ was estimated based on the concentration of the protein that bound 50% of the input DNA.

2.4. Inhibition Assay

KpSSB at a concentration of 625 nM was subjected to incubation with both dT30 and the extract of *P. cauliflora* at various concentrations (0, 7.8, 15.6, 31.2, 62.5, 125, 250, 500, and 1000 μg/mL). After EMSA, a titration curve was constructed based on the experimental data. The concentration of the *P. cauliflora* extract required achieving 50% inhibition (IC_{50}) of KpSSB activity that was directly determined from the graphical analysis.

2.5. Plant Material and Fruit Harvest

Fresh fruits of *P. cauliflora* (*M. cauliflora* O. Berg) were procured from a private farm located in Baoshan Township, Hsinchu County, Taiwan (24.7493° N, 121.0213° E). The ripe fruits (around 10 kg) were collected. The collected fruits were cleaned to remove impurities and inspected for damage, alterations, and signs of immaturity or overripeness. Of these collected fruits, 78 fruits of *P. cauliflora* (1000 g) were selected for further extraction. It took one day from the moment of harvesting the raw material to the point of conducting the analyses and obtaining the extract. The raw material was transported under controlled conditions at 4 °C.

2.6. Extract Preparation

The collected ripe fruits were then dried, pulverized into a fine powder, and subjected to extraction using 50% ethanol to explore their pharmacological potential. From 1000 g of these fruits, 27 g of dry powder were obtained after the lyophilization drying process (BTP-9ES, Benchtop Pro, Wolflabs, Pocklington, UK). In a liquid nitrogen atmosphere, a grinding mill (HR7629, Philips, Amsterdam, The Netherlands) was used for preparing a fine powder. The extraction process (one single extraction) involved placing plant powder (1 g) into a conical flask, adding the extraction solvent (100 mL), shaking the mixture on an orbital shaker for 5 h, and filtering the resulting extract using a 0.45 μm filter. The solvent (50% ethanol) in this extract was then removed via a hot air circulation oven at 40 °C. The *P. cauliflora* fruit extract was preserved at −80 °C until further use. To create a stock solution (20 mg/mL), the extract powder was dissolved in 20% DMSO. The stock solution was further diluted with the supplemented culture medium to the required assay concentrations for the anticancer cell assays. Incubation of B16F10 cells with the culture medium containing 0.2% DMSO served as a control group. In the case of EMSA, the stock solution was diluted with Buffer C to achieve the designated assay concentrations. Incubation of KpSSB with Buffer C plus 1% DMSO served as a control group for the EMSA experiments.

2.7. GC–MS Analysis

A Thermo Scientific TRACE 1300 Gas Chromatograph was used with a Thermo Scientific ISQ Single Quadrupole Mass Spectrometer system for GC–MS analysis. The conditions were: a Rtx-WAX column, 30 m × 0.25 mm i.d. × 0.25 μm film; the carrier gas, helium (a flow rate of 1 mL/min); the initial oven temperature, 50 °C (and increased to 250 °C at a rate of 10 °C/min); the compounds discharged from the column, a quadrupole mass detector with ions generated through the electron ionization method; and the mass range scanned, 29–650 amu. The relative mass fraction of each chemical component was determined using the peak area normalization method. By comparing the generated spectra with the NIST 2011 and Wiley 10th edition mass spectral libraries, compounds were tentatively identified.

2.8. Agar Well Diffusion Assay

The antibacterial activities of the *P. cauliflora* extract were assessed using an agar well diffusion assay [41]. A 0.1 McFarland standard suspension was prepared with the pathogens *Escherichia coli*, *Klebsiella pneumoniae*, *Pseudomonas aeruginosa*, and *Staphylococcus aureus*. These bacterial suspensions were then inoculated into sterile Petri dishes. The plates contained 60 mL of Muller–Hinton agar. The *P. cauliflora* extract was dissolved in 30% DMSO, and 10 mg of the extract was carefully transferred onto the agar plate. The plates were subsequently incubated at 37 °C for 12 h. The size (diameter) of the inhibition zone serves as an indicator of the antibacterial activity, with larger zones indicating higher potency of the extract. For comparison, ampicillin and 30% DMSO served as the positive and negative controls, respectively. The reported values show the mean standard deviation of at least three independent experiments.

2.9. Total Phenolic Content and Flavonoid Content

The total phenolic content (TPC) was assessed using the Folin–Ciocalteu method [42]. The developed blue color's absorbance was measured at 750 nm using a UV/VIS spectrophotometer (Hitachi U 3300, Hitachi High-Technologies, Tokyo, Japan). The obtained result was compared to the gallic acid standard curve and expressed as mg of gallic acid equivalent/g dry weight of plant. To determine the flavonoid content, the aluminum chloride colorimetric method was used [43]. The absorbance of both the extracts and standard solutions was measured at 510 nm with the same spectrophotometer. The results are expressed as mg of quercetin equivalent/g dry weight of plant. The individual standard of gallic acid and quercetin used was 0–500 μg/mL. Values show the mean standard deviation of at least three independent experiments.

2.10. Antioxidant Activity Analysis

We determined the antioxidant ability of the *P. cauliflora* extract using the 1,1-diphenyl-2-picrylhydrazyl (DPPH) radical scavenging assay [44]. The absorbance of the reaction was measured at 517 nm. DPPH free radical scavenging activity was calculated using the formula: %Radical scavenging activity = (Control OD − Sample OD)/Control OD × 100. Values show the mean standard deviation of at least three independent experiments. In comparison, L-ascorbic acid as a positive control had an IC_{50} value of 28.6 ± 0.2 μg/mL.

2.11. Trypan Blue Cytotoxicity Assay

We evaluated cell death using the trypan blue cytotoxicity assay [45]. In this assay, B16F10 cells (1×10^4) were prepared to incubate with the extract of *P. cauliflora* in a 100 µL volume. After 24 h, the cytotoxic activity of the extract was estimated via trypan blue dye.

2.12. Chromatin Condensation Assay

We examined apoptosis in cancer cells using the Hoechst 33342 staining method [46]. B16F10 cells were seeded in 96-well plates at a density of 5×10^3 cells per well. The cells were allowed to adhere for 16 h. These resultant cells were then incubated with the *P. cauliflora* extract for 24 h. Subsequently, the cells were washed with PBS and stained with Hoechst dye (1 µg/mL) in the dark for 10 min. ImageXpress Pico (Molecular Devices, Silicon Valley, CA, USA) was used to capture images of the stained cells. Image acquisition was carried out on the DAPI filter cubes. The obtained images were analyzed using the CellReporterXpress Version 2 software.

2.13. Clonogenic Formation Assay

We evaluated the inhibition of B16F10 cell proliferation using a clonogenic formation assay [47]. In this assay, B16F10 cells were seeded at a density of 1×10^3 cells per well in 6-well plates. These cells were then allowed to incubate overnight for attachment. Following the treatment with the *P. cauliflora* extract for a period of 5–7 days, the cells were washed with PBS, and the formed colonies were fixed with methanol. After staining with 0.5% crystal violet for 20 min, the number of colonies was counted under a light microscope to determine the extent of proliferation inhibition.

2.14. Wound-Healing Assay

We investigated the inhibition of B16F10 cell migration using the wound-healing assay [48]. The wound-healing assay is a well-established in vitro technique for assessing collective cell migration in two dimensions. In this assay, a cell-free area is generated within a confluent cell monolayer by removing the cells from the designated area. The presence of the cell-free area prompts the remaining cells to migrate and close the gap. Initially, B16F10 cells were seeded and grown in a serum-reduced medium for 6 h. Afterward, a linear wound was created across the well using a pipette tip. Subsequently, the cells were washed twice with the serum-reduced medium. After treatment with the *P. cauliflora* extract for 24 h, their migration ability was assessed.

2.15. MOE-Dock Analysis

The binding analysis was carried out using MOE-Dock. The binding capacities in comparison to each other were calculated. The structure of KpSSB was used for docking analysis. Any water molecules present in this crystal structure were removed using MOE. To ensure accuracy, a 3D protonation step followed by energy minimization was applied to add hydrogen atoms to the protein structure. The highest-ranked conformations were generated. The binding modes were predicted and visualized directly through the MOE-Dock tool.

2.16. Biorad Protein (Bradford) Assay

The dye reagent was prepared by diluting 1-part Dye Reagent Concentrate with 4 parts deionized water and filtering it through a 0.25 µm filter to remove any particulates. A bovine serum albumin (BSA) protein standard was prepared in concentrations ranging from 0.1 to 0.5 mg/mL and assayed in duplicate. For each standard and sample solution, 10 µL was pipetted into separate wells of a microtiter plate, and 200 µL of the prepared diluted dye reagent was added. The solutions were mixed by pipetting up and down using a multichannel pipette. The plate was then incubated at room temperature for at least 5 min before measuring the absorbance at 595 nm.

2.17. Statistical Analysis

Data were created at least 3 times and expressed as mean ± standard deviation. One-way ANOVA was used to measure significant differences between means. The statistical significance was defined as a p-value < 0.01. These analyses were performed using SPSS software version 26.0 for Windows (IBM, Chicago, IL, USA).

3. Results

3.1. Inhibition of KpSSB by the Extract of Plinia cauliflora

Due to the importance in DNA replication, repair, and recombination, the inhibition of SSB protein by the fruit extract of *P. cauliflora* as a potential antimicrobial therapy was investigated. KpSSB was used in this demonstration. This extract of *P. cauliflora* was obtained by 50% ethanol (Figure 1). Recombinant KpSSB was purified by Ni^{2+}-affinity chromatography. To quantify the binding affinity, we conducted EMSA to study the interaction between KpSSB and ssDNA at various protein concentrations. A biotinylated deoxythymidine (dT) homopolymer dT30 was used as a standard assay for the binding analysis of KpSSB. Through a streptavidin–horseradish peroxidase conjugate for detecting the biotin-labeled ssDNA and the complex, incubation of dT30 with KpSSB produced a band shift (Figure 2A). Input dT30 incubated with KpSSB of 625 nM was completely bound (Figure 2A). Through the titration curve, the binding constant ($[\text{Protein}]_{50}$) was calculated to be 175 ± 12 nM (Figure 2B). The extract of *P. cauliflora* was used to inhibit KpSSB activity (Figure 2C). In this inhibition assay, the extract at concentrations of 7.8–1000 μg/mL was used. The EMSA result showed that the ssDNA binding activity of KpSSB (625 nM) was inhibited by this extract at concentrations of 62.5–1000 μg/mL (Figure 2D). At a concentration of 250 μg/mL, this extract could completely inhibit the binding of KpSSB to dT30. Through the titration curve, the extract of *P. cauliflora* exhibited an inhibition capacity against KpSSB with an IC_{50} value of 73 ± 8 μg/mL (Figure 2D). The tuber extract from *Sinningia bullata* obtained by 50% ethanol was also used for investigating the anti-KpSSB properties; however, the extract of *S. bullata* (1000 μg/mL) did not influence the binding of KpSSB. Accordingly, a certain compound in the extract of *P. cauliflora*, alone or in combination, could be a potential anti-KpSSB agent.

3.2. Inhibition of KpSSB by Rutin

The fruit extract of *P. cauliflora* contains several bioactive natural products, including kaempferol, myricetin, quercetin, and rutin [9,49]. To investigate their inhibitory effects on KpSSB, these commercially available chemicals, which were dissolved in 10% dimethyl sulfoxide (DMSO) at concentrations ranging from 0 to 300 μM, were individually included in the binding assay. Using EMSA, we evaluated the impact of each compound on the dT30 binding activity of KpSSB. The results revealed that rutin exhibited an inhibition with an IC_{50} value of 293 ± 16 μM. On the other hand, kaempferol, myricetin, and quercetin at a concentration of 300 μM did not affect the binding of KpSSB to dT30. As a result, we have identified rutin as a new inhibitor of KpSSB. Our current research involves obtaining crystals of the KpSSB–rutin complex to conduct a more comprehensive analysis of their interactions.

3.3. Antibacterial Activity of the Extract of P. cauliflora

The extract of *P. cauliflora* could inhibit SSB (Figure 2), the essential protein for bacterial DNA replication. We further investigated the antibacterial activities of this extract using the agar well diffusion method (Figure 3). The zone of inhibition is a circular area around the spot of the extract in which the bacteria colonies do not grow. This extract (10 mg) exhibited antibacterial activities against the human pathogens *Escherichia coli*, *Klebsiella pneumoniae*, *Pseudomonas aeruginosa*, and *Staphylococcus aureus* used for this analysis and produced the inhibition zones of 19 ± 1, 19 ± 2, 18 ± 2, and 22 ± 2 mm, respectively. Among these bacteria, the extract of *P. cauliflora* had the highest antibacterial activity against *S. aureus*.

Thus, the extract of *P. cauliflora* was also effective for antipathogen chemotherapies. This extract may inhibit bacterial DNA replication, resulting in antibacterial activities.

3.4. TPC, TFC, and Antioxidant Ability of P. cauliflora

Many polyphenols possessing antioxidant activity can be developed as drug candidates [1,3]. Accordingly, we determined the total phenolic content (TPC), total flavonoid content (TFC), and the antioxidant activity of the extract of *P. cauliflora*. By using the modified Folin–Ciocalteu method [42] and aluminum chloride colorimetric method [43], the TPC and TFC of this extract were 113.2 ± 1.8 mg gallic acid/g and 102.8 ± 1.2 mg quercetin/g, respectively. By using a 1,1-diphenyl-2-picrylhydrazyl (DPPH) radical scavenging assay, the antioxidant ability of this extract was evaluated and quantified by IC_{50} value (Figure 4). Through the DPPH titration curve, the extract of *P. cauliflora* had an IC_{50} value of 46.5 ± 0.5 μg/mL. In comparison, L-ascorbic acid as a positive control had an IC_{50} value of 28.6 ± 0.2 μg/mL.

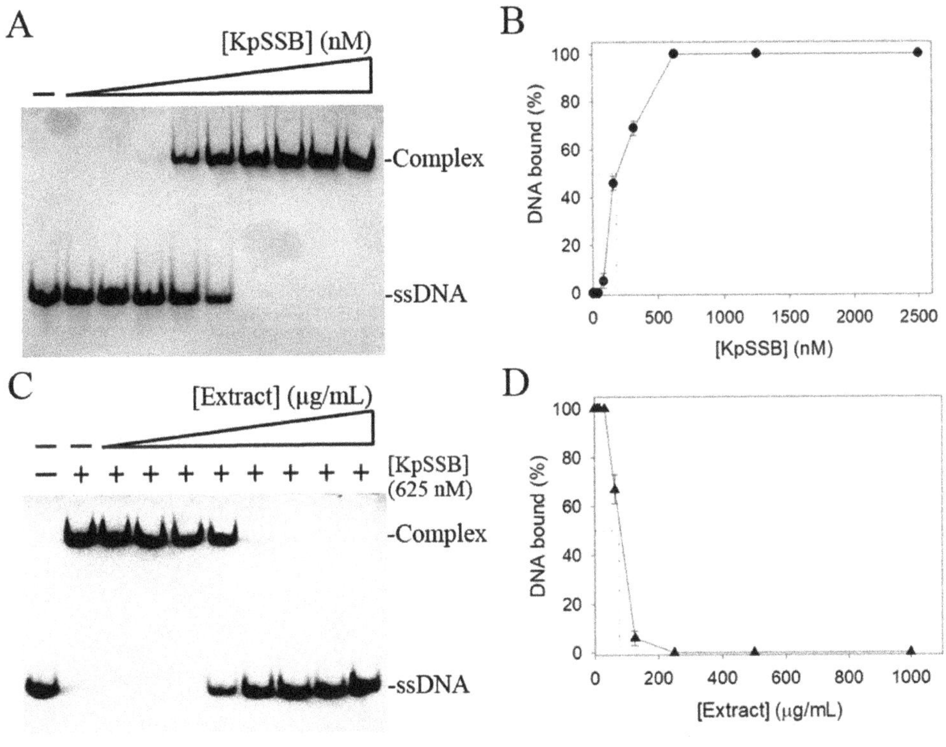

Figure 2. Inhibition of KpSSB by the fruit extract of *P. cauliflora*. (**A**) ssDNA binding of KpSSB. KpSSB (0, 20, 39, 78, 156, 312, 625, 1250, 2500, and 5000 nM) was incubated with dT30. A streptavidin–horseradish peroxidase conjugate was used for detecting this ssDNA and the complex. (**B**) ssDNA-binding ability of KpSSB. The $[Protein]_{50}$ of KpSSB was quantified based on the protein concentrations. The errors are standard deviation determined at 3 measurements. (**C**) Inhibition of the ssDNA-binding activity of KpSSB by the extract of *P. cauliflora*. KpSSB (625 nM) was incubated with this extract (0, 7.8, 15.6, 31.2, 62.5, 125, 250, 500, and 1000 μg/mL) and the binding activity was analyzed. (**D**) Inhibition ability of the extract of *P. cauliflora*. IC_{50} for KpSSB was determined through the titration curve. The errors are standard deviation determined at 3 measurements.

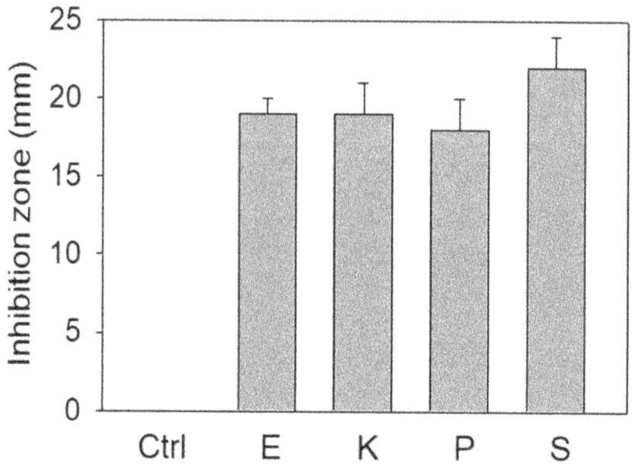

Figure 3. Agar well diffusion assay. The antibacterial capacities of the extract of *P. cauliflora* (10 mg) were quantified by the zone of inhibition. The zone of inhibition is a circular area around the spot of the extract in which the bacteria colonies do not grow. This extract exhibited antibacterial activities against human pathogens *Escherichia coli* (E), *Klebsiella pneumoniae* (K), *Pseudomonas aeruginosa* (P), and *Staphylococcus aureus* (S) used for this analysis and produced the inhibition zones of 19 ± 1, 19 ± 2, 18 ± 2, and 22 ± 2 mm, respectively. DMSO at 30% was used as a negative control.

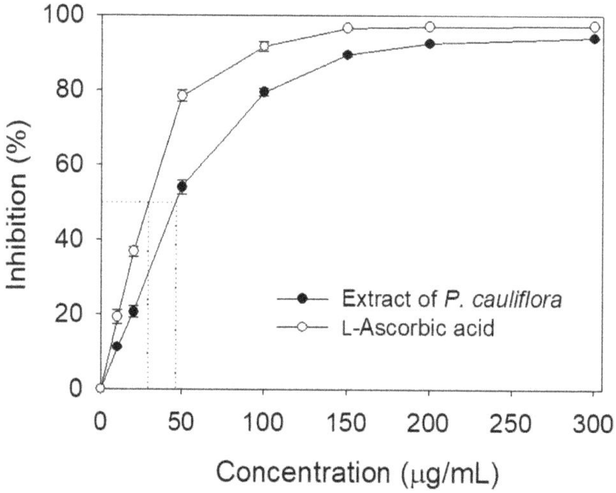

Figure 4. Antioxidant ability of the extract of *P. cauliflora*. Through the DPPH titration curve, the extract of *P. cauliflora* had an IC_{50} value of 46.5 ± 0.5 μg/mL. L-Ascorbic acid as a positive control had an IC_{50} value of 28.6 ± 0.2 μg/mL. The errors are standard deviation determined at 3 measurements.

3.5. GC–MS Analysis

Through GC–MS analysis, the most abundant compounds present in this extract were tentatively determined (Table 1). The top 15 compounds (>0.7%) were as follows: 5-hydroxymethylfurfural, 2,3-dihydro-3,5-dihydroxy-6-methyl-4h-pyran-4-one, 2,5-diformy lfuran, furfural, acetic acid, citraconic anhydride, formic acid, ethyl 4-hydroxy-3-methylbut-2-enoate, furfuryl alcohol, furyl hydroxymethyl ketone, 3-acetyl-3-hydroxyoxolane-2-one, 2,3-dihydro-5-hydroxy-2-(hydroxymethyl)-4H-pyran-4-one, 2(3H)-furanone, dihydro-4-hydroxy-, hydroxyacetone, and 1-hydroxybut-3-en-2-one. Possibly, certain compounds in this extract were capable of inhibiting KpSSB and thus attractive inhibitors.

Table 1. Compounds tentatively identified by the GC–MS analysis.

Peak No.	RT (min)	Name of Compounds	MF	CS	MW	Area (%)
1	19.13	5-Hydroxymethylfurfural	$C_6H_6O_3$	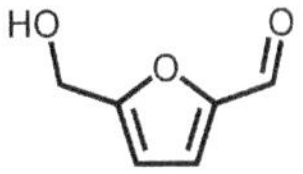	126	53.73%
2	17.14	2,3-Dihydro-3,5-dihydroxy-6-methyl-4h-pyran-4-one	$C_6H_8O_4$	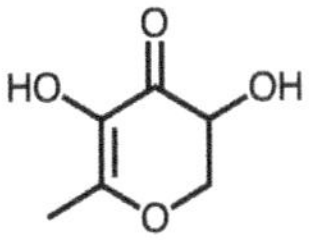	144	14.43%
3	14.36	2,5-Diformylfuran	$C_6H_4O_3$		124	6.06%
4	8.52	Furfural	$C_5H_4O_2$	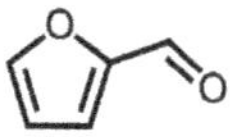	96	5.73%
5	8.28	Acetic acid	$C_2H_4O_2$	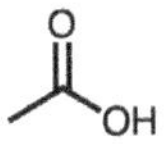	60	4.80%
6	11.46	Citraconic anhydride	$C_5H_4O_3$	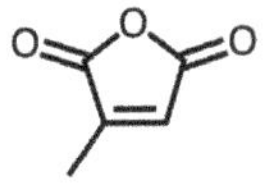	112	2.98%
7	8.92	Formic acid	CH_2O_2	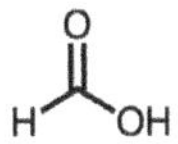	46	2.24%
8	20.39	Ethyl 4-hydroxy-3-methylbut-2-enoate	$C_7H_{12}O_3$	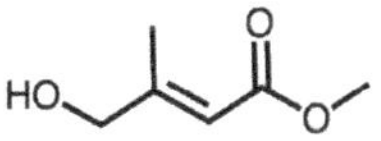	144	2.04%
9	10.86	Furfuryl alcohol	$C_5H_6O_2$	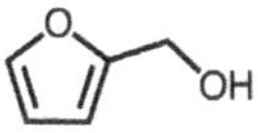	98	1.53%
10	14.68	Furyl hydroxymethyl ketone	$C_6H_6O_3$	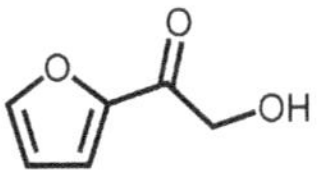	126	1.49%
11	17.65	3-Acetyl-3-hydroxyoxolane-2-one	$C_6H_8O_4$	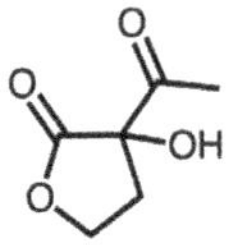	144	1.27%
12	21.26	2,3-Dihydro-5-hydroxy-2-(hydroxymethyl)-4H-pyran-4-one	$C_6H_8O_4$	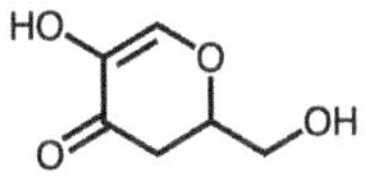	144	1.20%
13	19.83	2(3H)-Furanone, dihydro-4-hydroxy-	$C_4H_6O_3$	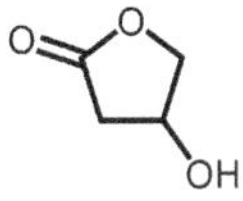	102	0.87%
14	6.59	Hydroxyacetone	$C_3H_6O_2$	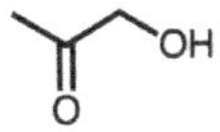	74	0.86%
15	7.85	1-Hydroxybut-3-en-2-one	$C_4H_6O_2$	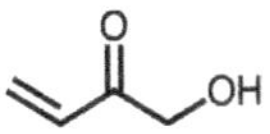	86	0.77%

3.6. Cytotoxic Effect of the Extract of *P. cauliflora*

Cancer remains a highly life-threatening disease on a global scale, and its mortality rates have been steadily increasing, making it a major contributor to human mortality [50,51]. Many natural products have demonstrated anticancer properties, especially concerning skin cancers [52]. Accordingly, we investigated the cytotoxic effect of the extract of *P. cauliflora*. B16F10 melanoma cells and a trypan blue staining assay (Figure 5) were used in this investigation. The B16F10 cell monolayers in 96-well microtitration plates were exposed to the extract at different concentrations per well. To achieve the indicated assay concentrations, the extract stock (20 mg/mL) was diluted using the culture medium. Incubation of B16F10 cells with the resulting extract solutions or the culture medium with 0.2% DMSO served as the treatment and control groups, respectively. Upon incubation, the extract at concentrations of 400, 800, 1000, and 1500 μg/mL caused cell death in B16F10 cells at rates of 2%, 17%, 72%, and 98%, respectively. Incubation of B16F10 cells with the extract (1500 μg/mL) resulted in nearly complete cell death. These findings indicate that the *P. cauliflora* extract exhibits anti-SSB activity and possesses anti-cancer potential, making it a promising candidate for further medical applications.

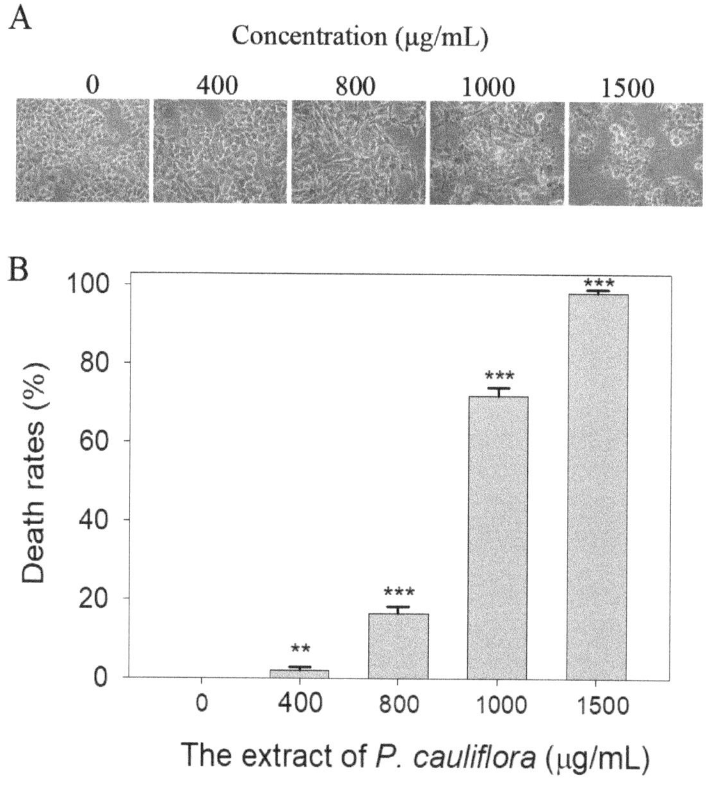

Figure 5. Cytotoxic effect of the extract of *P. cauliflora*. (**A**) The cytotoxic effect of the extract of *P. cauliflora* was estimated using trypan blue staining assay. The direct identification of live (unstained) and dead (blue) B16F10 cells was allowed by using this assay. The incubation of B16F10 cells with the extract at a concentration of 1500 μg/mL resulted in nearly complete cell death. (**B**) The death rates of B16F10 cells. Incubation with 400, 800, 1000, and 1500 μg/mL of the extract of *P. cauliflora* caused the deaths at rates of 2, 17, 72, and 98%, respectively. B16F10 cells were incubated with the culture medium with 0.2% DMSO as a control group. ** $p < 0.01$ and *** $p < 0.001$ compared with the control group.

3.7. The Extract of P. cauliflora Induced Apoptosis of B16F10 Cells

SSB plays a crucial role in various cellular processes, including DNA damage repair, maintenance of genome stability, and the regulation of cell apoptosis [19,53]. Given that the extract of *P. cauliflora* had anti-SSB properties, we investigated whether this extract could induce apoptosis of B16F10 cells using the Hoechst staining assay. Based on the results, we found that the extract of *P. cauliflora* could induce apoptosis of B16F10 cells (Figure 6A). The incubation of B16F10 cells with 400, 800, 1000, and 1500 μg/mL of this extract caused 4, 23, 75, and 98% DNA fragmentation (Figure 6B).

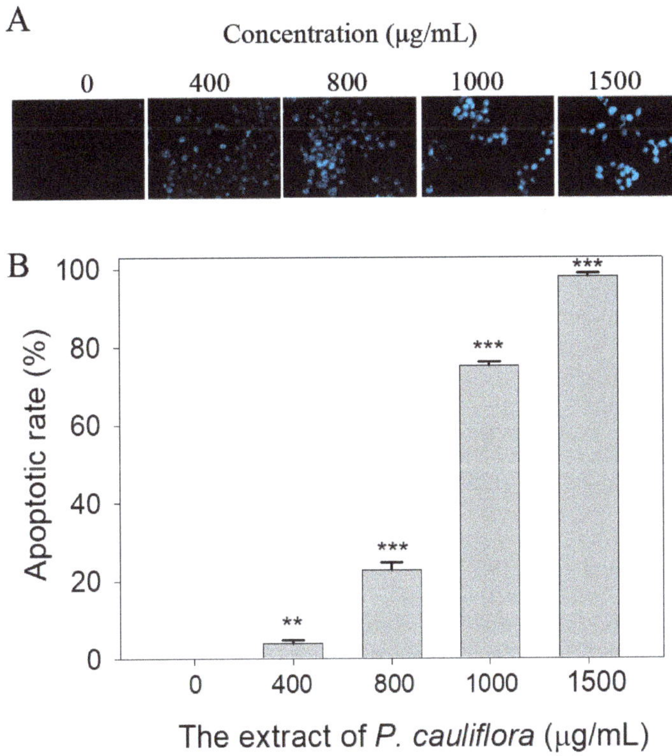

Figure 6. The extract of *P. cauliflora* induced apoptosis of the cells. (**A**) The apoptotic rate in B16F10 cells was estimated using the Hoechst staining assay. Based on our results, the extract of *P. cauliflora* could induce dose-dependent apoptosis of B16F10 cells. (**B**) The apoptotic rate. Incubation of B16F10 cells with 400, 800, 1000, and 1500 μg/mL of this extract caused 4, 23, 75, and 98% DNA fragmentation. The errors are the standard deviation as determined over 3 measurements. B16F10 cells were incubated with the culture medium with 0.2% DMSO as a control group. ** $p < 0.01$ and *** $p < 0.001$ compared with the control group.

3.8. The Extract of P. cauliflora Inhibited the Migration of B16F10 Cells

We also observed the extract of *P. cauliflora*'s ability to inhibit the cell migration using the wound healing assay (Figure 7). The wound healing assay is a well-established in vitro technique for assessing collective cell migration in two dimensions (Figure 7A). In this assay, a cell-free area is generated within a confluent cell monolayer by removing the cells from the designated area. The presence of the cell-free area prompts the remaining cells to migrate and close the gap. We observed that after 24 h of treatment, the *P. cauliflora* extract at concentrations of 400, 800, 1000, and 1500 μg/mL reduced B16F10 cell migration by 15%,

36%, 75%, and 100%, respectively (Figure 7B). These findings indicate that the *P. cauliflora* extract is capable of effectively inhibiting the migration of B16F10 cells.

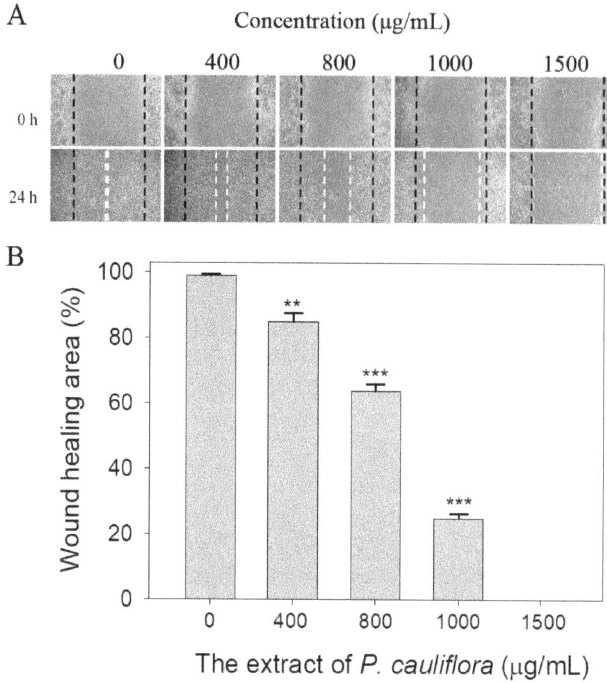

Figure 7. The extract of *P. cauliflora* inhibited the cell migration. (**A**) B16F10 cell migration was analyzed after a 24-h incubation with the *P. cauliflora* extract using the wound healing assay. The dashed lines indicate the wound edges. The *P. cauliflora* extract (1500 μg/mL) completely inhibited B16F10 cell migration. (**B**) The wound healing area was quantified, and the results showed that after 24 h of incubation, the *P. cauliflora* extract at concentrations of 400, 800, 1000, and 1500 μg/mL reduced B16F10 cell migration by 15%, 36%, 75%, and 100%, respectively. The error bars represent the standard deviation, calculated from three independent measurements. As a control group, B16F10 cells were incubated with the culture medium containing 0.2% DMSO. ** $p < 0.01$ and *** $p < 0.001$ compared with the control group.

3.9. The Extract of P. cauliflora Inhibited the Proliferation of B16F10 Cells

Through the clonogenic assay, we observed that the *P. cauliflora* extract effectively inhibited the cell proliferation (Figure 8A). The clonogenic assay is a well-established in vitro cell survival test, which assesses the capacity for a single cell to form a colony. When pre-treated with the *P. cauliflora* extract at concentrations of 400, 800, 1000, and 1500 μg/mL, the colony formation of B16F10 cells were suppressed by 8%, 29%, 69%, and 96%, respectively (Figure 8B). These results indicate that the *P. cauliflora* extract has a significant dose-dependent effect in suppressing the growth and proliferation of B16F10 cells.

3.10. Molecular Docking

Due to the high inhibitory capability, the compounds abundant in the extract of *P. cauliflora* were analyzed to determine the possible binding/inhibition modes via the docking studies. Based on the GC–MS analysis (Table 1), the top 3 compounds (>6.0%), namely 5-hydroxymethylfurfural, 2,3-dihydro-3,5-dihydroxy-6-methyl-4h-pyran-4-one, and 2,5-diformylfuran, were used for docking analysis. The binding mode of these compounds to

KpSSB was elucidated using the MOE Dock tool (Figure 9). Previously, we solved the crystal structure of KpSSB. The structure of KpSSB (Figure 9A) was therefore used for this docking analysis. KpSSB is a homotetramer. Given that the ssDNA-complexed structure of KpSSB is unavailable, the structure of *P. aeruginosa* SSB (PaSSB) bound by ssDNA dT25 (PDB ID 6IRQ) was used for analysis (Figure 9B). Bound ssDNA in PaSSB was found to adopt an S-shaped conformation (Figure 9B). Considering their structural similarity, the ssDNA-binding mode between KpSSB and PaSSB may be similar (Figure 9C). According to the docking results (Figure 9C), 5-hydroxymethylfurfural (Figure 9D), 2,3-dihydro-3,5-dihydroxy-6-methyl-4h-pyran-4-one (Figure 9E), and 2,5-diformylfuran (Figure 9F) occupy the ssDNA-binding sites of KpSSB. Thus, these compounds may influence the ssDNA-binding activity of KpSSB by this inhibition mechanism. The binding energies of these compounds were also calculated by MOE (Table 2). The predicted binding affinities for these KpSSB complexes were calculated for all potential binding geometries. Their docking scores revealed the following order for binding affinity (the S score): 5-hydroxymethylfurfural > 2,5-diformylfuran > 2,3-dihydro-3,5-dihydroxy-6-methyl-4h-pyran-4-one. The compound 5-hydroxymethylfurfural may possess the greatest binding affinity for KpSSB among the selected compounds, as indicated by the highest S score (−4.3646). However, this conjecture requires further experimental and structural investigations to be confirmed.

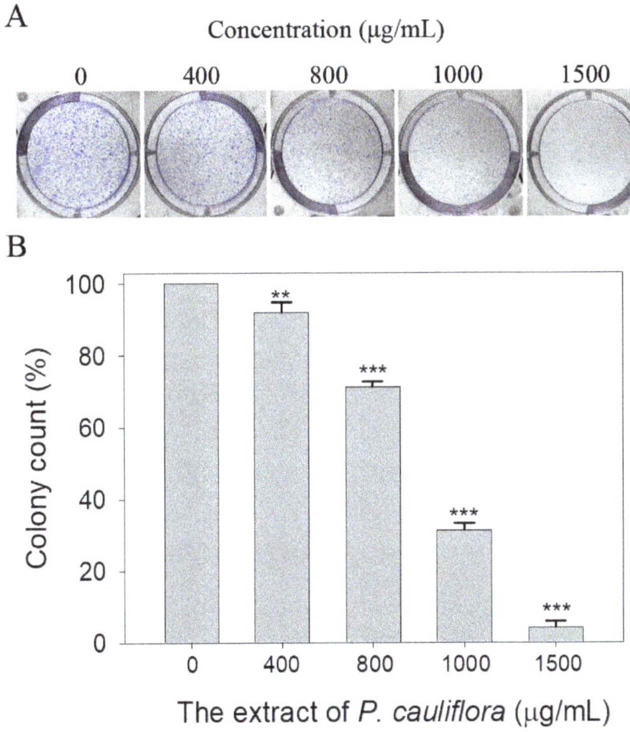

Figure 8. The extract of *P. cauliflora* inhibited the cell proliferation. (**A**) Estimation of the proliferation of B16F10 cells using the clonogenic assay. The *P. cauliflora* extract (1500 μg/mL) almost completely suppressed the colony formation. (**B**) The quantified results of the proliferation of B16F10 cells are presented. The *P. cauliflora* extract at concentrations of 400, 800, 1000, and 1500 μg/mL significantly inhibited the proliferation of B16F10 cells by 8%, 29%, 69%, and 96%, respectively. The error bars represent the standard deviation from 3 independent measurements. As a control group, B16F10 cells were incubated with the culture medium containing 0.2% DMSO. ** $p < 0.01$ and *** $p < 0.001$ compared with the control group.

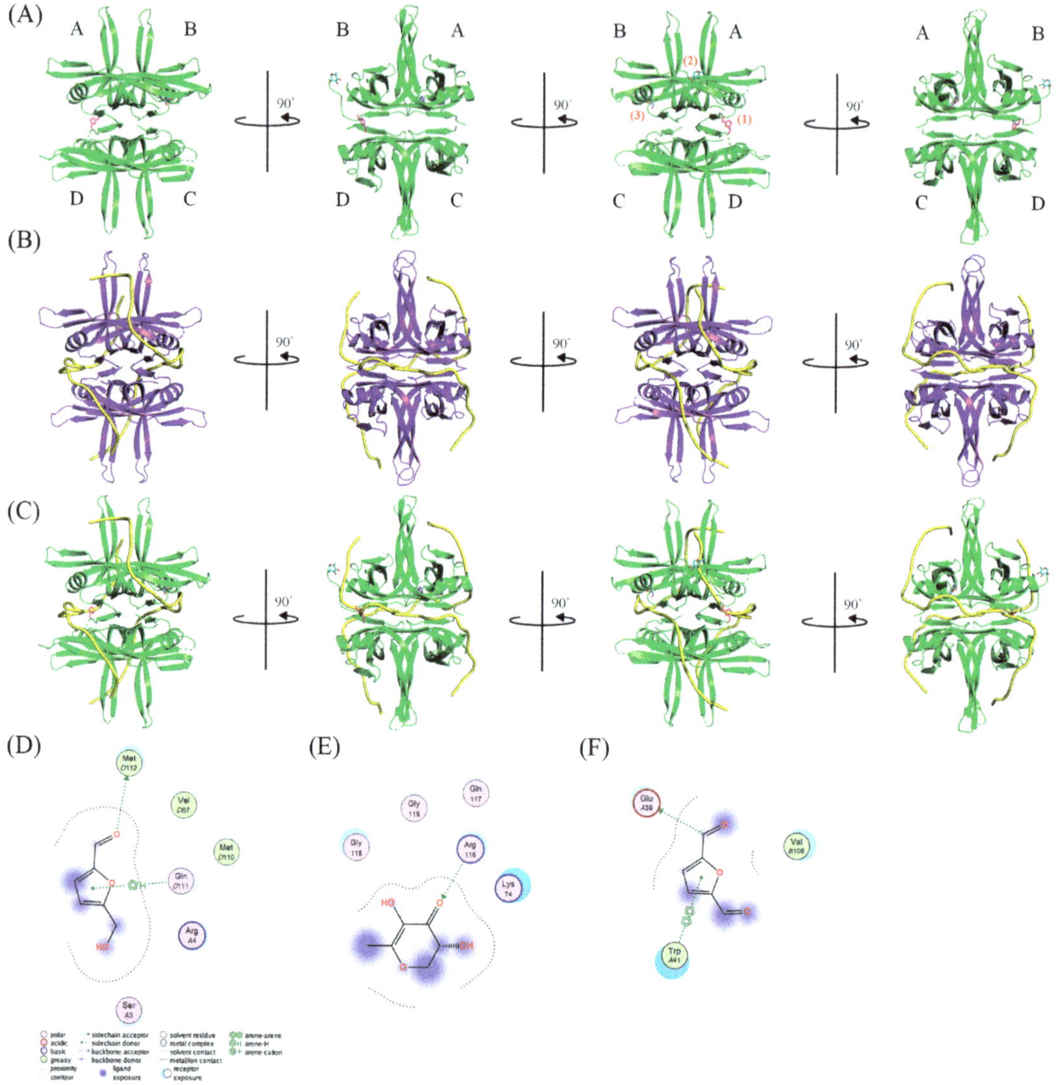

Figure 9. The docking studies. (**A**) Crystal structure of KpSSB. KpSSB is a homotetramer with subunits A, B, C, and D. The binding mode of (1) 5-hydroxymethylfurfural (light magenta), (2) 2,3-dihydro-3,5-dihydroxy-6-methyl-4h-pyran-4-one (cyan), and (3) 2,5-diformylfuran (slate) to KpSSB was elucidated using the MOE Dock tool. (**B**) The structure of PaSSB bound by ssDNA dT25. The complexed crystal structure of a PaSSB tetramer contains two ssDNA molecules (yellow). Given that the ssDNA-complexed structure of KpSSB is unavailable, the complexed structure of PaSSB was used for analyzing the ssDNA-binding mode of KpSSB. (**C**) The superimposed structures. Superimposing the ssDNA bound by PaSSB and the docked compounds bound by KpSSB indicates that these three compounds are likely to be situated at the sites of KpSSB for ssDNA binding. The binding modes of (**D**) 5-hydroxymethylfurfural, (**E**) 2,3-dihydro-3,5-dihydroxy-6-methyl-4h-pyran-4-one, and (**F**) 2,5-diformylfuran to KpSSB are shown. The interaction analysis is depicted through two-dimensional diagrams.

Table 2. Molecular docking analysis.

	S Score	Receptor Residue	Interaction	Distance (Å)	E (kcal/mol)
(1) 5-Hydroxymethylfurfural	−4.3646	(D) M112	H-donor	2.97	−1.5
		(D) Q111	pi–H	4.56	−0.5
		(D) Q111	pi–H	4.18	−0.5
(2) 2,3-Dihydro-3,5-dihydroxy-6-methyl-4h-pyran-4-one	−3.1840	(B) R116	H-acceptor	2.94	−4.0
(3) 2,5-Diformylfuran	−3.5429	(A) E39	H-donor	3.34	−0.6
		(A) W41	pi–pi	3.59	−0.0

4. Discussion

P. cauliflora has been associated with numerous ethnobotanical applications in treating various conditions such as diarrhea, labyrinthitis, skin irritations, flu, genitourinary problems, and asthma [10]. The commercial potential of *P. cauliflora* is significant, as its fruit is highly appreciated for both fresh consumption and for its use in making jelly, juice, ice cream, fermented beverages, and liqueurs [10]. This fruit is considered a functional food due to its richness in anthocyanins, exhibiting antioxidant and anti-inflammatory properties, making it sought after by the pharmaceutical industry for its nutraceutical characteristics. In our study, we investigated the extract of *P. cauliflora* obtained using 50% ethanol and found that it has the capability to inhibit the activity of KpSSB (Figure 2), an essential DNA replication protein in *K. pneumoniae. K. pneumoniae* is classified as one of the dangerous ESKAPE organisms and can effectively "escape" the biocidal action of antibiotics [31]. Given its association with life-threatening infections, it is important to combat this pathogen in some effective ways [34,35]. SSB plays a vital role in cellular survival, making it a potential target for pharmacological inhibition to combat bacterial pathogens. Inhibiting DNA replication and repair has a long-standing history as an antimicrobial strategy in antibiotic development. Notably, antibiotics such as quinolones and aminocoumarins were successfully developed for clinical applications by targeting bacterial DNA gyrase and topoisomerase IV [54,55]. As plants have been a rich source of pharmaceutical products with antibacterial properties, it is worthwhile to explore different plant extracts and continue the search for inhibitors against SSBs. The findings from our study provide promising insights into the potential therapeutic applications of *P. cauliflora* and its extract in combating bacterial pathogens by targeting essential DNA replication proteins.

According to the GC–MS analysis (Table 1), the top 15 compounds (>0.7%) in the fruit extract of *P. cauliflora* were tentatively identified as follows: 5-hydroxymethylfurfural, 2,3-dihydro-3,5-dihydroxy-6-methyl-4h-pyran-4-one, 2,5-diformylfuran, furfural, acetic acid, citraconic anhydride, formic acid, ethyl 4-hydroxy-3-methylbut-2-enoate, furfuryl alcohol, furyl hydroxymethyl ketone, 3-acetyl-3-hydroxyoxolane-2-one, 2,3-dihydro-5-hydroxy-2-(hydroxymethyl)-4H-pyran-4-one, 2(3H)-furanone, dihydro-4-hydroxy-, hydroxyacetone, and 1-hydroxybut-3-en-2-one. This extract may contain compounds that could potentially act as KpSSB inhibitors. However, it is essential to note that other compounds, such as catechin, epicatechin, quercetin, and rutin, have also been identified in various extracts from *P. cauliflora* [9]. In our study, we observed that rutin exhibited an inhibitory effect on the DNA-binding activity of KpSSB with an IC_{50} value of 293 ± 16 μM. Rutin, also known as vitamin P, is abundantly found in various vegetables and fruits and is known to possess antibacterial, anti-inflammatory, antitumor, and antiviral properties. Its anticancer potential has been associated with the regulation of multiple signaling pathways, including NF-κB, PI3K-Akt-mTOR, Nrf2, ERK, JNK, and p38 MAPK [56]. Given the inhibition capacity against the DNA replication protein SSB, it is worth continuing the search for inhibitors against SSB and for finding possible DNA-binding targets in cancer cells.

SSB typically recognizes ssDNA [57,58] through a highly conserved oligonucleotide/oligosaccharide-binding (OB) fold. The OB fold structure consists of a five-stranded β-barrel capped by an α-helix [59]. SSB is comprised of two domains: an N-terminal domain for ssDNA-binding and oligomerization (SSBn) and a C-terminal domain for protein–protein interaction (SSBc). To date, the structure of SSBc has not been observed. Recently, the structure of the six additional residues in SSBc is revealed by our crystal structure of KpSSB. These residues, 114-GGRQGG-119 in SSBc, can interact with SSBn but are released when SSBn binds to ssDNA [60]. Thus, this motif in SSB may be a switch in regulating the ssDNA binding [60]. In our study, the docking results indicated that R116 in KpSSB plays a crucial role in forming hydrogen bonds and interacting with the potential inhibitor 2,3-dihydro-3,5-dihydroxy-6-methyl-4h-pyran-4-one (Figure 8E). Consequently, this compound may act as an inhibitor, potentially affecting the regulatory switch for SSBn or ssDNA binding. Given the relationship between the GGRQ motif and the ssDNA binding sites, further investigations are needed to understand whether this motif also impacts the SSB_{35}/SSB_{65} distribution [24]. We are currently working on obtaining crystals of the KpSSB-inhibitor complex to conduct a more comprehensive analysis of these interactions.

P. cauliflora extracts have been previously recognized for their anti-proliferative activities and potential anti-carcinogenic effects on various cancer cell lines, including breast [14] and lung cancer cells [15]. The peel extracts exhibited inhibitory effects against *Bacillus subtilis*, *S. aureus*, *P. aeruginosa*, and *E. coli* bacteria [61]. In alignment with these findings, our study revealed the antibacterial activities (Figure 3), as well as the cytotoxic (Figure 5) and antiproliferative effects (Figure 8) of the *P. cauliflora* extract on the B16F10 cells. Additionally, we also found that the extract of *P. cauliflora* could inhibit the cell migration and induce apoptosis. These preliminary results suggest that the extract of *P. cauliflora* holds promise as a potential natural alternative for melanoma cancer, warranting further exploration for therapeutic applications.

By using a DPPH assay, the IC_{50} of an ethanol extract of *P. cauliflora* seeds and peel had been reported to be 27 [62] and 37 μg/mL [63], respectively. In this study, the fruit extract obtained using 50% ethanol had an IC_{50} value of 46.5 ± 0.5 μg/mL (Figure 4). Regardless of the cost of alcohol, ethanol could be a better solvent for extractions. In conclusion, this study represents the first identification that the fruit extract from *P. cauliflora* has the anti-KpSSB property. Rutin is a new inhibitor against KpSSB. Furthermore, the cytotoxic activities against melanoma cells were also investigated. Through GC–MS analysis, the top 15 compounds in this extract were detected, providing insights into the potential active components and their potential synergistic effects in these biological activities.

Author Contributions: E.-S.L. performed the experiments; E.-S.L. and C.-Y.H. analyzed the data; E.-S.L. and C.-Y.H. contributed to the study design and manuscript writing. All authors reviewed the results, contributed to the data interpretation, and approved the final version of the manuscript. All authors have read and agreed to the published version of the manuscript.

Funding: This research was supported by a grant from the National Science and Technology Council, Taiwan (NSTC 111-2622-E-025-002 to E.-S.L.).

Institutional Review Board Statement: Not applicable.

Informed Consent Statement: Not applicable.

Data Availability Statement: Not applicable.

Conflicts of Interest: The authors declare no conflict of interest.

References

1. Roszkowski, S. Application of Polyphenols and Flavonoids in Oncological Therapy. *Molecules* **2023**, *28*, 4080. [CrossRef] [PubMed]
2. Lobiuc, A.; Paválă, N.E.; Mangalagiu, I.I.; Gheorghiță, R.; Teliban, G.C.; Amăriucăi-Mantu, D.; Stoleru, V. Future Antimicrobials: Natural and Functionalized Phenolics. *Molecules* **2023**, *28*, 1114. [CrossRef] [PubMed]
3. Jakobušić Brala, C.; Karković Marković, A.; Kugić, A.; Torić, J.; Barbarić, M. Combination Chemotherapy with Selected Polyphenols in Preclinical and Clinical Studies-An Update Overview. *Molecules* **2023**, *28*, 3746. [CrossRef] [PubMed]
4. Atanasov, A.G.; Zotchev, S.B.; Dirsch, V.M.; Supuran, C.T. Natural products in drug discovery: Advances and opportunities. *Nat. Rev. Drug Discov.* **2021**, *20*, 200–216. [CrossRef] [PubMed]

5. WHO. Traditional medicine growing needs and potential. *WHO Policy Perspect. Med.* **2002**, *2*, 1–6.
6. Rufino-Palomares, E.E.; Pérez-Jiménez, A.; García-Salguero, L.; Mokhtari, K.; Reyes-Zurita, F.J.; Peragón-Sánchez, J.; Lupiáñez, J.A. Nutraceutical Role of Polyphenols and Triterpenes Present in the Extracts of Fruits and Leaves of Olea europaea as Antioxidants, Anti-Infectives and Anticancer Agents on Healthy Growth. *Molecules* **2022**, *27*, 2341. [CrossRef]
7. Khan, A.W.; Farooq, M.; Haseeb, M.; Choi, S. Role of Plant-Derived Active Constituents in Cancer Treatment and Their Mechanisms of Action. *Cells* **2022**, *11*, 1326. [CrossRef] [PubMed]
8. Gligorov, J.; Richard, S. Breast cancer: Weekly paclitaxel—Still preferred first-line taxane for mBC. *Nat. Rev. Clin. Oncol.* **2015**, *12*, 508–509. [CrossRef]
9. da Veiga Correia, V.T.; da Silva, P.R.; Ribeiro, C.M.S.; Ramos, A.; Mazzinghy, A.; Silva, V.D.M.; Júnior, A.H.O.; Nunes, B.V.; Vieira, A.L.S.; Ribeiro, L.V.; et al. An Integrative Review on the Main Flavonoids Found in Some Species of the Myrtaceae Family: Phytochemical Characterization, Health Benefits and Development of Products. *Plants* **2022**, *11*, 2796. [CrossRef]
10. Gasparotto Junior, A.; de Souza, P.; Lívero, F. *Plinia cauliflora* (Mart.) Kausel: A comprehensive ethnopharmacological review of a genuinely Brazilian species. *J. Ethnopharmacol.* **2019**, *245*, 112169. [CrossRef]
11. Chang, S.K.; Alasalvar, C.; Shahidi, F. Superfruits: Phytochemicals, antioxidant efficacies, and health effects—A comprehensive review. *Crit. Rev. Food Sci. Nutr.* **2019**, *59*, 1580–1604. [CrossRef]
12. Inada, K.O.P.; Leite, I.B.; Martins, A.B.N.; Fialho, E.; Tomás-Barberán, F.A.; Perrone, D.; Monteiro, M. Jaboticaba berry: A comprehensive review on its polyphenol composition, health effects, metabolism, and the development of food products. *Food Res. Int.* **2021**, *147*, 110518. [CrossRef]
13. Wu, S.B.; Dastmalchi, K.; Long, C.; Kennelly, E.J. Metabolite profiling of jaboticaba (*Myrciaria cauliflora*) and other dark-colored fruit juices. *J. Agric. Food Chem.* **2012**, *60*, 7513–7525. [CrossRef]
14. Reguengo, L.M.; Nascimento, R.P.D.; Machado, A.; Marostica Junior, M.R. Signaling pathways and the potential anticarcinogenic effect of native Brazilian fruits on breast cancer. *Food Res. Int.* **2022**, *155*, 111117. [CrossRef]
15. Machado, A.; Alves, M.D.R.; Nascimento, R.P.D.; Reguengo, L.M.; Marostica Junior, M.R. Antiproliferative effects and main molecular mechanisms of Brazilian native fruits and their by-products on lung cancer. *Food Res. Int.* **2022**, *162*, 111953. [CrossRef]
16. Filho, A.V.; Avila, L.B.; Lacorte, D.H.; Martiny, T.R.; Rosseto, V.; Moraes, C.C.; Dotto, G.L.; Carreno, N.L.V.; da Rosa, G.S. Brazilian Agroindustrial Wastes as a Potential Resource of Bioative Compounds and Their Antimicrobial and Antioxidant Activities. *Molecules* **2022**, *27*, 6876. [CrossRef]
17. Avila, L.B.; Barreto, E.R.C.; Souza, P.K.; Silva, B.Z.; Martiny, T.R.; Moraes, C.C.; Morais, M.M.; Raghavan, V.; Rosa, G.S.D. Carrageenan-Based Films Incorporated with Jaboticaba Peel Extract: An Innovative Material for Active Food Packaging. *Molecules* **2020**, *25*, 5563. [CrossRef] [PubMed]
18. Meyer, R.R.; Laine, P.S. The single-stranded DNA-binding protein of Escherichia coli. *Microbiol. Rev.* **1990**, *54*, 342–380. [CrossRef] [PubMed]
19. Bianco, P.R. The Biochemical Mechanism of Fork Regression in Prokaryotes and Eukaryotes-A Single Molecule Comparison. *Int. J. Mol. Sci.* **2022**, *23*, 8613. [CrossRef] [PubMed]
20. Antony, E.; Lohman, T.M. Dynamics of *E. coli* single stranded DNA binding (SSB) protein-DNA complexes. *Semin. Cell Dev. Biol.* **2019**, *86*, 102–111. [CrossRef] [PubMed]
21. Shereda, R.D.; Kozlov, A.G.; Lohman, T.M.; Cox, M.M.; Keck, J.L. SSB as an organizer/mobilizer of genome maintenance complexes. *Crit. Rev. Biochem. Mol. Biol.* **2008**, *43*, 289–318. [CrossRef] [PubMed]
22. Heller, R.C.; Marians, K.J. Replisome assembly and the direct restart of stalled replication forks. *Nat. Rev. Mol. Cell Biol.* **2006**, *7*, 932–943. [CrossRef] [PubMed]
23. Marians, K.J. PriA-directed replication fork restart in *Escherichia coli*. *Trends Biochem. Sci.* **2000**, *25*, 185–189. [CrossRef] [PubMed]
24. Lohman, T.M.; Ferrari, M.E. *Escherichia coli* single-stranded DNA-binding protein: Multiple DNA-binding modes and cooperativities. *Annu. Rev. Biochem.* **1994**, *63*, 527–570. [CrossRef]
25. Bianco, P.R. The mechanism of action of the SSB interactome reveals it is the first OB-fold family of genome guardians in prokaryotes. *Protein Sci.* **2021**, *30*, 1757–1775. [CrossRef]
26. Voter, A.F.; Killoran, M.P.; Ananiev, G.E.; Wildman, S.A.; Hoffmann, F.M.; Keck, J.L. A High-Throughput Screening Strategy to Identify Inhibitors of SSB Protein-Protein Interactions in an Academic Screening Facility. *SLAS Discov.* **2017**, *23*, 94–101. [CrossRef]
27. Glanzer, J.G.; Endres, J.L.; Byrne, B.M.; Liu, S.; Bayles, K.W.; Oakley, G.G. Identification of inhibitors for single-stranded DNA-binding proteins in eubacteria. *J. Antimicrob. Chemother.* **2016**, *71*, 3432–3440. [CrossRef]
28. Marceau, A.H.; Bernstein, D.A.; Walsh, B.W.; Shapiro, W.; Simmons, L.A.; Keck, J.L. Protein interactions in genome maintenance as novel antibacterial targets. *PLoS ONE* **2013**, *8*, e58765. [CrossRef]
29. Iftode, C.; Daniely, Y.; Borowiec, J.A. Replication protein A (RPA): The eukaryotic SSB. *Crit. Rev. Biochem. Mol. Biol.* **1999**, *34*, 141–180. [CrossRef]
30. Byrne, B.M.; Oakley, G.G. Replication protein A, the laxative that keeps DNA regular: The importance of RPA phosphorylation in maintaining genome stability. *Semin. Cell Dev. Biol.* **2019**, *86*, 112–120. [CrossRef]
31. Tommasi, R.; Brown, D.G.; Walkup, G.K.; Manchester, J.I.; Miller, A.A. ESKAPEing the labyrinth of antibacterial discovery. *Nat. Rev. Drug Discov.* **2015**, *14*, 529–542. [CrossRef]
32. Darby, E.M.; Trampari, E.; Siasat, P.; Gaya, M.S.; Alav, I.; Webber, M.A.; Blair, J.M.A. Molecular mechanisms of antibiotic resistance revisited. *Nat. Rev. Microbiol.* **2023**, *21*, 280–295. [CrossRef] [PubMed]

33. Du Toit, A. The cost of resistance. *Nat. Rev. Microbiol.* **2022**, *20*, 317. [CrossRef] [PubMed]
34. Dong, N.; Yang, X.; Chan, E.W.; Zhang, R.; Chen, S. Klebsiella species: Taxonomy, hypervirulence and multidrug resistance. *eBioMedicine* **2022**, *79*, 103998. [CrossRef]
35. Arato, V.; Raso, M.M.; Gasperini, G.; Berlanda Scorza, F.; Micoli, F. Prophylaxis and Treatment against *Klebsiella pneumoniae*: Current Insights on This Emerging Anti-Microbial Resistant Global Threat. *Int. J. Mol. Sci.* **2021**, *22*, 4042. [CrossRef] [PubMed]
36. Franzolin, M.R.; Courrol, D.D.S.; Silva, F.R.O.; Courrol, L.C. Antimicrobial Activity of Silver and Gold Nanoparticles Prepared by Photoreduction Process with Leaves and Fruit Extracts of *Plinia cauliflora* and *Punica granatum*. *Molecules* **2022**, *27*, 6860. [CrossRef]
37. Machado, G.H.A.; Marques, T.R.; de Carvalho, T.C.L.; Duarte, A.C.; de Oliveira, F.C.; Gonçalves, M.C.; Piccoli, R.H.; Corrêa, A.D. Antibacterial activity and in vivo wound healing potential of phenolic extracts from jaboticaba skin. *Chem. Biol. Drug Des.* **2018**, *92*, 1333–1343. [CrossRef]
38. Mott, M.L.; Berger, J.M. DNA replication initiation: Mechanisms and regulation in bacteria. *Nat. Rev. Microbiol.* **2007**, *5*, 343–354. [CrossRef]
39. Nižnanský, Ľ.; Osinová, D.; Kuruc, R.; Hengerics Szabó, A.; Szórádová, A.; Masár, M.; Nižnanská, Ž. Natural Taxanes: From Plant Composition to Human Pharmacology and Toxicity. *Int. J. Mol. Sci.* **2022**, *23*, 15619. [CrossRef]
40. Merecz-Sadowska, A.; Sitarek, P.; Kowalczyk, T.; Zajdel, K.; Kucharska, E.; Zajdel, R. The Modulation of Melanogenesis in B16 Cells Upon Treatment with Plant Extracts and Isolated Plant Compounds. *Molecules* **2022**, *27*, 4360. [CrossRef]
41. Balouiri, M.; Sadiki, M.; Ibnsouda, S.K. Methods for in vitro evaluating antimicrobial activity: A review. *J. Pharm. Anal.* **2016**, *6*, 71–79. [CrossRef] [PubMed]
42. Singleton, V.L.; Orthofer, R.; Lamuela-Raventos, R.M. Analysis of total phenols and other oxidation substrates and antioxidants by means of folin-ciocalteu reagent. *Methods Enzymol.* **1999**, *299*, 152–178.
43. Chang, C.C.; Yang, M.H.; Wen, H.M.; Chern, J.C. Estimation of total flavonoid content in propolis by two complementary colorimetric methods. *J. Food Drug Anal.* **2002**, *10*, 178–182.
44. Brand-Williams, W.; Cuvelier, M.E.; Berset, C. Use of a free radical method to evaluate antioxidant activity. *LWT Food Sci. Technol.* **1995**, *28*, 25–30. [CrossRef]
45. Strober, W. Trypan blue exclusion test of cell viability. *Curr. Protoc. Immunol.* **2001**, *21*, A-3B. [CrossRef]
46. Larsson, R.; Nygren, P. A rapid fluorometric method for semiautomated determination of cytotoxicity and cellular proliferation of human tumor cell lines in microculture. *Anticancer Res.* **1989**, *9*, 1111–1119.
47. Chen, M.H.; Yang, W.L.; Lin, K.T.; Liu, C.H.; Liu, Y.W.; Huang, K.W.; Chang, P.M.; Lai, J.M.; Hsu, C.N.; Chao, K.M.; et al. Gene expression-based chemical genomics identifies potential therapeutic drugs in hepatocellular carcinoma. *PLoS ONE* **2011**, *6*, e27186. [CrossRef]
48. Liang, C.C.; Park, A.Y.; Guan, J.L. In vitro scratch assay: A convenient and inexpensive method for analysis of cell migration in vitro. *Nat. Protoc.* **2007**, *2*, 329–333. [CrossRef]
49. Reynertson, K.A.; Yang, H.; Jiang, B.; Basile, M.J.; Kennelly, E.J. Quantitative analysis of antiradical phenolic constituents from fourteen edible Myrtaceae fruits. *Food Chem.* **2008**, *109*, 883–890. [CrossRef]
50. Dahiya, D.; Nigam, P.S. Biotherapy Using Probiotics as Therapeutic Agents to Restore the Gut Microbiota to Relieve Gastrointestinal Tract Inflammation, IBD, IBS and Prevent Induction of Cancer. *Int. J. Mol. Sci.* **2023**, *24*, 5748. [CrossRef]
51. Choi, J.R.; Kozalak, G.; di Bari, I.; Babar, Q.; Niknam, Z.; Rasmi, Y.; Yong, K.W. In Vitro Human Cancer Models for Biomedical Applications. *Cancers* **2022**, *14*, 2284. [CrossRef] [PubMed]
52. Dzialo, M.; Mierziak, J.; Korzun, U.; Preisner, M.; Szopa, J.; Kulma, A. The Potential of Plant Phenolics in Prevention and Therapy of Skin Disorders. *Int. J. Mol. Sci.* **2016**, *17*, 160. [CrossRef] [PubMed]
53. Xu, L.; Halma, M.T.J.; Wuite, G.J.L. Unravelling How Single-Stranded DNA Binding Protein Coordinates DNA Metabolism Using Single-Molecule Approaches. *Int. J. Mol. Sci.* **2023**, *24*, 2806. [CrossRef] [PubMed]
54. Sugino, A.; Peebles, C.L.; Kreuzer, K.N.; Cozzarelli, N.R. Mechanism of action of nalidixic acid: Purification of *Escherichia coli* nalA gene product and its relationship to DNA gyrase and a novel nicking-closing enzyme. *Proc. Natl. Acad. Sci. USA* **1977**, *74*, 4767–4771. [CrossRef]
55. Gellert, M.; O'Dea, M.H.; Itoh, T.; Tomizawa, J. Novobiocin and coumermycin inhibit DNA supercoiling catalyzed by DNA gyrase. *Proc. Natl. Acad. Sci. USA* **1976**, *73*, 4474–4478. [CrossRef]
56. Pandey, P.; Khan, F.; Qari, H.A.; Oves, M. Rutin (Bioflavonoid) as Cell Signaling Pathway Modulator: Prospects in Treatment and Chemoprevention. *Pharmaceuticals* **2021**, *14*, 1069. [CrossRef]
57. Richard, D.J.; Bolderson, E.; Cubeddu, L.; Wadsworth, R.I.; Savage, K.; Sharma, G.G.; Nicolette, M.L.; Tsvetanov, S.; McIlwraith, M.J.; Pandita, R.K.; et al. Single-stranded DNA-binding protein hSSB1 is critical for genomic stability. *Nature* **2008**, *453*, 677–681. [CrossRef]
58. Raghunathan, S.; Kozlov, A.G.; Lohman, T.M.; Waksman, G. Structure of the DNA binding domain of *E. coli* SSB bound to ssDNA. *Nat. Struct. Biol.* **2000**, *7*, 648–652. [CrossRef]
59. Dickey, T.H.; Altschuler, S.E.; Wuttke, D.S. Single-stranded DNA-binding proteins: Multiple domains for multiple functions. *Structure* **2013**, *21*, 1074–1084. [CrossRef]
60. Su, X.C.; Wang, Y.; Yagi, H.; Shishmarev, D.; Mason, C.E.; Smith, P.J.; Vandevenne, M.; Dixon, N.E.; Otting, G. Bound or free: Interaction of the C-terminal domain of *Escherichia coli* single-stranded DNA-binding protein (SSB) with the tetrameric core of SSB. *Biochemistry* **2014**, *53*, 1925–1934. [CrossRef]

61. Pinc, M.M.; Dalmagro, M.; da Cruz Alves Pereira, E.; Donadel, G.; Thomaz, R.T.; da Silva, C.; Macruz, P.D.; Jacomassi, E.; Gasparotto Junior, A.; Hoscheid, J.; et al. Extraction Methods, Chemical Characterization, and In Vitro Biological Activities of *Plinia cauliflora* (Mart.) Kausel Peels. *Pharmaceuticals* **2023**, *16*, 1173. [CrossRef] [PubMed]
62. Wang, W.H.; Tyan, Y.C.; Chen, Z.S.; Lin, C.G.; Yang, M.H.; Yuan, S.S.; Tsai, W.C. Evaluation of the antioxidant activity and antiproliferative effect of the jaboticaba (*Myrciaria cauliflora*) seed extracts in oral carcinoma cells. *Biomed. Res. Int.* **2014**, *2014*, 185946. [CrossRef]
63. Mendonça de Assis, P.; Cypriano Dutra, R.; Amarante, C.B.D.; Afonso Miranda Chaves, M.D.G.; Moreira, C.P.S.; Brandão, M.A.F.; Raposo, N.R.B. *Plinia cauliflora* (Mart.) Kausel: Toxicological assays, biological activities, and elemental analysis of organic compounds. *Nat. Prod. Res.* **2021**, *35*, 1727–1731. [CrossRef] [PubMed]

applied sciences

MDPI

Review

The Significance of Organic Horticulture in Mitigating Climate Change and Promoting the Production of Healthier Fruits and Vegetables

Awabo Manyaku [1], Hildegard Witbooi [1,*] and Charles Petrus Laubscher [2]

[1] Department of Agronomy, University of Fort Hare, Private Bag X1314, Alice 5700, South Africa; 201800568@ufh.ac.za

[2] Department of Horticultural Sciences, Cape Peninsula University of Technology, Bellville, Cape Town 7535, South Africa; laubscherc@cput.ac.za

* Correspondence: hwitbooi@ufh.ac.za; Tel.: +27-040-602-2585

Abstract: Organic horticulture is a holistic management system that follows good production practices and should be considered the cornerstone of mitigating climate change and producing healthier fruits and vegetables. This agroecosystem practice not only benefits the green economy but promotes and enhances soil biological activity, biodiversity, and other biological cycles in the sphere. The last decade has observed a rise in the production and consumption of organically certified agricultural products, and the biggest growth was registered in France (18%) due to its higher nutritional value of vitamin C (27%) and polyphenol content (72%), with a lowered risk of exposure to harmful chemicals of up to 70% and improved organoleptic properties. Between 2012 and 2020, the European Union's organic sector experienced significant growth, with a 56% expansion in organic land area, a 40% increase in organic producers, and a 114% increase in retail sales. The aim of this review was to evaluate the significant impact of organic horticulture on mitigating climate change and meeting consumer needs by examining key research areas, including Soil Health and Management, Pest and Disease Management, Climate Resilience and Adaptation, Carbon Sequestration and Climate Mitigation, Market and Consumer Preferences, and Policy and Institutional Support. The outcome of this review demonstrates that there are still numerous research studies required to evaluate how different farming systems and pedoclimatic conditions can contribute to more efficient horticultural practices.

Keywords: conventional agriculture; economic benefits; environmental effects; health benefits; nutrition; organic agriculture; traditional agriculture; vitamin C

1. Introduction

It is well-known that consumers are becoming more conscious of lifestyles that contribute to a low carbon footprint and prefer organic produce [1]. The top three countries that showed massive growth in the organic market in 2021/2022 are Canada (9.7%), Japan (8.4%), and Estonia (6.0%) [2]. The effect of global climate change is one of the greatest concerns faced by mankind in the 21st century. Organic farming provides an opportunity to counteract the effects of current farming practices, such as utilizing synthetic fertilizers, pesticides, and genetically modified organisms by minimizing pollution of the air, soil, and water, and enhancing the health and productivity of interdependent communities of plants, animals, and humans [1]. Organic farming, furthermore, eliminates both chronic and acute exposure of farm laborers, consumers [3], and surrounding aquatic and terrestrial ecosystems to toxic pesticides [4]. In addition, organic produce offers more nutritious products [5], which contain higher vitamin and mineral content [4]. Organic horticultural practices also have a downstream effect on the quality of the produce, which is perceived to have a better taste due to the higher sugar content and a longer shelf life as a result of metabolic integrity and superior cellular structure [4].

Modern-day horticulture provides many opportunities to incorporate more natural farming practices without affecting yield, with the ultimate aim of higher returns. Integrated and applied organic farming promotes healthy soils and soil microbiota [4], thereby enhancing the availability of nutrients to cultivated plants [6], which offers a more sustainable approach. Organic farming prevents genetic mutations and immune development in insects, an adverse effect associated with pesticide use, thereby potentially reducing pest outbreaks [7] organic farming practices are considered cheaper and more economically competitive as they reduce various input costs associated with conventional farming practices, such as the use of pesticides, herbicides, and synthetic fertilizers [3], whereas traditional agriculture refers to a type of farming that was practiced in the previous years. It is what people used to do to produce food before the discovery of new farming methods and chemicals [8,9]. Like traditional farming systems, closed organic farming systems clearly rely on resources found in nature and within agricultural bodies [10]. This allows for a more harmonious adjustment to nature and, as such, constitutes a privileged ethical strategy for humanity [11].

Climate is described by characteristic conditions such as temperature, precipitation, humidity, soil, snow, and ice cover. The climate is constantly changing due to various natural factors, of which human activity is the most recent and concerning factor that has had an increasing impact on the Earth's climate over the past 200 years. Its impact is defined by the greenhouse effect [12], with temperature changes increasing due to high levels of manufacturing and economic activity, including increases in major greenhouse gas emissions, such as carbon dioxide and methane [13]. Organic agriculture has further demonstrated its capacity to reduce greenhouse gas (GHG) emissions through carbon sequestration and employing fewer inputs. Consequently, in the area of climate change, the transition from conventional agriculture to organic practices is being acknowledged as a viable alternative farming system that holds promise for addressing both climate change mitigation and environmental protection objectives [14,15].

2. Organic Farming

Organic farming is a production system that mitigates or largely eliminates the need to use synthetic fertilizers, pesticides, growth regulators, and feed additives and relies essentially on aspects of crop rotation, crop residues, animal manure, legumes, cover crops, non-agricultural organic wastes, and biological pest control (cover crop and insects), which all contribute to maintaining optimal soil productivity, texture, and nutrients to the cultivated plants [16]. This farming technique has evolved into an agricultural system that focuses on resource conservation and recycling, with the aim of building a more sustainable production system [17].

Organically farmed produce is one of the most popular commodities that modern consumers are most interested in purchasing. The organic food market has grown significantly during the past few years (2016–2017), with several countries in the market showing double-digit growth according to the International Federation of Organic Agriculture Movements (IFOAM) and Research Institute of Organic Agriculture FIBL's most recent study, which supports this assertion. The countries with the largest increases were France (18%), Spain (16%), and Denmark (15%) [3,18].

In South Africa, organic farming has a long history, and the country is a founding member of the International Federation of Organic Movements (IFOAM) [19]. South Africa largely supports organic farming, with an estimated organic produce market of between ZAR 200-400 million in 2005 [20]; however, there is currently no comprehensive database to capture the exact number of organic farmers across the country [19]. However, organic certification bodies may be contracted to release up-to-date figures with a real-time breakdown of certified organic certification of fruit and vegetable production.

Organic foods are often perceived as being healthier and having better flavor than conventionally grown crops, as conventional farming practices are usually associated with unwanted residues from pesticides and herbicides. These may accumulate in the soil and certain crops, such as potato tubers, and, therefore, negatively impact organoleptic properties as well as human and animal health [21]. Higher concentrations of nitrates and lutein were found in conventionally farmed potato tubers when compared to organic potato tubers [18]. Despite lutein reportedly being a health benefit for eye health, it may become detrimental when consumed in high amounts, including the yellowing of human skin due to the nature of its pigments as a carotenoid. Furthermore, potatoes from organic production were, on average, richer in polyphenolic compounds, such as phenolic acids and flavonoids.

Organic cultivation practices are rapidly being adopted and are already being practiced in more than 120 countries throughout the world [22]. From a global perspective, organic horticulture is still a niche practice since less than 1% of worldwide farmland is cultivated naturally and since only a small percentage of the worldwide populace is consuming organic produce in noteworthy amounts. Since the generation yields from organic farming practices are moderately low, and the goals of organic farming, according to standards and measures, are not being achieved on most farms, these practices need improvements that are based on scientific evidence and better implementation [23].

Organic farming practices are not without their disadvantages; fruits and vegetables are typically grown in heated greenhouses and require higher energy inputs, contributing to high emissions. In addition, air freight is a large producer of greenhouse gases contributing to a higher carbon footprint, which is of particular concern as most of these products are transported by air. Organically farmed produce is also highly fragile and more prone to spoilage, resulting in large amounts of waste [24].

3. Economic Benefits of Organic Cultivation

Whilst organic farming is a system that eliminates the use of synthetic fertilizer on farms by improving agricultural practices that would conventionally utilize hormones, fertilizers, feed additives, and pesticides, it relies heavily on natural methods such as animal manure, crop residues, crop rotation, off-farm organic waste, and the utilization of biological systems for crop protection and nutrient mobilization [25]. This suggests a more hands-on approach to farming and requires more attention to the land being farmed. According to Das et al. [25], organic farming yields more nutritious food that is safe for consumption and reduces input costs at large, and due to the niche market, premium prices can be obtained for organic produce. Premium prices, however, exclude a majority of the global population, which could potentially be corrected earlier on in the production costs. Organic farming assists farmers in running sustainable homes [26], and, in turn, contributes to the overall combating of global climate change. In 2017, it was reported that the number of organic products worldwide was increasing significantly (Figure 1) [25], and in 2022, the statistics report showed that organic agriculture land in hectares (M = millions) in each continent of the world has also increased (Figure 2). Between 2012 and 2020, the organic sector in the European Union (EU) experienced notable growth, with a 56% expansion in organic land area, a 40% rise in the number of organic producers, and a noteworthy 114% increase in the monetary value of retail sales [27]. In addition to the premium prices that can be charged for organic produce, organic certification can indirectly be linked to other economic benefits since certified farmer organizations (or buyers) in developing countries typically provide services, such as price information, training, credit, or value addition, to assist farmers to meet certification criteria and produce the quality demanded in worldwide organic markets. As smallholder access to services is often limited, the additional services provided by certified organizations can assist with the improvement of the smallholder farms, resulting in increased revenue [1,25].

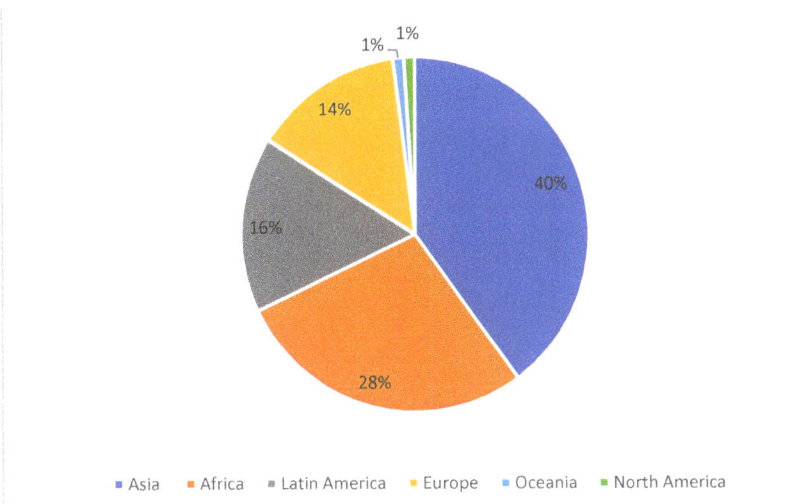

Figure 1. Percentage of organically farmed produce by each continent of the world in 2017, adapted with permission from [25].

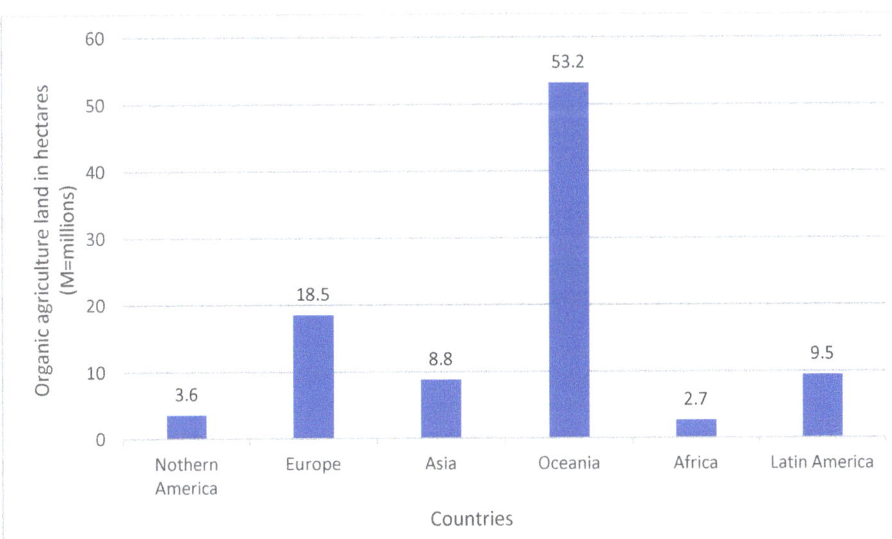

Figure 2. Organic agriculture land in hectares (M = millions) in each continent of the world in 2022, adapted with permission from [2].

4. Conventional vs. Organic Horticulture

Organic farming and conventional farming are two common management practices in agriculture, which use different methods and practices [26]. Conventional agricultural processes use synthetic pesticides and chemical fertilizers to produce higher yields and profits [25], while organic farming, in simple terms, means the cultivation of plants without the use of synthetic fertilizers or pesticides [26]. Organically produced plant-based foods are often thought to have better health-promoting properties than those produced in conventional or integrated production systems [28]. Organically grown foods, especially leafy greens, and tubers, have a high dry matter content compared to conventional cultivation methods [25]. Furthermore, conventionally produced plant-based foods and meats are becoming recognized risk factors for both their safety and quality [28].

Organic fortification with organic farming results in nitrogen gas (N_2) emissions on the same yield scale as non-organic farming systems. Strengthening organic farming will address climate change mitigation, as organic farming systems are reported to provide greater ecosystem services and social benefits in a larger context compared to non-organic systems. Additionally, organic farming promotes fair labor practices, supports local economies, and enhances social well-being. It also considers environmental, economic, and social equity aspects, promoting a holistic and resilient approach to food production [29].

5. Traditional vs. Organic Agriculture

Traditional agriculture, used by centuries of local farming practices [8,9,30], fosters resilience in the face of contemporary challenges and passes the knowledge to the next generation [31]. These traditional methods of farming are a priceless source of information, providing concepts and methods that can strengthen the ability of modern agricultural systems to withstand adverse weather conditions. Traditional farming mostly relies on the natural environment, which is subject to irregular weather patterns and global climate change [32]. The fact that many small farmers adapt to and even anticipate climate change by increasing the use of locally adapted drought-tolerant varieties, water harvesting, mixed cropping, agroforestry, soil conservation practices, and a host of other traditional methods, is the most significant finding regarding the relationships between climate change and peasant agriculture [33]. Crop diversification, maintaining regional genetic diversity, managing soil organic matter, conserving water, and harvesting methods are more strategies included in this list of methods [10].

There are many similarities between traditional agriculture and organic agriculture, as they both depend on natural materials. However, in traditional agriculture, farmers manage risk using experiential knowledge and locally available resources [33], whereas organic farmers lack strategies for risk management; for example, during the dry months, small-scale farmers cannot enter the market, a time when the growing season is for fruits and vegetables [31]. Traditional agricultural practices are usually restricted to small farmers [9]. Evidence has indicated that traditional knowledge is crucial for managing natural resources, maintaining ecosystems, protecting cultural landscapes [32], and managing agricultural productivity [33]. Numerous international conservation programs have emphasized the value of traditional knowledge, which is widely acknowledged. Traditional vegetable knowledge is seriously threatened by factors such as habitat loss, the introduction of new varieties, historical policies, the stigma associated with using traditional vegetables, and altered lifestyles. For example, traditional vegetables grow well in drought-prone locations [9]. The current situation demands that traditional and organic agriculture be integrated.

As already mentioned, organic inputs can reduce GHG emissions [34,35]; also, in traditional agriculture systems, one of the salient features is low energy input [9], which contributes positively to GHG emissions. Farmers who preserve crop genetic resources rely on their traditional agricultural expertise as an essential component of their production process and a method of surviving [36]. Because of their agroecological characteristics, traditional agricultural techniques have the ability to adapt to and reduce the effects of climate change. They enhance the resilience and agrobiodiversity of agroecosystems. They are also inexpensive, energy efficient, and dependent on locally accessible resources. The expertise of traditional agriculture is preserved by indigenous people. In light of climate change, traditional agriculture can be used as a substitute technique for sustainable food production. Traditional agriculture contributes to energy conservation, human health safety, natural resource management, and socio-ecological integrity, in addition to reducing climate change [9].

6. Potential Health Benefits of Organically Grown Fruits and Vegetables

Various health benefits are associated with organic farming, such as reduced occupational health risks, as both farmers and farm employees are less likely to be exposed to toxic pesticides and residues [1]. Organic produce appears to have higher levels of health-promoting phytonutrients, vitamins, and minerals. Plant foods with higher concentrations of polyphenols have been shown to be produced through organic cultivation practices and to contain phytochemicals with anticarcinogenic, anti-inflammatory, antibacterial, antioxidant, antihypertensive, immunological regulating, cardioprotective, vasodilatory, and analgesic properties, according to numerous scientific reports [37,38].

Several high-profile meta-analyses published in recent years have challenged the notion that organic foods are healthier, whilst others have found differences between organic and conventional produce that may translate to better health outcomes for consumers, such as higher levels of antioxidants, like vitamin C and polyphenols, lower levels of cadmium (Cd) and pesticide contamination, a lower incidence of antibiotic-resistant bacteria, and less water content (greater dry matter per unit fresh weight) in organic produce [39,40]. Studies have shown that vitamin C was significantly higher in organically versus conventionally produced plant foods [38,41]. Citrus, strawberries, bell peppers, and potatoes are some of the crops that are reportedly high in naturally occurring vitamin C. Tables 1 and 2 show organically produced vegetables and fruits with vitamin C contents. It would, however, be beneficial for tuber crops, like potatoes, to be actively considered for more natural and organic production since it is a global staple food.

Table 1. Vitamin C content (mg/100 g) of organically and conventionally produced vegetables.

Vegetable	Organic Produce	Reference	Conventional Produce	References
Potatoes	68.8	[42]	13.50	[43]
Cabbage	107	[42]	27.0	[44]
Spinach	150	[45,46]	28.1	[44]
Beetroot	20.36	[47]	10.21	[48]
Carrot	5.3	[49]	2.9	[50]

Table 2. Vitamin C content (mg/100 g) of organically and conventionally produced fruit.

Fruits	Organic Produce	References	Conventional Produce	References
Banana	8.71	[51]	8.7	[44]
Oranges	58.30	[52]	53.2	[44]
Raspberry	174.90	[53]	26.2	[44]
Strawberry	69	[54]	58.8	[44]
Mango	92.8	[55]	36.4	[44]

A growing number of recent studies have demonstrated the beneficial effects of crop-based foods on human health. The consumption of fruits and vegetables, in particular, has been demonstrated to be useful in preventing cardiovascular disease [56] and more so due to the higher levels of flavonoid expression in organic cultivation. Cereals, legumes, and berries have also been shown to aid in maintaining human health. Their impact can be attributed to a variety of factors. Firstly, crop-based foods are often low in sugar and fat, especially in processed forms, but they are also high in nutritionally beneficial components, such as vitamins [4,57], antioxidants, and bioactive chemicals [36]. Fruit and berry extracts have been shown in vitro to decrease cancer cell proliferation. Thus, elements from fruits, particularly berries, have been shown to inhibit cellular processes associated with tumor growth. Furthermore, extracts from strawberries were found to reduce the rate of cancer cell growth, with organically grown strawberry extracts showing a higher inhibition of cancer cell growth compared to commercially produced strawberry extracts [36,58].

7. Strength, Weakness, Opportunities, and Threats

Conventional and organic horticulture both have unique benefits and difficulties when it comes to producing fruits and vegetables. A SWOT analysis has been carried out to assess the strength, weakness, opportunities, and threats related to each approach in order to provide a thorough comparison. This analysis, which is shown in Table 3, provides insightful information on how organic and conventional horticulture compete with one another by identifying important variables that affect their efficacy, viability, and market potential.

Table 3. A SWOT analysis depicting the strengths, weaknesses, opportunities, and threats of organic horticulture compared to conventional horticulture production of fruit and vegetables.

Factor	Strengths	Weaknesses	Opportunities	Threats
Cultivation methods	The exclusion of chemical pesticides, herbicides, and fertilizers leads to lower environmental impacts [58–63]	Requires more land to produce a similar yield to conventional, and it demands more labor [58]	Positive environmental impacts [58]	Adaptation of organic cultivation at a large scale may lead to land use change [64]
Product quality	More nutritional, healthier, and safer produce [25,28,37,38,41]	Lower yields compared to conventional production [58,60]	Highly marketable produce [18,27]	Insect, pest, and disease damage to produce [24,65,66]
Production yields	Lower yields with improved nutritional value [25]	Lower yields compared to conventional production [58,60]	Lower yields demand higher prices [1,25]	Some consumers cannot afford to buy at higher prices [1,25]
Soil fertility	High soil organic matter content, encouraging soil biological activity [67]	Degraded soil can take time to recover [68]	Reduction in environmental degradation [65]	Face similar environmental problems (increased wind, downstream sedimentation, and losses in biodiversity, water quality, and habitat loss) [69]
Organic fertilizer	Use of naturally occurring products, such as manure, crop rotation, and crop residues [16]	High prices for organic fertilizers and a lack of organic means [69]	Farmers can use Indigenous knowledge to produce organic fertilizer [68]	The limited supply and variation of organic products in specialized stores [65]
Economy	It is highly profitable and increases employment year round [64]	Sometimes a lack of access to financial facilities for organic farming [65]	Government and policymaker involvement can boost organic agriculture's hope for the future [66]	A decrease in organic farmers in certain countries due to the drop in profit during the first years of converting from conventional to organic farming [70]
Education	Use of Indigenous knowledge [66]	Low level of farmer literacy [65]	Applying and executing scientific achievements [65]	The weakness of educational and promotional planning [65]
Environmental and climate	In organic farming, drought-tolerant crops can be used [32,71]	Reduction in soil organic matter when there is little rainfall [67]	Less rainfall is good for harvesting season [69]	Soil becomes more acidic (salinization) when rainfall decreases [67]

8. Organic Agriculture, Conventional Agriculture, and Traditional Agriculture

The agricultural industry includes a range of farming practices, each requiring different regulatory compliances and methods. The Table 4 below gives a thorough rundown of three common agricultural methods organic, conventional, and, traditional. It draws attention to the essential distinctions and parallels between methods, input use, environmental effects, and conformity to rules and laws.

Table 4. Comparisons of traditional, organic, and conventional farming practices and compliances.

Practices	Compliances	Traditional Cultivation	Organic Cultivation	Conventional Cultivation
Historical	Knowledge preservation and development or the advancement of farming practices	Historic farming practices preserving cultural and Indigenous knowledge [8,9,30]	Natural cultivation through crop diversification, genetic diversity, soil organic management, watering conservation, and harvesting techniques [72]	Advancement of modern practices to advance large-scale commercial agriculture
Environmental	Legislation and policies	Few to no documented reports	Organic certification in place [1,25]	Food health and safety legislation [73]
	CO_2 footprint reduction	Very low CO_2 footprint [8]	Low CO_2 footprint [74]	High CO_2 footprint [74]
	Fewer chemical residues	Natural and traditional remedies and concoctions [32]	Alternative and organic pesticides and fungicides registered usage	Integrated pest and disease management [75]
Cultivation	Meeting food security	Often linked with self-sufficiency and subsistence farming [31]	Reported as alternative cultivation practices and limited large-scale production [76]	Aimed at meeting food security, often overproduction and wastage
	Land use	Smaller land use areas often sustainable self-sufficient farming [31]	More land but smaller cultivation areas [58]	More intensive cultivation, more or less land use [59,61]
	Volume of production	Lower production [8]	Low-volume production [59,61]	Higher volume of production [59,61,73]
	Speed of production	Reliant environmental [32]	Subjected to environmental conditions [3]	Subjected to environmental conditions for open ground or controlled undercover production
	Adaptable to various seasons	Open field [32]	Open field or undercover [3]	Open field or undercover [59]
Propagation	Accepted or approved methods used	Mainly self-harvested organic seed and traditional vegetative materials and methods used [77,78]	Organic seed and asexual propagation using organic plant material [77,78]	Seed and vegetative propagation methods using commercial seed in conventional farming [77,78]
Soil preparation	Minimize destruction of soil profiles	Traditional farming animals (cows and horses) are used to prepare the soil, with plant residues [79]	Reduced tillage with permanent plant cover to establish improved and nurtured soils [80]	Conventional tillage defines soil management practices with mouldboard plowing, followed by secondary cultivation to create a seedbed often degradation of soils [80]
Fertilizing and nutrition	Chemical usage regulation	Organic manure used on cropland and traditional fertilizers broadcasted by hand [81]	Animal manure, crop residues, green manure, bio-fertilizers, and bio-solids from agro-industries and food processing wastes [82–84]	Chemical fertilizers are used to provide nutrients to the plants, and they are applied using heavy machinery, boom sprayers, etc. [25,85]
Irrigation	Water management	Traditional irrigation systems of flood irrigation and hand watering with the utilization of available sources of water, rainwater harvest, wastewater, and sewage as fertilizer [86–89]	Applications of in-organic agriculture use drip irrigation and overhead irrigation with water conservation strategies through mulching [90]	Conventional agricultural systems use large quantities of irrigation fresh water and fertilizers [91]
Weed control	Weed management strategies	Nonchemical weed strategies used in traditional farming, such as crop rotations, cover crops, intercropping, and hand weeding [92]	Increasing seeding rate suppresses the development of weeds through the use of mulch and long-term weed seedbank management [93,94]	Intensive mechanical weeding and chemical weed control [95]

Table 4. *Cont.*

Practices	Compliances	Traditional Cultivation	Organic Cultivation	Conventional Cultivation
	Chemical use application	No chemical use [72]	No chemical use [1]	More agrochemicals used [59]
Pest management		An eradicative method is used in traditional farming [96]	Organic farming employs plant-derived pest control compounds, microbial-based insecticides, and biopesticides derived from organic material for effective pest management [97,98]	Chemical pest control is used in conventional agriculture [99]
Disease management		Field sanitation and the use of clean seeds to reduce the pathogen population size [100]	In organic farming, natural amendments are used, and it also uses biological remedies, physical weeding, and crop rotation to combat disease [101]	Synthetic amendments (fungicides, pesticides) are used in conventional farming [101]
Storage		Traditional storage methods, including tree shade, small huts, straw, leaves, and sorghum stalks, are used for insulation and sun protection, sometimes selling produce immediately after harvest [102]	The storage system utilizes vapor barrier materials, like polyethylene liner, bags, sacks, cloth-coated boxes, and recyclable packaging, for low temperature and high humidity conditions [102]	Storages in organic farming are characterized by low temperatures and high humidity [102]
Harvesting		Manual harvesting [102]	Manual harvesting and machinery use [59]	Manual harvesting and machinery use [59]
Packaging and handling		Packing harvested produce in paper, cardboard, plastic, or recycled bags [103]	Containers like bags, crates, hampers, baskets, cartons, bulk bins, and palletized containers are used [104]	Containers like bags, crates, hampers, baskets, cartons, bulk bins, and palletized containers are used [104]
Transport		Less transportation sometimes is required when selling produce immediately after harvest [102]	Lower transport requirements. Products are often sold at local markets [104]	Large volumes of produce are transported by large trucks [104]
Consumers and product marketing	Knowledge	It is safe for human health [31]	Promote a healthier lifestyle. Improve taste [25]	Its products are recognized as risk factors for their safety and quality [28]
	Attitude	Not perfect products [31]	Less perfect product (environmental and insect damage) [25]	Perfect graded products [28]
	Affordability	Low costs [31]	More expensive [1,25]	Expensive

9. Greenhouse Gas Emissions (GHGs)

The greenhouse effect is defined as the increase in the Earth's surface temperature (lower layers of the atmosphere) caused by the accumulation of greenhouse gases. As a result of this phenomenon, temperatures are higher than normal, causing irreversible effects, such as climate change and global warming [13,61]. Greenhouse gases include water vapor, carbon dioxide (CO_2), methane (CH_4), and ozone (O_3). Carbon dioxide (CO_2) is the main contributor to the greenhouse effect (up to 72%) globally. The next most important component is CO_2 (9–26%), with CH_4 and O_3 contributing 4–9% and 3–7%, respectively [13]. The challenge of agricultural development today is not only to meet the demand for food but also to be more environmentally conscious. The global demand for food is expected to increase in the future as the population increases; however, increased food demand in the form of food crops, horticulture, plantations, and animal husbandry has increased productivity, leading to increased GHG emissions, specifically from the agricultural sector [64]. Combined emissions from the agriculture sector account for an estimated one-fifth of the annual increase in GHG emissions. Agricultural production and biomass burning also contribute to CO_2 emissions, as do land-use changes, such as deforestation. Agriculture is also considered a major source of CH_4 and nitrous oxide (N_2O) emissions, estimated to account for about 50% and 70% of total anthropogenic emissions, respectively [35].

In order to simultaneously achieve increased crop yields and reduce greenhouse gas emissions in small-scale homestead crop cultivation, a variety of approaches are needed, as no single approach can adequately address the complexity of crop production and the challenge of greenhouse gas emissions simultaneously. Adopting different approaches can result in positive synergistic effects that exceed the additive effects of each approach used separately. Nonetheless, due to the lack of scientific data in the field, further efforts are urgently needed, including research and field demonstrations, to identify optimal combinations of different approaches [62].

Carbon dioxide and other greenhouse gases produced by natural processes were present in the atmosphere long before humans played a significant role. These gases have kept the Earth warmer (about 32 °C) than it would have been in their absence by partially blocking the infrared radiation emitted by the Earth [63].

The food system contributes up to 30% of GHG emissions, with agriculture accounting for the largest share of this percentage. In agriculture, livestock is the largest source of GHG emissions, while outdoor horticulture can also have a significant impact. The method by which food is transported, packaged, prepared, and stored can also result in high GHG emissions [24]. Urban food production, such as urban agriculture, community gardens, allotment gardens, and home gardens, is an important part of urban ecosystems since they not only contribute to improving the physical, mental, and social health of people but also the health of the environment. The local food movement has seen growing interest in urban gardens as an alternative to mainstream food systems. Urban food production has the potential to produce food with reduced transport, energy, and packaging intensity and greater carbon sequestration compared to conventional food systems, thus reducing net GHG emissions [70].

Fertilizer discharge and leaching into the environment also lead to environmental problems, such as eutrophication [96]. Several studies have focused on slow-release fertilizers as a solution to reducing groundwater pollution, reducing greenhouse gas emissions, and mitigating the impacts of climate change [105]. Organic inputs, in particular slow-release nutrient pellets and liquid guano, have become imminent in all farming practices due to their risk to GHG emissions.

10. Mitigating and Adapting to Climate Change

Some studies report that if crop yields decline due to climate change, the effects on the well-being of subsistence farming households will be severe, particularly if the subsistence component of production declines [41]. Organic agriculture can help with adaptation by increasing water infiltration, which allows soils to absorb most of the rainwater, even during heavy rainfall events by improving water-holding capacity, which promotes plant survival during drought periods. At the same time, organic agriculture can lower fossil fuel emissions by up to 60% compared to conventional agriculture and limit the long-term use of fertilizer and agrochemicals by 20% [41]. However, the greatest contribution that organic agriculture has to offer to climate change mitigation is from carbon sequestration. On average, 79%, that is, 0.1–0.5 tons of organic carbon, can be captured per hectare of land in humid temperate conditions. The most important aspect of the relationship between climate change and organic farming is the realization that many small farmers cope with and even prepare for climate change by increasing the use of drought-tolerant local varieties, water harvesting, mixed cropping, agroforestry, and soil organic practices [41].

Organic systems may significantly lower GHG emissions and hence contribute to mitigating global warming by increasing diversity at the farm level and being included in high-diversity landscapes. Organic farmers who do not utilize agrochemical inputs and, instead, diversify their traditional farms and rural landscapes, could significantly cut

GHG emissions from pesticide and fertilizer production [41]. Organic systems, on average, include more soil organic carbon than conventional systems, notwithstanding the wide range of organic farming practices and underlying concepts. Soil organic carbon benefits from organic farming may only be realized if this management system is combined with larger carbon inputs than in conventional treatments [65].

11. Effect of Soil, Water, and Environmental Impact: Organic Farming vs. Conventional

Agricultural practices incur substantial environmental costs, encompassing soil degradation, freshwater contamination, eutrophication, and loss of biodiversity and habitats [102]. In response to these challenges, organic agriculture has emerged as a touted sustainable alternative, characterized by lower environmental impacts compared to conventional high-input agriculture [68,76,85,106,107]. Extensive research indicates that organic farming not only mitigates these environmental costs but also holds potential benefits for soil health. Studies demonstrate that organic farming can lead to higher soil organic matter content and stable aggregation, contributing to improved soil water supply capacity and potentially higher water-limited crop yields compared to conventional farming [85,108]. The positive impact of organic agriculture on soil properties, such as increased organic matter content, enhanced aggregate stability, and reduced bulk density, suggests a correlation with improved ecosystem water relations, emphasizing the multifaceted environmental advantages associated with the adoption of organic farming practices [109].

The environmental consequences of agricultural practices significantly contribute to global climate change, marking agriculture both a contributor and a victim in this complex dynamic. Agricultural activities release substantial amounts of greenhouse gases, making them a contributor to climate change, while simultaneously suffering the consequences of climate-induced impacts on production [110,111]. The adverse effects of climate change on agriculture manifest in various forms, including salinity, drought, heat stress, cold stress, and other abiotic stresses experienced by plants due to climatic factors [112,113]. Climate change exacerbates challenges, such as water scarcity, soil fertility loss, and increased pest infestations, resulting in significant undesirable impacts on crops [112]. Table 5 provides a concise summary of the anticipated effects of individual climate change variables on soil processes. Conventional agriculture is identified as a major contributor to climate change, whereas organic farming is recognized for its comparatively lower environmental and climate impacts when measured per unit of land, though not necessarily per unit of output [1]. This underscores the complex relationship between agricultural practices, climate change, and environmental sustainability.

Table 5. Summary of the expected effects of individual climate change variables on soil processes [110].

Increasing temperature	Loss of soil organic matter Reduction in the labile pool of soil organic matter Reduction in moisture content increase in the mineralization rate Loss of soil structure Increase in soil the respiration rate
Increasing carbon dioxide concentration	Increase in soil organic matter Increase in water use efficiency More availability of carbon to soil microorganisms Accelerated nutrient cycling
Increasing rainfall	Increase in soil moisture or soil wetness and enhanced surface runoff and erosion Increase in soil organic matter Nutrient leaching Increased reduction in ions and nitrates Increased volatilization loss of nitrogen Increase in productivity in arid regions
Reduction in rainfall	Reduction in soil organic matter Soil salinization Reduction in nutrient availability

12. Challenges and Opportunities for the Development of Plant Organic Farming

Several studies have compared organic and conventional systems and have found that organic systems have lower yields, depending on the crop type and agroecological condition [106,107]. The issue of lower yields in organic farming may be caused by many factors, such as the type of seed used, the environmental conditions of the plants, and many more. There is no availability of certified seeds and other plant materials suitable for organic and low-input farming in many countries, and farmers tend to use inorganic seeds and then follow the organic method to produce their products. Hence, those plants produce less yield because, in nature, they require high inputs to be able to produce good yield. Environmental conditions play a vital role in plant yield, and it is important for the farmers to understand the requirements of the plant and the type of seed they will use, whether or not it is drought-resistant, and many other characteristics before choosing the environment to produce their product in to avoid loss of yield.

One of the main problems facing developing countries is unemployment, particularly for a significant majority of the less-skilled population. Organic farming creates jobs in rural areas since it requires over 15% more labor than conventional farming. Some of the most widely used organic farming methods, like strip farming, nonchemical weeding, and the creation, gathering, and delivery of organic supplements, all call for a substantial amount of labor. The adoption of organic farming in developed countries and for struggling with finances farmers in developing countries may be delayed by a shortage of labor and associated costs. However, labor and its associated costs are not a restriction for countries such as India. Instead, a large portion of rural areas can benefit from organic farming by creating job opportunities [114].

Therefore, differences in the types of work and planting and harvesting timetables may offer more rural women employment options, as well as more evenly distributed and stable employment prospects for males working in agriculture. All year long, farmers and farm laborers are kept busy with crops and mechanical weed control. Conventional farming only offers part-time employment opportunities because more labor is needed in the spring and autumn. As a result, organic farming also takes cost effectiveness into consideration [114].

The progression of organic farming encounters disruptions in its structural dynamics, as evidenced by the cessation of organic practices in some farms and the adoption of such practices in others [115]. In certain countries, the number of organic farms is declining faster than new ones are being established, leading to a reduction in the overall area dedicated to organic crops. For example, in the European Union, the discontinuation rate of organic farming in 2005 was reported to be 7.3% [116]. This highlights the existence of barriers impeding the seamless growth of organic farming, and their identification is crucial for understanding the challenges that may significantly influence future organic farming policies. The formulation of effective policies promoting organic farming necessitates not only the recognition of factors facilitating the adoption of organic production methods but also an understanding of the factors leading to the discontinuation of such methods. One significant prevention for farmers considering organic production patterns is the conversion period, which demands a swift restructuring of their farms. This process involves incurring additional costs related to investments and information access, while concurrently leading to periodic drops in incomes attributable to factors like diminished yields and the inability to command a price premium [117].

13. Conclusions

The need for growth in the organic horticulture production systems worldwide is evident. It is not only a requirement for the mitigation of climate change but produces food that is more acceptable for consumers. Organic horticulture can have a remarkable impact on mitigating climate change due to the reduction in carbon sequestration by not using chemical fertilizers that release large quantities of GHGs into the atmosphere. This is notably advantageous to the environment due to no leaching of nutrients and chemicals that contaminate water bodies in the soil. In addition, it is the second farming system after the traditional farming system that is safe for both the farm workers and the environment since it requires a higher rate of biologically active and safe chemicals than the rate of synthetic chemicals currently applied in conventional horticulture. Many researchers have found that organic horticulture produces healthier fruits and vegetables compared to conventional cultivation. While acknowledging this nutritional superiority, it is essential to recognize the trade-off in yield quantity, with organic farming generally producing lower yields than conventional methods. Organic farmers receive compensation in the form of premium prices for their products, which is a reflection of their dedication to sustainable and health-conscious farming methods. Because of its many advantages, organic horticulture is seen as an essential and long-term solution to the problems that face the environment and modern agriculture. Organic farming offers promising environmental benefits and job opportunities, and addressing challenges related to yield, socio-economic factors, and barriers to adoption is crucial for its sustainable development. Furthermore, the integration of organic farming systems and traditional farming systems would be more advantageous to farmers, the environment, and climate change since these farming systems share many similarities.

The creation of supporting policies that take into account the complexity of organic agriculture and promote the adoption of environmentally friendly methods should be the main priorities of government support for research funding and policymakers. Suggestions are, therefore, made to add a curriculum that includes organic agriculture for primary and secondary schools. Additionally, support should be given to specialized institutes that teach organic agriculture. Higher education in organic agriculture systems should be developed so that organic agriculture can be understood more and obtain the attention and required resources to be at the same level as conventional agriculture in terms of research knowledge and information. Consumers' and growers' education and awareness about the drawbacks of conventional agriculture should be actively promoted. By developing a deeper understanding of conventional agriculture, consumers can make informed choices, contributing towards a more sustainable and responsible food production system. In addition, specifically, African governments must acknowledge the variety of interests represented in the organic sector, making sure that each is fairly taken into account and giving underprivileged groups more consideration. On all levels, governments need to take a proactive role in promoting organic agriculture, and data on organic agriculture need to be recorded yearly in each country throughout the African continent so that the growth and level of it can be known to identify the gaps and opportunities for further actions. There are other developing countries around the world that are succeeding in organic agriculture, such as Turkey, and according to the 2024 statistics, the top three organic producers are India, Uganda, and Thailand. Extensive agricultural systems and the minimal use of agrochemicals make it easy to transition to organic farming, reducing chemical pollution and promoting sustainable practices. Experiences with organic agriculture can be used for defining and improving policy programs, market development, exports, extension services, and research activities in other similar countries.

Author Contributions: Writing—original draft preparation, A.M.; conceptualization, writing—review and editing, visualization, supervision, and funding acquisition, H.W. writing—review and editing, C.P.L. All authors have read and agreed to the published version of the manuscript.

Funding: This research was funded by the Govan Mbeki Research and Development Centre (GMRDC), grant number C461.

Conflicts of Interest: The authors declare no conflicts of interest.

References

1. Meemken, E.; Qaim, M. Organic Agriculture, Food Security, and the Environment. *Annu. Rev. Resour. Econ.* **2018**, *10*, 39–63. [CrossRef]
2. Helga, W.; Jan, T.; Bernhard, S. *The World of Organic Agriculture Statistics and Emerging Trends 2024*; IFOAM—Organics International: Bonn, Germany, 2024.

3. Mie, A.; Andersen, H.R.; Gunnarsson, S.; Kahl, J.; Kesse-guyot, E.; Rembia, E.; Quaglio, G.; Grandjean, P. Human Health Implications of Organic Food and Organic Agriculture: A Comprehensive Review. *Environ. Health Glob. Access Sci.* **2017**, *16*, 1–22. [CrossRef] [PubMed]
4. Tal, A. Making Conventional Agriculture Environmentally Friendly: Moving beyond the Glorification of Organic Agriculture and the Demonization of Conventional Agriculture. *Sustainability* **2018**, *10*, 1078. [CrossRef]
5. Baranski, M.; Srednicka-Tober, D.; Volakakis, N.; Seal, C.; Sanderson, R.; Dominika, S.; Stewart, G.B.; Benbrook, C.; Biavati, B.; Markellou, E.; et al. Higher Antioxidant and Lower Cadmium Concentrations and Lower Incidence of Pesticide Residues in Organically Grown Crops: A Systematic Literature Review and Meta-Analyses. *Br. J. Nutr.* **2014**, *112*, 794–811. [CrossRef] [PubMed]
6. Bender, S.F.; Van Der Heijden, M.G.A. Soil Biota Enhance Agricultural Sustainability by Improving Crop Yield, Nutrient Uptake and Reducing Nitrogen Leaching Losses. *J. Appl. Ecol.* **2015**, *1*, 228–239. [CrossRef]
7. Borel, B. When the pesticides run out. *Nature* **2017**, *543*, 302–304. [CrossRef] [PubMed]
8. Singh, R.; Singh, G.S. Traditional agriculture: A climate-smart approach for sustainable food production. *Eng. Ecol. Environ.* **2017**, *2*, 296–316. [CrossRef]
9. Ba, Q.X.; Lu, D.J.; Kuo, W.H.J.; Lai, P.H. Traditional farming and sustainable development of an indigenous community in the mountain area—A case study of Wutai Village in Taiwan. *Sustainability* **2018**, *10*, 3370. [CrossRef]
10. Kwon, C.T. Trait Improvement of Solanaceae Fruit Crops for Vertical Farming by Genome Editing. *J. Plant. Biol.* **2023**, *66*, 1–14. [CrossRef]
11. Crowder, D.W.; Reganold, J.P. Financial Competitiveness of Organic Agriculture on a Global Scale. *Proc. Natl. Acad. Sci. USA* **2015**, *112*, 7611–7616. [CrossRef]
12. Alirezaei, M.; Onat, N.C.; Tatari, O.; Abdelaty, M. The Climate Change-Road Safety-Economy Nexus: A System Dynamics Approach to Understanding Complex Interdependencies. *Systems* **2017**, *5*, 6. [CrossRef]
13. Mikhaylov, A.; Moiseev, N.; Aleshin, K.; Burkhardt, T. Global Climate Change and Greenhouse Effect. *Entrep. Sustain.* **2020**, *7*, 2897–2913. [CrossRef] [PubMed]
14. Syamsiyah, J.; Ariyanto, D.P.; Herawati, A.; Herdiansya, G.; Dwisetio, P.K. Rice paddy field on Gentungan, Karanganyar: Soil C Humic Acid, Fulvic Acid, and Stock as Affected by Period of Organic Practices. *IOP Conf. Ser. Earth Environ. Sci.* **2023**, *1165*, 1–8. [CrossRef]
15. Patle, G.T.; Badyopadhyay, K.K.; Kumar, M. An overview of organic agriculture: A potential strategy for climate change mitigation. *J. Appl. Nat. Sci.* **2014**, *6*, 872–879. [CrossRef]
16. Yadav, M. Towards A Healthier Nation: Organic Farming and Government Policies in India. *Int. J. Adv. Res. Dev.* **2017**, *2*, 153–159.
17. Smith, L.G.; Williams, A.G.; Pearce, B.D. The Energy Efficiency of Organic Agriculture: A Review. *Renew. Agric. Food Syst.* **2015**, *30*, 280–301. [CrossRef]
18. Kazimierczak, R.; Srednicka-Tober, D.; Hallmann, E.; Kopczynska, K.; Zarzynska, K. The Impact of Organic vs. Conventional Agricultural Practices on Selected Quality Features of Eight Potato Cultivars. *Agronomy* **2019**, *9*, 799. [CrossRef]
19. Uhunamure, S.E.; Kom, Z.; Shale, K.; Nethengwe, N.; Steyn, J. Perceptions of Smallholder Farmers towards Organic Farming in South Africa. *Agriculture* **2021**, *11*, 1157. [CrossRef]
20. Naidoo, V.; Ramatsetse, M.K. Assessment of the Consumer Purchase Intentions of Organic Food at the Hazel Food Market in Pretoria, South Africa. *Environ. Econ.* **2016**, *7*, 81–88. [CrossRef]
21. Djaman, K.; Sanogo, S.; Koudahe, K.; Allen, S.; Saibou, A.; Essah, S. Characteristics of Organically Grown Compared to Conventionally Grown Potato and the Processed Products: A Review. *Sustainability* **2021**, *13*, 6289. [CrossRef]
22. Mgbenka, R.N.; Amechi, O.E.; Ike, E.C. Organic Farming in Nigeria: Need for Popularization and Policy Organic Farming in Nigeria: Need for Popularization and Policy. *World J. Agric. Sci.* **2015**, *11*, 346–355.
23. Rahmann, G.; Ardakani, M.R.; Bàrberi, P.; Boehm, H.; Canali, S.; Chander, M.; David, W.; Dengel, L.; Erisman, J.W.; Galvis-Martinez, A.C.; et al. Organic Agriculture 3.0 Is Innovation with Research. *Org. Agric.* **2017**, *7*, 169–197. [CrossRef]
24. Garnett, T.; Smith, P.; Nicholson, W.; Finch, J. Food Systems and Greenhouse Gas Emissions. *Foodsource* **2016**, 1–31.
25. Das, S.; Chatterjee, A.; Pal, T.K. Organic Farming in India: A Vision towards a Healthy Nation. *Food Qual. Saf.* **2020**, *4*, 69–76. [CrossRef]
26. Mariappan, K.; Zhou, D. A Threat of Farmers' Suicide and the Opportunity in Organic Farming for Sustainable Agricultural Development in India. *Sustainability* **2019**, *11*, 2400. [CrossRef]
27. Moreno-Pérez, O.M.; Blázquez-Soriano, A. What future for organic farming? Foresight for a smallholder Mediterranean agricultural system. *Agric. Food Econ.* **2023**, *11*, 34. [CrossRef]
28. Alsanius, B.W.; Von Essen, E.; Hartmann, R.; Vagsholm, I.; Doyle, O.; Schmutz, U.; Stützel, H.; Fricke, A.; Dorais, M. The "One Health"-Concept and Organic Production of Vegetables and Fruits. *Acta Hortic.* **2019**, *1242*, 1–14. [CrossRef]
29. Skinner, C.; Gattinger, A.; Krauss, M.; Krause, H.; Mayer, J.; Heijden, M.G.A.V.D.; Mäder, P. The Impact of Long-Term Organic Farming on Soil-Derived Greenhouse Gas Emissions. *Sci. Rep.* **2019**, *9*, 1702. [CrossRef] [PubMed]
30. Orlove, B.; Roncoli, C.; Kabugo, M.; Majugu, A. Indigenous climate knowledge in southern Uganda: The multiple components of a dynamic regional system. *Clim. Change* **2010**, *100*, 243–265. [CrossRef]
31. Jiao, W.; Yang, X.; Li, Y. Traditional knowledge's impact on soil and water conservation in mountain agricultural systems: A case study of Shexian Dryland stone terraced System, China. *Ecol. Indic.* **2024**, *159*, 111742. [CrossRef]

32. Altieri, M.A.; Nicholls, C.I. The adaptation and mitigation potential of traditional agriculture in a changing climate. *Clim. Change* **2017**, *140*, 33–45. [CrossRef]
33. Jouzi, Z.; Azadi, H.; Taheri, F.; Zarafshani, K.; Gebrehiwot, K.; Passel, S.V.; Lebailly, P. Organic Farming and Small-Scale Farmers: Main Opportunities and Challenges. *Ecol. Econ.* **2017**, *132*, 144–154. [CrossRef]
34. Allendorf, T.D.; Brandt, J.S.; Yang, J.M. Local perceptions of Tibetan village sacred forests in northwest Yunnan. *Biol. Conserv.* **2014**, *169*, 303–310. [CrossRef]
35. Marble, S.C.; Prior, S.A.; Runion, G.B.; Torbert, H.A.; Gilliam, C.H.; Fain, G.B. The Importance of Determining Carbon Sequestration and Greenhouse Gas Mitigation Potential in Ornamental Horticulture. *Hort. Sci.* **2011**, *46*, 240–244. [CrossRef]
36. Bourn, D.; Prescott, J. A Comparison of the Nutritional Value, Sensory Qualities, and Food Safety of Organically and Conventionally Produced Foods. *Crit. Rev. Food Sci.* **2002**, *42*, 1–34. [CrossRef]
37. Aksoy, Z.; Öz, O. Protection of traditional agricultural knowledge and rethinking agricultural research from farmers' perspective: A case from Turkey. *J. Rural Stud.* **2020**, *80*, 291–301. [CrossRef]
38. Faller, A.L.K.; Fialho, E. Polyphenol content and antioxidant capacity in organic and conventional plant foods. *J. Food Compos. Anal.* **2010**, *23*, 561–568. [CrossRef]
39. Popa, M.E.; Mitelut, A.C.; Popa, E.E.; Stan, A.; Popa, V.I. Organic Foods Contribution to Nutritional Quality and Value. *Trends Food Sci. Technol.* **2019**, *84*, 15–18. [CrossRef]
40. Reeve, J.R.; Hoagland, L.A.; Villalba, J.J.; Carr, P.M.; Atucha, A.; Cambardella, C.; Davis, D.R.; Delate, K. Organic Farming Soil. Health, and Food Quality: Considering Possible Links. *Adv. Agron.* **2016**, *137*, 319–368.
41. Chhabra, R.; Kolli, S.; Bauer, J.H. Organically Grown Food Provides Health Benefits to *Drosophila melanogaster*. *PLoS ONE* **2013**, *8*, e52988. [CrossRef]
42. Hajslová, J.; Schulzová, V.; Slanina, P.; Janné, K.; Hellenäs, K.E.; Andersson, C. Quality of organically and conventionally grown potatoes: Four-year study of micronutrients, metals, secondary metabolites, enzymic browning and organoleptic properties. *Food. Addit. Contam.* **2005**, *22*, 514–534.
43. Ellong, E.N.; Bellard, S.; Adenet, C.; Rochefort, K. Polyphenols, Carotenoids, Vitamin C Content in Tropical Fruits and Vegetables and Impact of Processing Methods. *JFNS* **2015**, *6*, 299–313. [CrossRef]
44. Mieszczakowska-Frąc, M.; Celejewska, K.; Płocharski, W. Impact of innovative technologies on the content of vitamin C and its bioavailability from processed fruit and vegetable products. *Antioxidants* **2021**, *10*, 54. [CrossRef] [PubMed]
45. Ismail, A.; Cheah, S.F. Determination of Vitamin C, b-carotene and Riboflavin Contents in Five Green Vegetables Organically and Conventionally Grown. *Malays. J. Nutr.* **2003**, *9*, 31–39. [PubMed]
46. Sharma, J.; Agarwal, S. Impact of Organic Fertilizers on Growth, Yield and Quality of Spinach. *Indian J. Plant Sci.* **2014**, *3*, 2319–382437.
47. Yoon, Y.E.; Kuppusamy, S.; Cho, K.M.; Kim, P.J.; Kwack, Y.B.; Lee, Y.B. Influence of cold stress on contents of soluble sugars, vitamin C and free amino acids including gamma-aminobutyric acid (GABA) in spinach (*Spinacia oleracea*). *Food Chem.* **2017**, *215*, 185–192. [CrossRef] [PubMed]
48. Pavlović, N.; Zdravković, M.; Mladenović, J.; Štrbanović, R.; Zdravković, J. Analysis of fresh and processed carrots and beets from organic and conventional production for the content of nutrients and antioxidant activity. *Acta Agric. Serb.* **2020**, *25*, 171–177. [CrossRef]
49. Bender, I.; Edesi, L.; Hiiesalu, I.; Ingver, A.; Kaat, T.; Kaldmäe, H.; Talve, T.; Tamm, I.; Luik, A. Organic carrot (*Daucus carota* L.) production has an advantage over conventional in quantity as well as in quality. *Agronomy* **2020**, *10*, 1420.
50. Kazimierczak, R.; Siłakiewicz, A.; Hallmann, E.; Średnicka-Tober, D.; Rembialkowska, E. Chemical composition of selected beetroot juices in relation to beetroot production system and processing technology. *Not. Bot. Horti Agrobot. Cluj. Napoca* **2016**, *44*, 491–498. [CrossRef]
51. Yusuf, E.; Tkacz, K.; Turkiewicz, I.P.; Wojdyło, A.; Nowicka, P. Analysis of chemical compounds' content in different varieties of carrots, including qualification and quantification of sugars, organic acids, minerals, and bioactive compounds by UPLC. *Eur. Food Res. Technol.* **2021**, *247*, 3053–3062. [CrossRef]
52. Hasanah, R.; Daningsih, E.; Titin, T. The analysis of nutrient and fiber content of banana (*Musa paradisiaca*) sold in Pontianak, Indonesia. *Biofarm. J. Nat. Prod. Biochem.* **2017**, *15*, 21–25. [CrossRef]
53. Fatin Najwa, R.; Azrina, A. Comparison of vitamin C content in citrus fruits by titration and high-performance liquid chromatography (HPLC) methods. *Int. Food Res. J.* **2017**, *24*, 726–733.
54. Frías-Moreno, M.N.; Parra-Quezada, R.A.; González-Aguilar, G.; Ruíz-Canizales, J.; Molina-Corral, F.J.; Sepulveda, D.R.; Salas-Salazar, N.; Olivas, G.I. Quality, bioactive compounds, antioxidant capacity, and enzymes of raspberries at different maturity stages, effects of organic vs. Conventional fertilization. *Foods* **2021**, *10*, 953. [CrossRef]
55. Kilic, N.; Burgut, A.; Gündesli, M.A.; Nogay, G.; Ercisli, S.; Kafkas, N.E.; Ekiert, H.; Elansary, H.O.; Szopa, A. The effect of organic, inorganic fertilizers and their combinations on fruit quality parameters in strawberry. *J. Hortic.* **2021**, *7*, 354. [CrossRef]
56. Maldonado-Celis, M.E.; Yahia, E.M.; Bedoya, R.; Landázuri, P.; Loango, N.; Aguillón, J.; Restrepo, B.; Guerrero Ospina, J.C. Chemical Composition of Mango (*Mangifera indica* L.) Fruit: Nutritional and Phytochemical Compounds. *Front. Plant Sci.* **2019**, *10*, 1073. [CrossRef]

57. Zhou, D.; Luo, M.; Shang, A.; Mao, Q.; Li, B.; Gan, R.; Li, H. Antioxidant Food Components for the Prevention and Treatment of Cardiovascular Diseases: Effects, Mechanisms, and Clinical Studies. *Oxid. Med. Cell. Longev.* **2021**, *2021*, 1–17. [CrossRef] [PubMed]
58. Johansson, E.; Hussain, A.; Kuktaite, R.; Andersson, S.C.; Olsson, M.E. Contribution of Organically Grown Crops to Human Health. *Int. J. Environ. Res. Public. Health* **2014**, *11*, 3870–3893. [CrossRef] [PubMed]
59. Foteinis, S.; Efthalia, C. Life cycle assessment of organic versus conventional agriculture. A case study of lettuce cultivation in Greece. *J. Clean. Prod.* **2016**, *112*, 2462–2471. [CrossRef]
60. Manganaris, G.; Goulas, V.; Vicente, A.; Terry, L. Berry antioxidants: Small fruits providing large benefit. *J. Sci. Food Agric.* **2013**, *94*, 825–833. [CrossRef] [PubMed]
61. Huang, S.; Kuo, L.; Chou, K. The Applicability of Marginal Abatement Cost Approach: A Comprehensive Review. *J. Clean. Prod.* **2016**, *127*, 59–71. [CrossRef]
62. Kim, D.; Grieco, E.; Bombelli, A.; Hickman, J.E.; Sanz-Cobena, A. Challenges and Opportunities for Enhancing Food Security and Greenhouse Gas Mitigation in Smallholder Farming in Sub-Saharan Africa. A Review. *Food Secur.* **2021**, *13*, 457–476. [CrossRef]
63. Azhar, M.; Zahir, M.; Zaman, K.; Naz, L. Global Estimates of Energy Consumption and Greenhouse Gas Emissions. *Renew. Sustain. Energy Rev.* **2014**, *29*, 336–344. [CrossRef]
64. Prastiyo, S.E.; Irham, I.; Hardyastuti, S.; Jamhari, J. The Impact of Regional Policy and Population Growth on Environmental Kuznets Curve for Agricultural Sector in Indonesia: A Provincial Dynamic Panel Data Analysis. *Curr. World Environ.* **2020**, *15*, 364–370. [CrossRef]
65. Aguilera, E.; Lassaletta, L.; Gattinger, A.; Gimeno, B.S. Managing Soil Carbon for Climate Change Mitigation and Adaptation in Mediterranean Cropping Systems: A Meta-Analysis. *Agric. Ecosyst. Environ.* **2013**, *168*, 25–36. [CrossRef]
66. Aghasafari, H.; Karbasi, A.; Mohammadi, H.; Calisti, R. Determination of the best strategies for development of organic farming: A SWOT—Fuzzy Analytic Network Process approach. *J. Clean. Prod.* **2020**, *277*, 124039. [CrossRef]
67. Namboothiripad, P. An Impact Study on Organic Agriculture—SWOC Analysis. *Indian J.Pure Appl. Biosci.* **2021**, *9*, 92–98. [CrossRef]
68. Krause, J.; Machek, O. A comparative analysis of organic and conventional. *Agric. Econ.* **2018**, *64*, 1–8.
69. Obbineni, J.; Kandasamy, I.; Vasantha, W.B.; Smarandache, F. Combining SWOT analysis and neutrosophic cognitive maps for multi-criteria decision making: A case study of organic agriculture in India. *Soft Comput.* **2023**, *27*, 18311–18332. [CrossRef] [PubMed]
70. Cleveland, D.A.; Phares, N.; Nightingale, K.D.; Weatherby, R.L.; Radis, W.; Ballard, J.; Campagna, M.; Kurtz, D.; Livingston, K.; Riechers, G.; et al. The Potential for Urban Household Vegetable Gardens to Reduce Greenhouse Gas Emissions. *Landsc. Urban Plan.* **2017**, *157*, 365–374. [CrossRef]
71. Kodirkhonov, B.M.; Uljabaev, J.R.; Kholdorov, S.M. Organic agriculture in Uzbekistan: SWOT analysis. *IOP Conf. Ser. Earth Environ. Sci.* **2022**, *979*, 012140. [CrossRef]
72. Altieri, M.A.; Nicholls, C.I.; Henao, A.; Lana, M.A. Agroecology and the design of climate change-resilient farming systems. *Agron. Sustain. Dev.* **2015**, *35*, 869–890. [CrossRef]
73. Çakmakçı, S.; Çakmakçı, R. Quality and Nutritional Parameters of Food in Agri-Food Production Systems. *Foods* **2023**, *12*, 351. [CrossRef] [PubMed]
74. Holka, M.; Kowalska, J.; Jakubowska, M. Reducing Carbon Footprint of Agriculture—Can Organic Farming Help to Mitigate Climate Change? *Agriculture* **2022**, *12*, 1383. [CrossRef]
75. Merot, A.; Fermaud, M.; Gosme, M.; Smits, N. Effect of conversion to organic farming on pest and disease control in French vineyards. *Agronomy* **2022**, *10*, 1047. [CrossRef]
76. Seufert, V.; Ramankutty, N. 'Many shades of gray—The context-dependent performance of organic agriculture. *Sci. Adv.* **2017**, *3*, e1602638. [CrossRef] [PubMed]
77. Makena, I.M.; Matsiliza-Mlathi, B.; Kleynhans, R. Seed germination and vegetative propagation of *Helichrysum odoratissimum*. *S. Afr. J. Bot.* **2023**, *161*, 531–541. [CrossRef]
78. Nkongho, R.N.; Ndam, L.M.; Akoneh, N.N.; Tongwa, Q.M.; Njilar, R.M.; Agbor, D.T.; Sama, V.; Ojongakpa, O.T.; Ngone, A.M. Vegetative propagation of F1 tomato hybrid (*Solanum lycopersicum* L.) using different rooting media and stem-nodal cuttings. *J. Agric. Food Res.* **2023**, *11*, 100470. [CrossRef]
79. Gebregziabher, S.; Mouazen, A.M.; Van Brussel, H.; Ramon, H.; Nyssen, J.; Verplancke, H.; Behailu, M.; Deckers, J.; De Baerdemaeker, J. Animal drawn tillage, the Ethiopian ard plough, maresha: A review. *Soil. Tillage Res.* **2006**, *89*, 129–143. [CrossRef]
80. Kobierski, M.; Lemanowicz, J.; Wojewódzki, P.; Kondratowicz-Maciejewska, K. The effect of organic and conventional farming systems with different tillage on soil properties and enzymatic activity. *Agronomy* **2020**, *10*, 1809. [CrossRef]
81. Rahman, K.M.A.; Zhang, D. Effects of fertilizer broadcasting on the excessive use of inorganic fertilizers and environmental sustainability. *Sustainability* **2018**, *10*, 759. [CrossRef]
82. Singh, R.P.; Singh, R.K.; Singh, S.P.; Srivastava, V.; Bohra, J.S. Organic Farming: A need of sustainable agriculture. *Indian Farming* **2014**, *63*, 38–40.
83. Sacco, D.; Moretti, B.; Monaco, S.; Grignani, C. Six-year transition from conventional to organic farming: Effects on crop production and soil quality. *Eur. J. Agron.* **2015**, *69*, 10–20. [CrossRef]
84. Dubey, L.; Dubey, M.; Jain, P. Role of green manuring in organic farming. *Plant Arch.* **2015**, *15*, 23–26.

85. Hartmann, M.; Frey, B.; Mayer, J.; Ma, P. Distinct soil microbial diversity under long-term organic and conventional farming. *ISME J.* **2015**, *9*, 1177–1194. [CrossRef]
86. Mvungi, A.; Mashauri, D.; Madulu, N.F. Management of water for irrigation agriculture in semi-arid areas: Problems and prospects. *Phys. Chem. Earth* **2005**, *30*, 809–817. [CrossRef]
87. Habtu, S.; Yoshinobu, K. Traditional Irrigation Management in Betmera-Hiwane, Ethiopia: The Main Peculiarities for the Persistence of Irrigation Practices. *J. Mt. Sci.* **2006**, *3*, 139–146. [CrossRef]
88. Palakuru, M. An overview of indian traditional irrigation systems for sustainable agricultural practices. *Int. J. Mord. Agric.* **2020**, *9*, 12–22.
89. Chojnacka, N.; Wuttikhun, T.; Songkhum, P.; Khemthong, P.; Loahhasurayothin, K. Controlled release studies of boron and zinc from layered double hydroxides as the micronutrient hosts for agricultural application. *Appl. Clay Sci.* **2018**, *152*, 311–322.
90. Al-Hello, H.A.; Jubier, A.R. Evaluation and Determination of the Best Irrigation Methods Suitable for the Agricultural Lands of Abu Gharq in Babylon. *IOP Conf. Ser. J. Earth Environ. Sci.* **2023**, *1213*, 012094. [CrossRef]
91. Rahmad, I.F.; Nababan, E.B.; Tanti, L.; Triandi, B.; Ekadiansyah, E.; Fragastia, V.A. Application of the Alcohol Sensor MQ-303A to Detect Alcohol Levels on Car Driver. In Proceedings of the 2019 7th International Conference on Cyber and IT Service Management, CITSM 2019, Jakarta, Indonesia, 6–8 November 2019; pp. 247–255.
92. Pannacci, E.; Lartanzi, B.; Tei, F. Non-chemical weed management strategies in minor crops: A review. *Crop Prot.* **2017**, *96*, 44–58. [CrossRef]
93. Kovács, E.B.; Dorner, Z.; Csík, D.; Zalai, M. Effect of Environmental, Soil and Management Factors on Weed Flora of Field Pea in South-East Hungary. *Agronomy* **2023**, *13*, 1864. [CrossRef]
94. Brown, B.; Gallandt, E.R. To each their own: Case studies of four successful, small-scale organic vegetable farmers with distinct weed management strategies. *Renew. Agric. Food Syst.* **2019**, *34*, 373–379. [CrossRef]
95. Saile, M.; Spaeth, M.; Schwarz, J.; Bahrs, E.; Claß-Mahler, I.; Gerhards, R. Weed control in a pesticide-free farming system with mineral fertilizers. *Weed Res.* **2023**, *63*, 196–206. [CrossRef]
96. Ha, T.M. A Review on the Development of Integrated Pest Management and Its Integration in Modern Agriculture. *Asian J. Agric. Food Sci.* **2014**, *2*, 336–340.
97. Bravo, A.; Soberón, M. Can microbial-based insecticides replace chemical pesticides in agricultural production? *J. Microbiol. Biotechnol.* **2023**, *16*, 2011–2014. [CrossRef]
98. Ali, S.; Ahmad, N.; Dar, M.A.; Manan, S.; Rani, A.; Alghanem, S.M.S.; Khan, K.A.; Sethupathy, S.; Elboughdiri, N.; Mostafa, Y.S.; et al. Nano-Agrochemicals as Substitutes for Pesticides: Prospects and Risks. *Plants* **2023**, *13*, 109. [CrossRef] [PubMed]
99. Shennan, C.; Krupnik, T.J.; Baird, G.; Cohen, H.; Forbush, K.; Lovell, R.J.; Olimpi, E.M. Organic and Conventional Agriculture: A Useful Framing? *Annu. Rev. Environ.* **2017**, *42*, 317–346. [CrossRef]
100. Aduo, B.C.; Nwadili, C.O. A survey of indigenous farming systems and practices in Ebonyi state Nigeria: In search of sustainable plant disease management options. *Global Sci. J.* **2019**, *7*, 184–190.
101. Khatri, S.; Dubey, S.; Jelsbak, S.; Sharma, S. Organic farming induces changes in bacterial community and disease suppressiveness against fungal phytopathogens. *Appl. Soil. Ecol.* **2023**, *181*, 104658. [CrossRef]
102. Kasso, M.; Bekele, A. Post-harvest loss and quality deterioration of horticultural crops in Dire Dawa Region, Ethiopia. *J. Saudi Soc. Agric. Sci.* **2018**, *17*, 88–96. [CrossRef]
103. Faqeerzada, M.A.; Rahman, A.; Joshi, R.; Park, E.; Cho, B.-K. Postharvest technologies for fruits and vegetables in South Asian countries: A review. *Agric. Sci. Korean J. Agric. Sci.* **2018**, *45*, 325–353.
104. Munasinghe, S.; Jayasuriya, N. Packing and Transporting Procedures Used by Sri Lankan Farmers to Reduce Post-Harvest. Losses. 2023. Available online: https://www.researchgate.net/publication/371377929 (accessed on 30 May 2024).
105. Benlamlih, F.Z.; Lamhamedi, M.S.; Pepin, S.; Benomar, L.; Messaddeq, Y.; Delgado, M.; Mesa, S.; Bedmar, E. Evaluation of a New Generation of Coated Fertilizers to Reduce the Leaching of Mineral Nutrients and Greenhouse Gas (N_2O) Emissions. *Agronomy* **2021**, *11*, 1129. [CrossRef]
106. Dietrich, L.; Scharer, M.; Kundel, D.; Mader, P.; Kahmen, A. Reduced plant water use can explain higher soil moisture in organic compared to conventional farming systems. *Agric. Ecosyst. Environ.* **2022**, *332*, 107915.
107. Meier, M.S.; Stoessel, F.; Jungbluth, N.; Juraske, R.; Schader, C.; Stolze, M. Environmental impacts of organic and conventional agricultural products—Are the differences captured by life cycle assessment ? *J. Environ. Manag.* **2015**, *149*, 193–208. [CrossRef] [PubMed]
108. Crittenden, S.J.; Poot, N.; Heinen, M.; Van Balen, D.J.M.; Pulleman, M.M. Soil & Tillage Research Soil physical quality in contrasting tillage systems in organic and conventional farming. *Soil. Tillage Res.* **2015**, *154*, 136–144.
109. van Tuyll, A.; Boedijn, A.; Brunsting, M.; Barbagli, T.; Blok, C.; Stanghellini, C. Quantification of Material Flows: A First Step towards Integrating Tomato Greenhouse Horticulture into a Circular Economy. *J. Clean. Prod.* **2022**, *379*, 379. [CrossRef]
110. Pareek, N. Climate Change Impact on Soils: Adaptation and Mitigation Climate Change Impact on Soils: Adaptation and Mitigation. *MOJ Ecol.* **2018**, *2*, 136–139.
111. Sikder, R.; Xiaoying, J. Climate Change Impact and Agriculture of Bangladesh. *J. Earth Syst. Sci.* **2014**, *4*, 35–40.
112. Malhi, G.S.; Kaur, M.; Kaushik, P. Impact of Climate Change on Agriculture and Its Mitigation Strategies. *Sustainability* **2021**, *13*, 1318. [CrossRef]

113. Arora, N.K. Impact of climate change on agriculture production and its sustainable solutions. *J. Environ. Sustain.* **2019**, *2*, 95–96. [CrossRef]
114. Pandey, J.; Singh, A. India is mainly an agricultural country, where agriculture contributes to about 14.6 percent in gross domestic product (GDP) and support over 58 percent of nation's population for livelihood (GOI, 2010). *J. Sci. Res.* **2010**, *56*, 47–72.
115. Läpple, D.; Kelley, H. Understanding the uptake of organic farming: Accounting for heterogeneities among Irish farmers. *Ecol. Econ.* **2013**, *88*, 11–19. [CrossRef]
116. Othieno, L.; Shinyekwa, I. This Document Is Discoverable and Free to Researchers Across the Globe Due to the Work of AgEcon Search. Trade, Revenue and Welfare Effects of the East African Community Customs Union Principle of Asymmetry on Uganda. 2011 [Preprint]. Available online: http://ageconsearch.umn.edu (accessed on 30 May 2024).
117. Łuczka, W.; Kalinowski, S. Barriers to the development of organic farming: A polish case study. *Agriculture* **2020**, *10*, 536. [CrossRef]

MDPI AG
Grosspeteranlage 5
4052 Basel
Switzerland
Tel.: +41 61 683 77 34

Applied Sciences Editorial Office
E-mail: applsci@mdpi.com
www.mdpi.com/journal/applsci

www.ingramcontent.com/pod-product-compliance
Lightning Source LLC
LaVergne TN
LVHW070452061126
851858LV00011B/218